New Digital Work II

Ulrike Schmuntzsch · Alexandra Shajek ·
Ernst Andreas Hartmann
Editors

New Digital Work II

Digital Sovereignty of Companies and Organizations

 Springer

Editors
Ulrike Schmuntzsch
Institut für Innovation und Technik
VDI/VDE Innovation + Technik GmbH
Berlin, Germany

Alexandra Shajek
Institute for Innovation and Technology (iit)
VDI/VDE Innovation + Technik GmbH
Berlin, Germany

Ernst Andreas Hartmann
Institut für Innovation und Technik
VDI/VDE Innovation + Technik GmbH
Berlin, Germany

ISBN 978-3-031-69993-1   ISBN 978-3-031-69994-8 (eBook)
https://doi.org/10.1007/978-3-031-69994-8

Contents

About the Editors

Dr. Ulrike Schmuntzsch studied 'Business Psychology' at the former University of Applied Science in Lüneburg and 'Human Factors' at the Technical University (TU) Berlin. Focusing on Work and Engineering Psychology, she worked as a research associate at the Chair of Human-Machine Systems at the TU Berlin from 2011 until 2022. As part of several application-oriented research projects with various industry partners, she was responsible for human-centred evaluation and design. Her doctoral thesis, which she completed in 2014, focuses on the development and evaluation of a warning glove as a means of user support during maintenance work in industrial applications. Since 2022, she has been a Research Associate at the Institute for Innovation and Technology (iit) and a consultant at VDI/VDE Innovation + Technik GmbH. In this role, she is part of the project team 'Digital Sovereignty in Business'.

Dr. Alexandra Shajek is a Research Associate at the iit and Team Leader of the 'Education and Work' group at VDI/VDE Innovation + Technik GmbH. Currently, she manages the project 'Federal Report on Young Academics 2025' for a German Federal Ministry. At the iit, she worked on projects such as 'Case studies on the effects of the COVID-19 pandemic on operational transformation processes in organizations' among others. Previously, she worked as a visiting researcher at the German Institute for Economic Research (DIW) and, in addition, completed her doctorate in the field of innovation research at Humboldt University of Berlin.

Dr. Ernst Andreas Hartmann after studying psychology, is specialising in work and organisational psychology—obtained his doctorate as Dr. rer. nat. at the Faculty of Mechanical Engineering at RWTH Aachen University in 1995. In the 1990s, he worked at the Hochschuldidaktisches Zentrum/Lehrstuhl Informatik im Maschinenbau (University Teaching Centre/Chair of Information Technology in Mechanical Engineering) at RWTH Aachen. In this context, he engaged in projects on academic reform and took part in the development of new forms of academic teaching/learning. Furthermore, he carried out research on the design of man–machine systems and issues of industrial work organisation. In the mid-1990s, he was an internal consultant for organisation and process development at John Deere Werke Mannheim. In

2002, he qualified as lecturer (habilitation) in psychology and received the 'venia legendi' for Work and Organisational Psychology; since then, he has been a private lecturer for work systems and process design at RWTH Aachen. From 2001 to 2004, he was responsible for the scientific coordination of the programme 'Lernkultur Kompetenzentwicklung' ('competence development and learning cultures') of the German Federal Ministry of Education and Research at the 'Arbeitsgemeinschaft Betriebliche Weiterbildungsforschung ABWF e.V.' ('Association for Research in Continuing Education').

From 2004 to 2016, Ernst Andreas Hartmann was head of the Socio-economic Department at VDI/VDE Innovation + Technik GmbH in Berlin; since 2016, he has been head of the Education, Science and Humanities Department. Since 2007, he has functioned as one of the directors of the Institute for Innovation and Technology (iit).

Digital Sovereignty of Companies and Other Organizations—An Introduction

Ernst Andreas Hartmann⊙, **Alexandra Shajek**⊙, and **Ulrike Schmuntzsch**⊙

Abstract This volume is the fourth in a series of books on Digital Sovereignty in economy. In this volume, Digital Sovereignty of companies and other organizations is the focus. Digital Sovereignty at the organizational level is described through the concepts of human, structural, and relational capital, which represent key elements of an organization's capacity to address (digital) challenges. Human capital refers to the knowledge and competences of the individual employees within the company. The structures and processes enabling the organization to access, combine, and develop its knowledge base constitute structural capital. Relational capital relates to knowledge available through the external relations of the organization. These three types of capital are integrated with two key aspects of control to form a core pillar of sovereignty: confidence in action—the degree to which an organization can trust that the selected actions will achieve the intended results—and freedom of action—the variety of options available for the organization to pursue its aims. The three kinds of capital combined with the two aspects of control yield a matrix with six cells, which can be used to analyze and design digital sovereignty for organizations. In this book, contributions from computer science, economics, social sciences, organizational psychology, work psychology and human factors, mechanical and industrial engineering, and law describe interdisciplinary building blocks for Digital Sovereignty. These contributions are presented with respect to the six cells of the analysis and design matrix, highlighting their relevance for individual aspects of Digital Sovereignty.

Keywords Digital sovereignty · Digital transformation · Sociotechnical analysis · Sociotechnical design

E. A. Hartmann (✉) · A. Shajek · U. Schmuntzsch
Institute for Innovation and Technology (iit), Steinplatz 1, Berlin, Germany
e-mail: hartmann@iit-berlin.de

U. Schmuntzsch et al. (eds.), *New Digital Work II*,
https://doi.org/10.1007/978-3-031-69994-8_1

1 Digital Sovereignty in Economic Contexts

This volume which covers Digital Sovereignty[1] of companies and organizations is the fourth—and, for the time being, final—in a series of the Institute for Innovation and Technology (iit)'s publications with respect to Digital Sovereignty in economic contexts. Within these economic contexts, two pillars of Digital Sovereignty are distinguished: Digital Sovereignty of individual employees, at their workplace, and Digital Sovereignty of companies or other organizations.

The first two publications, in German, contain contributions to both pillars (Hartmann 2021b, 2022). A third book, in English, focused on the first pillar, Digital Sovereignty of individual employees at their workplace (Shajek and Hartmann 2023). Consequently, this edition focuses on Digital Sovereignty of companies and other organizations.

In the following text, a matrix describing the Digital Sovereignty of companies and organizations will be presented (below, Sect. 2). Based on this descriptive structure, all contributions to this volume will be presented with respect to this matrix structure (below, Sects. 3, 4, 5).

2 A Matrix Describing Digital Sovereignty of Companies and Other Organizations

At its core, sovereignty refers to entities—people, organizations, states—having control over their situation, and their environments (Couture and Toupin 2019; Hartmann 2021a). Similarly, Digital Sovereignty can be defined as control within digitalized contexts (Couture and Toupin 2019; Hartmann 2021a). To further define control, concepts from psychological control theory (Luchman and González-Morales 2013; Oesterreich 1981), action regulation theory (Hacker 2005), and sociotechnical systems theory (Brandt et al. 1999; Cherns 1976; Mühlbradt et al. 2022; Mumford 2006; Trist and Bamforth 1951) were utilized.

This was approach was taken for various reasons:

- … all three approaches—control theory, action regulation theory, and sociotechnical systems theory—fit together under the conceptual roof of (sociotechnical) systems theory (Hartmann 2005),
- … in this combination, they allow a systemic perspective—including people, organization, and technology—which is specifically useful for assessing Digital Sovereignty in industrial application contexts (Hofmann et al. 2023),

---

[1] In this chapter, the term 'Digital Sovereignty' is used. In other contributions and publications, 'Sociodigital Sovereignty' is used instead, e.g. in the contribution of Schmuntzsch and Hartmann (2025) to this volume. For the time being, we consider both terms as equivalent and use them interchangeably. The future will show which term will become the one which will be predominantly used.

- ... these concepts have been used for decades to design digital work environments (Hacker 1987).

For the purpose of this book, it is assumed that organizations—in a similar, but not identical way as individuals—can be perceived as subjects, and acting entities, selecting courses of action to achieve specific goals (Coleman 1986; Hacker 2005; Hartmann 2005, 2021a; Luhmann 1982).

Following Rainer Oesterreich (Oesterreich 1981), two aspects constitute control within an action-related framework as referred to above:

- **Confidence of action** (in the German Original: *Effizienz*): When selecting a specific course of action, the acting entity—individual or organization—can be confident that the selected action will bring about the intended result(s).
- **Freedom of action** (in the German original: *Divergenz*): When pursuing a specific goal, the acting entity—individual or organization—may choose from a broad set of actions with discretion.

These two aspects—confidence of action and freedom of action—constitute one dimension of Digital Sovereignty of companies and organizations (Hartmann 2021a; Hofmann et al. 2023) (Fig. 1).

The other dimension of Digital Sovereignty of companies and organizations draws on concepts from the methodology of knowledge balance sheets (Hartmann et al. 2014; Mertins 2005). In knowledge-based economies, knowledge is the main asset of organizations, and knowledge—its protection and development—is a core issue of Digital Sovereignty when applied to organizations, as many contributions to this edition will show (Schlinkert et al. 2025; Straub 2025).

In knowledge balance sheets, three forms of intellectual capital are distinguished (Alwert et al. 2005; Hartmann et al. 2014; Mertins 2005):

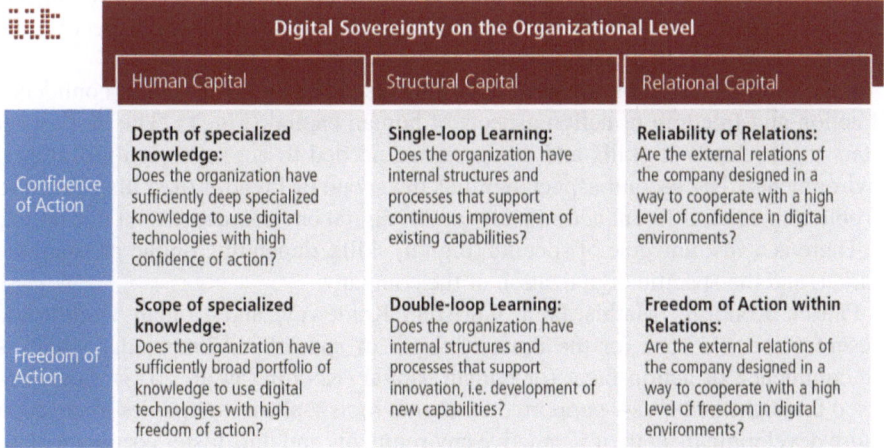

Digital Sovereignty on the Organizational Level		
Human Capital	**Structural Capital**	**Relational Capital**
Confidence of Action — **Depth of specialized knowledge:** Does the organization have sufficiently deep specialized knowledge to use digital technologies with high confidence of action?	**Single-loop Learning:** Does the organization have internal structures and processes that support continuous improvement of existing capabilities?	**Reliability of Relations:** Are the external relations of the company designed in a way to cooperate with a high level of confidence in digital environments?
Freedom of Action — **Scope of specialized knowledge:** Does the organization have a sufficiently broad portfolio of knowledge to use digital technologies with high freedom of action?	**Double-loop Learning:** Does the organization have internal structures and processes that support innovation, i.e. development of new capabilities?	**Freedom of Action within Relations:** Are the external relations of the company designed in a way to cooperate with a high level of freedom in digital environments?

Fig. 1 Facets of digital sovereignty of organizations (own illustration, Hartmann 2021a)

- **Human capital**: The knowledge of the organization's members, i.e. the company's employees
- **Structural capital**: The knowledge embedded in the (organizational or technological) structures and processes within the organization, or the organization's capability, based on these structures and processes, to access and (re-)combine its own knowledge base
- **Relational capital**: The knowledge embedded in the external relations of the organization, or the organization's capability, based on these relations, to access knowledge available in the organization's environment.

Combining the two aspects of control—confidence of action and freedom of action—with these three forms of intellectual capital—human—structural, and relational capital—yields the two-by-three matrix of Digital Sovereignty of companies and organizations, describing six facets of Digital Sovereignty (Fig. 1). For each facet, a leading question is given to illustrate the meaning.

In the following, the contributions to this edition will be presented according to their relations to the three forms of intellectual capital—human, structural, and relational capital.

3   Human Capital

In the previous edited volume in this series, dedicated to Digital Sovereignty of individual employees at their workplaces (Shajek and Hartmann 2023), the domain was structured according to the three pillars of sociotechnical systems: People, organization, and technology (Hartmann 2005; Mühlbradt et al. 2022). Regarding the people pillar, competence development is arguably the most prominent aspect. Accordingly, several contributions to that volume focused on learning and competences (Kanyane 2023; Senderek 2023; Windelband 2023). From the organization's perspective, the same issues constitute human capital.

Regarding human capital, the matrix of facets distinguishes between confidence of action and freedom of action aspects of human capital (Fig. 1). The first aspect refers to the depth of skills and competences needed to act sovereignly in digital environments. The second aspect signifies the scope or breadth of skills needed to be able to pursue different action pathways in digital environments with discretion.

There is a vast universe of specific (digital) skills, depending on the product, the market, and the specific organization of the company.

Patrick Ködding, Mathis Jahn, Christian Koldewey, and Roman Dumitrescu present a specific case for the issue of depth of specialized knowledge,which is the confidence of action facet for human capital (Ködding et al. 2025). Scenario-based foresight methods—supported by digital tools—allow companies to envision future developments in their respective environments and thus foster corporate planning. These methodologies are, however, rather demanding, and Patrick Ködding and his co-authors find in their expert interviews that insufficient expertise is a major hurdle for the application of scenario-based foresight methods, especially in small and medium-sized enterprises (SMEs). They present a total of seven key messages about the current situation in practice, and offer a range of suggestions of how improvements can be made by harnessing advanced digital technologies, including generative AI.

When considering the breadth or scope of specialized knowledge within a company—the freedom of action facet for human capital—the interrelations and functional interdependencies between different skill sets are of specific interest. Role concepts describe such different skill sets and their interactions and interdependencies. René Wöstmann, Lukas Schulte, Florian Meierhofer, Gunter Beitinger, and Jochen Deuse describe data-driven problem-solving in producing companies, relying on a set of corresponding roles—IT experts, domain experts, data scientists, management, and the orchestrator, also known as the citizen data scientist. For each of these roles, a corresponding skill set is described. Taken together, the functional interplay of these roles—supported by complementary skills—enables these companies to choose from a range of possible courses of action, providing a solid base for their Digital Sovereignty (Wöstmann et al. 2025). This role concept may at the same time be seen as an aspect of structural capital, the structures, and processes within an organization enabling the organization to access and combine knowledge and skill resources within the organization's boundaries in order to cope with new challenges.

Furthermore, Ulrike Schmuntzsch, and Ernst Hartmann (Schmuntzsch and Hartmann 2025) combine, in an industrial case study (Schmuntzsch and Hartmann 2023), the workplace-centered approach of the previous edited volume on New Digital Work (Shajek and Hartmann 2023) with the organization-centered perspective taken in this present volume. They support analysis and design issues during the development and implementation of a new software environment which transforms core business processes of the company in question. In doing so, they use the nine facets of Digital Sovereignty at the workplace (Schmuntzsch and Hartmann 2023, 2025) as well as the six facets of Digital Sovereignty at the organizational level (Fig. 1). Focal points of their analysis concern preserving and developing the knowledge of employees (human capital) as well as evoking a change of mindset within the company from solitary expertism towards a culture of common knowledge sharing (structural capital).

More examples regarding structural capital will be discussed in the next section.

4 Structural Capital

As with human capital, the matrix of facets distinguishes between confidence of action and freedom of action aspects also for structural capital (Fig. 1).

Structural capital—as defined above—describes the knowledge embedded in the structures and processes within the organization; in other words, the organization's capability, based on these structures and processes, to access and (re-)combine its own knowledge base.

These structures and processes may be organizational or technical in nature (Hartmann 2005). In the following, one contribution with a more technical focus will be presented for the confidence of action aspect, followed by two contributions with an organizational focus for the freedom of action aspect.

Industry 4.0 relies heavily on data-driven approaches, in general, and also specifically in the domain of predictive maintenance. Christopher Braun and Marco F. Huber show that the prognostic and diagnostic performance of data-driven machine learning systems in this domain can be significantly improved when combined with a knowledge-based component, integrating prior knowledge into the data processing, thus bringing about a hybrid approach (Braun and Huber 2025). This prior knowledge can be theoretical knowledge, e.g. relating to laws of physics, or—even more interesting in the context of Digital Sovereignty—experiential knowledge held by employees of the company. This approach allows the company to become more effective and efficient—and thus more confident—in their already existing processes, consequently being an example of single-loop learning (Argyris 1976; Argyris and Schön 1996).

Double-loop learning, on the other hand, does not relate to improving existing structures and processes, but rather to the generation of new ones, thus increasing the freedom of action aspect of Digital Sovereignty (Argyris 1976; Argyris and Schön 1996). Organizational development aims at generating those new structures and processes. Simone Kauffeld and Ann-Kathleen Berg, (Kauffeld and Berg 2025) describe a training program designed to educate change makers in companies, being able to support organizational change and business transformation in dynamically developing environments increasingly incorporating digital technologies as core elements. This concept relies heavily on work-integrated learning, referring to the VDI/VDE guideline 7100 on "work design conducive to learning"[2] as a helpful tool for companies to develop more learning-intensive work environments. Work-integrated learning is complemented by inputs from outside the organization, and transfer projects build the bridge between individual learning and organizational development.

Another connection between individual and organizational learning—which also relates to double-loop learning, here in the context of innovation—is explored by Gina Glock (Glock 2025). She asks whether autonomy on the individual or group level, as a feature of work (system) design, can be considered a prerequisite of innovative capabilities on the company level. In doing so, she takes up research by Edward Lorenz and Antoine Valeyre (Lorenz and Valeyre 2005) on the effect of work design on innovative capabilities of companies and, eventually, innovation itself, as well as former work within iit on measuring innovative capabilities (Hartmann et al. 2014). In this context of individual autonomy and innovative capabilities of the organization, Gina Glock (Glock 2025) analyses data from the German BIBB/ BAuA employment survey. This representative survey is conducted every six years, carried out by the Federal Institute for Vocational Education and Training (BIBB)

---

[2] www.vdi.de/richtlinien/details/vdivde-mt-7100-lernfoerderliche-arbeitsgestaltung-ziele-nutzen-begriffe.

and the Federal Institute for Occupational Safety and Health (BAuA). The current survey dates from 2018[3] and includes 20,012 respondents, aged 15 and above who are in paid employment for at least 10 h per week.

Based on these data she identifies three clusters of employees:

- Cluster I—*The potentials*: There are clear innovation-friendly resources—like the necessity to react to new tasks and problems—for this group, but these are hardly used in everyday life due to little autonomous work and organizational requirements including high performance pressure.
- Cluster II—*The passive balanced*: This group shows a considerably higher level of autonomy. On the other hand, they encounter—compared to the other groups—less frequently new topics or tasks and have less often to try out new things. Also, independent problem-solving is less common among them as compared to the other clusters.
- Cluster III—*The autonomous front runners*: This group is characterized by intense learning necessities as well as high autonomy. They often make difficult decisions and the details of their work are not prescribed. Relatively often they encounter new tasks and need a high level of problem-solving skills. On the other hand, these employees encounter exceedingly high organizational demands, and have to resolve conflicts frequently.

Gina Glock (Glock 2025) discusses opportunities to develop the potentials of these three clusters of employees in a way to maximize their impact on corporate innovation.

5   Relational Capital

Relational capital refers to companies or other organizations using their external relations—to other companies, research and education institutions, associations, clusters, public bodies or other organizations/institutions—to make the knowledge of these external partners internally available, and also to produce new knowledge together with these partners, as e.g. in collaborative research and development partnerships.

As Digital Sovereignty of companies relies heavily on relations with other entities—e.g. providers of cloud platforms, partners collaborating in digital value chains—it is not surprising that a significant part of the contributions to this volume is related to relational capital.

Sebastian Straub (Straub 2025) analyses the European Data Act and its implications for Digital Sovereignty, specifically regarding rights and obligations of different partners in business relations. The overarching aim of this regulation is to improve data availability to bring forward the exploitation of economic potentials resting in rich and diverse pools of data. In the European Data Act, this is pursued by creating

---

[3] https://www.baua.de/DE/Themen/Monitoring-Evaluation/Zahlen-Daten-Fakten/BIBB-BAuA/BIBB-BAuA-2018.html.

data access rights for users of specific products, third parties and public bodies, thus limiting the data ownership of the original manufacturers of the respective products. On the other hand, measures are taken to protect the interests of the manufacturers, especially regarding sensitive business information concerning intellectual property rights or trade secrets. In this way, the European Data Act is designed to balance a greater amount of freedom of action for users, third parties, and public bodies against confidence of action for manufacturers regarding their intellectual property rights.

A crucial factor for confidence of action in external relations is trust regarding these external actors, their commercial and legal environments, and their products. Ulla Coester and Norbert Pohlmann (Coester and Pohlmann 2025) discuss aspects of trust and trustworthiness with a special emphasis on providers and users of AI-based or algorithmic systems. They point out that adherence to legal regulations is not enough, rather, AI providers need to convince their customers that they are actively caring to respect the interests of their customers, even if this would restrict their own freedom of action. Ulla Coester and Norbert Pohlmann develop a model of trustworthiness regarding specific products, specific providers, and the domains the providers are embedded in, in terms of commercial, ethical, and legal structures or orientations. Finally, they argue that trust and trustworthiness are crucial cornerstones of digital transformation.

Data sovereignty is another—and strongly related—pillar of Digital Sovereignty. For data sovereignty to exist in digital ecosystems, two conditions need to be met: the collection, processing, and use of data must be transparent for the data providers, and data providers must be able to actively control the disclosure and use of their data. In this perspective, data sovereignty—quite similar to Digital Sovereignty as a whole—equals transparency plus self-determination. Denis Feth, Christian Jung, and Andreas Eitel use this perspective to describe requirements and possible solutions for data sovereignty (not only) in industrial digital ecosystems (Feth et al. 2025). Specifically, they describe data cockpits, data usage control, and data trustees as possible designs to implement data sovereignty. In terms of the confidence of action and freedom of action aspects, this contribution relates strongly to the confidence of action aspect, since data sovereignty puts data providers in a situation where they can be confident that their actions within the ecosystem will only lead to data collection, processing, and use which they intended to occur. Additionally, also freedom of action is touched, as data disclosure and usage can be actively and selectively controlled for different sets of data, thus providing new opportunities to act.

In a similar way regarding the goals, but with a different technical approach, Anne Mareike Schlinkert, Leonhard Kunczik, Orlando Hohmeier, and Michael Kuehne-Schlinkert discuss Vertical Federated Learning as a promising way to increase Digital Sovereignty, especially for SMEs (Schlinkert et al. 2025). Their core issue is to resolve the paradox of data usage: Data needs to be shared—or at least combined (see below)—in order to take advantage of powerful data analysis and processing tools, while at the same time intellectual property, which is embedded in this data, needs to be protected and kept private for the intellectual property owners. Their proposed solution employs Federated Machine Learning—more specifically: Vertical Federated Learning—as a means to dissolve this paradox. In Federated Learning, algorithms are trained across decentral sets of data without exchanging any data. The authors demonstrate in two case studies that the performance of their Vertical Federated Learning approach shows a performance similar to traditional central learning models. Thus, companies can—e.g. along value chains—combine their data for powerful data analyses, without actually giving any of their data away. Like in the case of the contribution by Denis Feth and co-authors (Feth et al. 2025), this approach fosters confidence of action as well as freedom of action.

Federated learning is also regarded by Nils Jahnke, Marieke Rohde, and Tom Kraus in the broader context of edge computing (Jahnke et al. 2025). The core idea of edge computing is to move data processing and storage closer to the point of data creation, offering advantages such as reduced latency in data analytics and local analysis of confidential or privacy-relevant data. Despite these advantages, specifically SMEs still find it difficult to take up edge computing for their purposes.

With reference to application in various industries—agriculture and food, healthcare, manufacturing and processing, smart living, and water management—they describe potentials and challenges for the implementation of edge computing applications. By supporting digital transformation, edge computing fosters the optimization of existing processes as well as the invention of new ones, thus enabling single-loop as well as double-loop learning (Argyris 1976; Argyris and Schön 1996).

Among the challenges of introducing edge computing, the difficulties of efficient management and orchestration of edge devices, and missing standards for interoperability and portability are mentioned, among others.

Regarding the relational capital aspects of Digital Sovereignty, edge computing supports freedom of action, as it opens up completely new ways of data processing and storage for application areas where this was not possible before, e.g. because of latency or confidentiality/privacy reasons. Because data sovereignty is also an issue of edge computing, confidence of action is also enhanced, because companies can rather be confident about what will—and will not—happen with their data.

Hyperscalers, powerful providers of cloud infrastructure, contributed strongly to a 'platformization' of digital markets, imposing substantial risks especially for SMEs. These risks include vendor lock-in effects due to the power and capabilities of hyperscalers to make it extremely unattractive for users to use any other (cloud) resources than the hyperscalers'. As a consequence, user companies may lose their own innovative capabilities by over-reliance on the services of the hyperscaler. Due to hyperscalers' controlling their customers' access to their own data, data security objectives may be jeopardized. Finally, resilience of the customers' operations is endangered when, for some reason or another, the services of the hyperscaler break down and cannot (immediately) be replaced by other providers' services.

As a remedy for these risks, Peter Ganten, Miriam Seyffarth, and Nils Kuhlmann propose using data infrastructures based on open-source technologies. They also emphasize the importance of standardization to provide a level playing field for all (Ganten et al. 2025). By providing a way out of vendor lock-in situations, this approach specifically enhances freedom of action in digital networks and ecosystems.

As relational capital directly rests on the structures and dynamics of networks and ecosystems, a closer look at these structures and dynamics is helpful for analyzing the context while designing new digital solutions. Juliane Balder and Rainer Stark pursue this approach by setting up a stakeholder analysis in the context of their project to develop an open innovation (OI) platform in the furniture and furnishing industry (Balder and Stark 2025). An important aspect of this stakeholder analysis was to understand which stakeholders hold which power over the OI platform to be developed, and how a good balance of power and a preferably equal standing of the participants could be achieved.

Furthermore, Juliane Balder and Rainer Stark (Balder and Stark 2025) devise a set of criteria for developing OI platforms according to Digital Sovereignty. These criteria refer to the categories Data, Distribution and Federation, Interfaces, Software for Technical Platform, Hardware, Technical Control, Competencies, and Jurisdiction. Some of these criteria contribute predominantly to the confidence of action within the relations aspect, like e.g. Data, which would be completely fulfilled when users have full data control, incl. reading, modifying, deleting, and choosing storage location. Other criteria contribute rather to the freedom of action within the relations aspect, like e.g. Interfaces, which would be completely fulfilled when users have access to all data and functions via open, freely usable interfaces with open-source reference implementation.

A very strong point for relational capital is made by Katharina Dassel and Maxie Lutze, based on an example of cross-organizational Digital Sovereignty in healthcare (Dassel and Lutze 2025). In the context of patient-centered digital care pathways encompassing prevention, outpatient and inpatient treatment, rehabilitation, and care, they argue that Digital Sovereignty cannot be achieved by single organizations alone, but rather by cooperating ecosystems of organizations—care providers—bringing about systems of shared Digital Sovereignty and responsibilities along these digital care pathways. In its core, this contribution refers to the confidence of action aspect of inter-organizational relations, because, as is argued here, Digital Sovereignty can only be achieved collectively. The topics to be addressed in these inter-organizational negotiation and design processes include (1) roles & responsibilities, (2) data exchange, access, and use, (3) IT infrastructure, hardware and software, interoperability, (4) IT & cybersecurity, (5) transformation capabilities, (6) ethics and responsible practices, trustworthy algorithms, and sustainability, (7) competencies.

6 A Unique, Complex, and Completely Different Case—Wikimedia

A final contribution can be seen to serve as a kind of 'contrast medium' for the rest of this volume. All the contributions presented so far are concerned with knowledge as intellectual property and how this can be utilized and protected—in digitalized environments—according to the goals and purposes of the company (or other organization). Consequently, this intellectual property—knowledge—is regarded as—human, structural, relational—*capital.*

Quite a different picture emerges when looking at one of the probably most knowledge-intensive and knowledge-rich organizations on this planet—Wikimedia (Klempert and Ménard 2025). Rather than gather and protect knowledge as private property—as *capital*—Wikimedia regards knowledge as a free resource to share with everyone—as *commons.*

Moreover, by providing free access to a tremendous wealth of knowledge, Wikimedia contributes significantly to the Digital Sovereignty of individuals and organizations worldwide.

Taken together, this very specific case may serve as a stimulus for reflection on the key assumptions of the other contributions of this volume, and on how things might as well be a bit different.

7 Concluding Remarks

This book is the fourth in a series of books on Digital Sovereignty in economic contexts, regarding Digital Sovereignty of individual employees at their workplaces as well as Digital Sovereignty of companies and other organizations. In this book, the last aspect, Digital Sovereignty of companies and other organizations, stands in focus.

To describe and analyze Digital Sovereignty on an organizational level, it was assumed that aspects of knowledge and competence (Škrinjarić 2022) are paramount for organizations to be able to master challenges, including challenges induced by digital technologies.

In this line of thinking, concepts from the context of knowledge balance sheets were employed, namely human, structural, and relational capital (Alwert et al. 2005). Human capital is the body of knowledge held by the individual employees. Structural capital refers to the structures and processes within the organization, and how these structures and processes allow to access, combine, and develop knowledge and competences within the organization. Relational capital, finally, encompasses the knowledge embedded in the external relations of the organization, to other organizations, research and educational institutions, public bodies, and others.

The third book of the series focused on Digital Sovereignty of individual employees at their workplaces (Shajek and Hartmann 2023). There, a socio-technical approach was used, distinguishing between human, organizational, and technological sub-systems. (Hartmann and Shajek 2023). The two approaches—socio-technical systems on the individual level and human, structural, and relational capital on the organizational level—are closely interrelated. Human capital refers to issues of individual competences, which is also a core aspect of the human sub-system of sociotechnical systems. The organizational sub-system contains structures and functions for cooperation, knowledge sharing and knowledge development, in a similar way as structural capital on the level of the organization. These two forms of capital—human and structural—relate to internal affairs of the organization and correspond, as described above, with two sub-systems of socio-technical systems. Relational capital, however, addresses phenomena outside the organization, which is not an issue for the socio-technical perspective on the individual workplace level.

Technology is a separate sub-system on the individual workplace level. On the organizational level, technology is rather seen as a cross-sectional structure, affecting, enabling—or restricting—human, structural, and relational capital alike.

These inter-relations between workplace and company level perspectives can be used within a holistic approach for analyzing and designing Digital Sovereignty in companies in industrial practice (Schmuntzsch and Hartmann 2023, 2025).

Because Digital Sovereignty is an issue relevant for both academia and industry, authors from both domains cooperated in this book.

Regarding disciplines and knowledge domains, authors cover topics from the IT and computer science domain, like open source (Ganten et al. 2025), edge computing (Jahnke et al. 2025), federated learning (Schlinkert et al. 2025), Hybrid AI/theory-guided data science (Braun and Huber, 2025), and data cockpits (Feth et al., 2025).

Economics, social sciences, organizational psychology, and organizational development (Glock, 2025; Kauffeld and Berg, 2025) are represented as well as work psychology and human factors (Schmuntzsch and Hartmann, 2025).

Moreover, knowledge domains also include mechanical and industrial engineering (Balder and Stark, 2025; Wöstmann et al., 2025) and law (Straub, 2025).

This variety of knowledge domains shows that Digital Sovereignty is a deeply interdisciplinary endeavor. This also means that the editors would like to encourage readers to adopt an interdisciplinary perspective, and to have a look not only at contributions from their own fields of competence, but also from other disciplinary perspectives. In this sense, we wish our readers a fruitful and enriching experience with this book.

References

Alwert, K., Heisig, P., Mertins, K.: Wissensbilanzen—Intellektuelles Kapital erfolgreich nutzen und entwickeln. In K. Mertins (Ed.), *Wissensbilanzen: Intellektuelles Kapital erfolgreich nutzen und entwickeln; mit 16 Tabellen* (pp. 1–17). Springer (2005). https://doi.org/10.1007/3-540-275 19-3_1

Argyris, C.: Single-loop and double-loop models in research on decision making. Adm. Sci. Quart. **21**(3), 363 (1976). https://doi.org/10.2307/2391848

Argyris, C., Schön, D.A.: Organizational learning II: theory, method, and practice. Addison-Wesley, Organization development series (1996)

Balder, J., Stark, R.: Development of an open innovation knowledge plat-form in the context of 446 digital sovereignty. In U. Schmuntzsch, A. Shajek, E.A. Hartmann (Eds.), *New Digital Work II: Digital Sovereignty of Companies and Organizations*. Springer (2025)

Brandt, D., Hartmann, E.A., Sander, C., Strina, G.: Designing and simulating sociotechnical systems: concepts and strategies. Human Factors and Ergonomics in Manufacturing & Service Industries **9**(3), 245–252 (1999). https://doi.org/10.1002/(SICI)1520-6564(199922)9:3%3c245::AID-HFM2%3e3.0.CO;2-X

Braun, C., Huber, M.F.: Hybrid AI-driven advances in prognostics and health management within manufacturing environments. In U. Schmuntzsch, A. Shajek, & E. A. Hartmann (Eds.), *New Digital Work II: Digital Sovereignty of Companies and Organizations*. Springer (2025)

Cherns, A.: The principles of sociotechnical design. Human Relat. **29**(8), 783–792 (1976). https://doi.org/10.1177/001872677602900806

Coleman, J.S.: Social theory, social research, and a theory of action. Am. J. Sociol. **91**(6), 1309–1335 (1986). https://doi.org/10.1086/228423

Couture, S., Toupin, S.: What does the notion of "sovereignty" mean when referring to the digital? New Media Soc. **21**(10), 2305–2322 (2019). https://doi.org/10.1177/1461444819865984

Dassel, K., Lutze, M.: Digital care-pathways and inter-organizational systems: a perspective on digital sovereignty along shared responsibilities. In U. Schmuntzsch, A. Shajek, & E. A. Hartmann (Eds.), *New Digital Work II: Digital Sovereignty of Companies and Organizations*. Springer (2025)

Feth, D., Jung, C., Eitel, A.: Concepts for data sovereignty in digital value Chains: Data Cockpits—Data Usage Control—Data Trustees. In U. Schmuntzsch, A. Shajek, E.A. Hartmann (Eds.), *New Digital Work II: Digital Sovereignty of Companies and Organizations*. Springer (2025)

Ganten, P., Seyffarth, M., Kuhlmann, N.: Successful digital transformation in economy and industry requires open source. In U. Schmuntzsch, A. Shajek, E.A. Hartmann (Eds.), *New Digital Work II: Digital Sovereignty of Companies and Organizations*. Springer (2025)

Glock, G.: Innovation capacity in manufacturing: a question of autonomy? In U. Schmuntzsch, A. Shajek, E.A. Hartmann (Eds.), *New Digital Work II: Digital Sovereignty of Companies and Organizations*. Springer (2025)

Hacker, W.: Software-ergonomie; Gestalten Rechnergestützter Geistiger Arbeit?! In W. Schönpflug & M. Wittstock (Eds.), *Berichte des German Chapter of the ACM. Software-Ergonomie '87 Nützen Informationssysteme dem Benutzer?* (pp. 31–54). Vieweg+Teubner Verlag (1987). https://doi.org/10.1007/978-3-322-82971-9_2

Hacker, W.: Allgemeine Arbeitspsychologie (2., vollständig überarbeitete und ergänzte Auflage). Huber (2005)

Hartmann, E.A.: *Arbeitssysteme und Arbeitsprozesse. Mensch—Technik—Organisation: Bd. 39*. vdf Hochschulverl. an der ETH (2005)

Hartmann, E.A.: Digitale Souveränität in der Wirtschaft—Gegenstandsbereiche, Konzepte und Merkmale. In E.A. Hartmann (Ed.), *Digitalisierung souverän gestalten: Innovative Impulse im Maschinenbau*. Springer (2021a). https://doi.org/10.1007/978-3-662-62377-0_1

Hartmann, E.A. (Ed.).: *Digitalisierung souverän gestalten: Innovative Impulse im Maschinenbau*. Springer (2021b). https://link.springer.com/book/https://doi.org/10.1007/978-3-662-62377-0 https://doi.org/10.1007/978-3-662-62377-0

Hartmann, E.A. (Ed.). *Digitalisierung souverän gestalten II: Handlungsspielräume in digitalen Wertschöpfungsnetzwerken*. Springer (2022). https://link.springer.com/book/https://doi.org/10.1007/978-3-662-64408-9 https://doi.org/10.1007/978-3-662-64408-9

Hartmann, E.A., Shajek, A.: New digital work and digital sovereignty at the workplace—an introduction. In A. Shajek, E.A. Hartmann (Eds.), *New Digital Work: Digital Sovereignty at the Workplace* (1st ed. 2023, pp. 1–15). Springer International Publishing; Imprint Springer (2023). https://doi.org/10.1007/978-3-031-26490-0_1

Hartmann, E.A., von Engelhardt, S., Hering, M., Wangler, L., Birner, N.: *Der iit-Innovationsfähigkeitsindikator: Ein neuer Blick auf die Voraussetzungen von Innovationen.* Institut für Innovation und Technik (iit) (2014). http://www.iit-berlin.de/de/indikator/downlo ads/iit_perspektive_innovationsfaehigkeitsindikator.pdf

Hofmann, A., Hartmann, E.A., Shajek, A.: Digitale Souveränität in soziotechnischen Systemen—KI-Nutzung und Krisenbewältigung. Gruppe. Interaktion. Organisation. Zeitschrift Für Angewandte Organisationspsychologie (GIO), **54**(1), 95–105 (2023). https://doi.org/10.1007/s11 612-023-00674-9

Jahnke, N., Rohde, M., Kraus, T.: Edge computing for digital sovereignty in the data economy. In U. Schmuntzsch, A. Shajek, & E. A. Hartmann (Eds.), *New Digital Work II: Digital Sovereignty of Companies and Organizations.* Springer (2025)

Kanyane, M.: Digital work—transforming the higher education landscape in South Africa. In A. Shajek & E. A. Hartmann (Eds.), *New Digital Work: Digital Sovereignty at the Workplace.* Springer (2023)

Kauffeld, S., Berg, A.-K.: Shaping transformation: becoming a Changemaker. In U. Schmuntzsch, A. Shajek, E.A. Hartmann (Eds.), *New Digital Work II: Digital Sovereignty of Companies and Organizations.* Springer (2025)

Klempert, A., Ménard, D.: Wikipedia's Atypical Organizational Model: Digital Sovereignty 20 Years in the Making. In U. Schmuntzsch, A. Shajek, E.A. Hartmann (Eds.), *New Digital Work II: Digital Sovereignty of Companies and Organizations.* Springer (2025)

Ködding, P., Jahn, M., Koldewey, C., Dumitrescu, R.: Challenges for Scenario-based Foresight and Potentials for Digital Technologies: Insights from Practice. In U. Schmuntzsch, A. Shajek, E.A. Hartmann (Eds.), *New Digital Work II: Digital Sovereignty of Companies and Organizations.* Springer (2025)

Lorenz, E., Valeyre, A.: Organisational innovation, human resource management and labour market structure: a comparison of the EU-15. J. Ind. Relat. **47**(4), 424–442 (2005). https://doi.org/10. 1111/j.1472-9296.2005.00183.x

Luchman, J.N., González-Morales, M.G.: Demands, control, and support: A meta-analytic review of work characteristics interrelationships. J. Occup. Health Psychol. **18**(1), 37–52 (2013). https:// doi.org/10.1037/a0030541

Luhmann, N.: Autopoiesis, Handlung und kommunikative Verständigung. Z. Soziol. **11**(4), 366–379 (1982). https://doi.org/10.1515/zfsoz-1982-0403

Mertins, K. (Ed.).: *Wissensbilanzen: Intellektuelles Kapital erfolgreich nutzen und entwickeln; mit 16 Tabellen.* Springer (2005). https://doi.org/10.1007/3-540-27519-3

Mühlbradt, T., Shajek, A., Hartmann, E.A.: Methoden der Analyse und Gestaltung komplexer soziotechnischer Systeme: Trends in der Forschung. In Gesellschaft für Arbeitswissenschaft (Chair), *Frühjahrskonferenz der Gesellschaft für Arbeitswissenschaft 2022,* Magdeburg (2022)

Mumford, E.: The story of socio-technical design: reflections on its successes, failures and potential. Inf. Syst. J. **16**(4), 317–342 (2006). https://doi.org/10.1111/j.1365-2575.2006.00221.x

Oesterreich, R.: *Handlungsregulation und Kontrolle.* Urban & Schwarzenberg (1981)

Schlinkert, A.M., Kunczik, L., Hohmeier, O., Kuehne-Schlinkert, M.: Preserving digital sovereignty in data-driven manufacturing networks. In U. Schmuntzsch, A. Shajek, E.A. Hartmann (Eds.), *New Digital Work II: Digital Sovereignty of Companies and Organizations.* Springer (2025)

Schmuntzsch, U., Hartmann, E.A.: Highly Automated and Master of the Situation?! Approach for a Human-Centered Evaluation of AI Systems for More Sociodigital Sovereignty. In C. Stephanidis, M. Antona, S. Ntoa, & G. Salvendy (Eds.), *Communications in Computer and Information Science. HCI International 2023 Posters* (Vol. 1832, pp. 494–501). Springer Nature Switzerland (2023). https://doi.org/10.1007/978-3-031-35989-7_63

Schmuntzsch, U., Hartmann, E.A.: Analyzing and developing sociodigital sovereignty on individual an organizational levels—a case study. In U. Schmuntzsch, A. Shajek, E.A. Hartmann (Eds.), *New Digital Work II: Digital Sovereignty of Companies and Organizations.* Springer (2025)

Senderek, R.: Work based Learning in the Mexican Automotive Sector. In A. Shajek, E.A. Hartmann (Eds.), *New Digital Work: Digital Sovereignty at the Workplace.* Springer (2023)

Shajek, A., Hartmann, E.A. (Eds.).: *New Digital Work: Digital Sovereignty at the Workplace.* Springer (2023)

Škrinjarić, B.: Competence-based approaches in organizational and individual context. Human. Soc. Sci. Commun. **9**(1) (2022). https://doi.org/10.1057/s41599-022-01047-1

Straub, S.: The European data act and its impact on corporate digital Sovereignty. In U. Schmuntzsch, A. Shajek, E.A. Hartmann (Eds.), *New Digital Work II: Digital Sovereignty of Companies and Organizations.* Springer (2025)

Trist, E.L., Bamforth, K.W.: Some social and psychological consequences of the longwall method of coal-getting. Human Relations **4**(1), 3–38 (1951). https://doi.org/10.1177/001872675100400101

Windelband, L.: Artificial intelligence and assistance systems for technical vocational education and training: opportunities and risks. In A. Shajek, E.A. Hartmann (Eds.), *New Digital Work: Digital Sovereignty at the Workplace.* Springer (2023)

Wöstmann, R., Schulte, L., Meierhofer, F., Beitinger, G., Deuse, J.: Future challenges of data-driven problem solving in producing companies in context of Sigital Sovereignty and lessons learned from electronics industry. In U. Schmuntzsch, A. Shajek, E.A. Hartmann (Eds.), *New Digital Work II: Digital Sovereignty of Companies and Organizations.* Springer (2025)

Dr. Ernst Andreas Hartmann after studying psychology, specialising in work and organisational psychology—obtained his doctorate as Dr. rer. nat. at the Faculty of Mechanical Engineering at RWTH Aachen University in 1995. In the 1990's, he worked at the Hochschuldidaktisches Zentrum/Lehrstuhl Informatik im Maschinenbau (University Teaching Centre/Chair of Information Technology in Mechanical Engineering) at RWTH Aachen. In this context, he engaged in projects on academic reform and took part in the development of new forms of academic teaching/learning. Furthermore, he carried out research on the design of man-machine systems, and issues of industrial work organisation. In the mid-1990's, Ernst A. Hartmann was an internal consultant for organisation and process development at John Deere Werke Mannheim. In 2002, he qualified as lecturer (habilitation) in psychology and received the 'venia legendi' for Work and Organisational Psychology; since then, he has been a private lecturer for work systems and process design at RWTH Aachen. From 2001 to 2004, he was responsible for the scientific coordination of the programme 'Lernkultur Kompetenzentwicklung' ('competence development and learning cultures') of the German Federal Ministry of Education and Research at the 'Arbeitsgemeinschaft Betriebliche Weiterbildungsforschung ABWF e.V.' ('Association for Research in Continuing Education').

From 2004 to 2016, Ernst A. Hartmann was head of the Socio-economic Department at VDI/VDE Innovation + Technik GmbH in Berlin; since 2016, he has been head of the Education, Science, and Humanities Department. Since 2007 he has functioned as one of the directors of the Institute for Innovation and Technology (iit).

Dr. Alexandra Shajek is a research associate at the iit and team leader of the 'Education and Work' group at VDI/VDE Innovation + Technik GmbH. Currently, she manages the project 'Federal Report on Young Academics 2025' for a German Federal Ministry. At the iit, she worked on projects such as 'Case studies on the effects of the Covid-19 pandemic on operational transformation processes in organizations' among others. Previously, Alexandra Shajek worked as a visiting researcher at the German Institute for Economic Research (DIW) and, in addition, completed her doctorate in the field of innovation research at Humboldt University of Berlin.

Dr. Ulrike Schmuntzsch studied 'Business Psychology' at the former University of Applied Science in Lüneburg and 'Human Factors' at the Technical University (TU) Berlin. Focusing on Work and Engineering Psychology, she worked as a research associate at the Chair of Human-Machine Systems at the TU Berlin from 2011 until 2022. As part of several application-oriented research projects with various industry partners, she was responsible for human-centered evaluation and design. Her doctoral thesis, which she completed in 2014, focuses on the development and evaluation of a warning glove as a means of user support during maintenance work in industrial applications. Since 2022, Ulrike Schmuntzsch has been a research associate at the Institute for Innovation and Technology (iit) and a consultant at VDI/VDE Innovation + Technik GmbH. In this role, she is part of the project team 'Digital Sovereignty in Business'.

Successful Digital Transformation in Economy and Industry Requires Open Source

Peter Ganten, Miriam Seyffarth, and Nils Kuhlmann

Abstract The immense transformation processes surrounding digitalization in business and industry are not only bringing about changes like the shift to the cloud, but are also raising questions about digital sovereignty. A key answer here is the use of open source and open standards. But, there is more to it such as cooperation instead of walling-off and the creation of reliably available open structures in the digital domain.

Keywords Digital sovereignty · Open source · Interoperability

1 The Fourth Industrial Revolution and Its Challenges

Economy and industry are currently undergoing a tremendous transformation process. This is because the fourth industrial revolution—i.e., the intelligent connection of machines and manufacturing processes with the help of information and communications technology—is in full swing and is revolutionizing all areas of life under the buzzword "Industry 4.0". At the same time, these upheavals in the age of data spaces and the platform economy are bringing questions regarding the digital sovereignty of the economy into focus: Are we and our companies still an active part of value creation? Who controls our digital infrastructures and the data flows? Can we still be innovative or is our creative scope already severely limited by our dependence on a few large service providers? Will innovation in the future only mean using

P. Ganten (✉) · M. Seyffarth
OSB Alliance, Pariser Platz 6a, 10117 Berlin, Germany
e-mail: ganten@osb-alliance.com

M. Seyffarth
e-mail: seyffarth@osb-alliance.com

N. Kuhlmann
Univention GmbH, Mary-Somerville-Str. 1, Bremen, Germany
e-mail: kuhlmann@univention.de

© The Author(s) 2025
U. Schmuntzsch et al. (eds.), *New Digital Work II*,
https://doi.org/10.1007/978-3-031-69994-8_2

and applying the inventions of others—as the big IT companies sometimes want us to believe?

While some large companies are already developing visions of an industry in which AI systems can be trained and entirely new business models can be developed with the help of huge amounts of data, many small and medium-sized companies (SMEs) have not even completed the third industrial revolution, i.e., the very fundamental introduction of digital processes into the manufacturing process. While some are working on the automated factory floor of the future, others are still putting pen to paper. This "digital gap" presents an additional challenge. Since the European economy and industry tend to be dominated by SMEs, the requirements and needs as well as the challenges of those need to be focused on in particular.

What both large and small companies have in common, however, is that they will increasingly have to shift their digital processes and the processing of accumulated data to the cloud if they want to survive in an ever changing digital market. This is because cloud solutions are simply the current state of the art, simplifying many processes or making them possible in the first place. In this transformation process, there is a risk that dependencies on individual software providers that already exist will be dramatically exacerbated. We therefore need to determine how we want to shape this change process and what course we need to set so that we are not just passive spectators of the fourth industrial revolution, but self-determined players, who can use this change to improve prosperity, security and our potential for the future.

2 What Kind of Digitalization Do We Want

Digitalization offers unimagined opportunities to elevate humankind's existence to a completely new level of development. We could manage the planet's energy and climate crises, and hopefully overcome them for good at some point, we could master the pressing challenges in medicine, and educate ourselves in a completely different way through advanced learning methods. We could also alleviate the labor shortages that exist in more and more industries with the help of automation and information technology. There are many more opportunities that one might not even imagine today, as technological developments in AI and Large Language Models (LLM), for example, are opening up new possibilities every day.

At the same time, as already mentioned, there is a risk that digitalization will lead us into irresolvable dependencies. More specifically, this means that our companies are in a worse bargaining position compared to already dominant software and service providers and are essentially open to extinction, that the infrastructures on which critical digital processes are based can no longer be considered secure, that we can no longer benefit from free markets, and that the data stored in the cloud or generated there can be used not only to our advantage but also against us. In short, we may lose the ability to control and shape our digital systems and our ability to innovate and compete.

If we want to profit from digitalization in the long term, we must find a way to deal with these dangers. The challenge, then, is to harness the opportunities of digitalization for the benefit of all while, at the same time, minimizing the risks.

3 Platforms, Monopolies and the Problem of Dependence on Hyperscalers

The shift of moving data and processes to the cloud offers numerous opportunities and positive development possibilities for business operations, especially with regard to scaling. For example, a wide variety of services can be ordered and used at any time at the click of a button, and cloud infrastructures provide access to quasi-endless computing power for diverse purposes, including for AI applications or for connecting to different networks from all industries and from all over the world.

At the same time, another transformation is taking place as well: for numerous companies, value is no longer necessarily created by manufacturing a hardware product, but by processing the data generated during the manufacturing and usage of the product. In addition, cloud providers are increasingly acting as platforms, thereby inserting themselves between manufacturing companies and their customers.

Particularly in the case of digital products or products with significant digital components, platforms are taking on the role of retailers. Thanks to their position and already existing dependencies on the customers' side, they can take quite different margins in the distribution of digital products. In addition, they collect data on production, usage, intervene in the complete communication between customers and suppliers, and thus give themselves yet another additional advantage.

Thus, innovation and value creation are shifting to these platforms where the data is being processed—and, eventually, away from the companies in commerce and manufacturing. A few large cloud service providers—mostly from the US—now dominate the market. This leads to an increasing dependence of our economy on these individual providers with negative impact on the ability to control and on freedom of scope, on competitiveness and innovation, and on supply chains, as well as IT security and data protection.

Therefore, the increasing "cloudification" also poses some significant risks for companies:

- **"Platformization" and vendor lock-in**: Most large cloud providers tend to develop into platforms to enable vertical and horizontal scaling of their assortment. From the perspective of a hyperscaler, i.e., a large and scalable cloud provider, customers should no longer have any reasons to use any other service outside their own cloud service production line. Customers should be able to cover all their needs from just one source. Hyperscalers use a wide variety of strategies to achieve this goal. These range from excellent integration of software applications and the most complete assortment possible for all needs, to closed interfaces and locked-in data that can only be used within the respective hyperscaler at reasonable prices, and costs charged for external network traffic. Once customers are in this vendor lock-in, they have no option but to accept all the cloud provider's terms and conditions, for example arbitrary price increases or changes in product design. They can no longer switch providers at short notice and must go along with all justified or unjustified license price increases.

- **Losing the ability to innovate and freedom of scope**: By becoming dependent on individual large providers, commerce and industry lose the ability to build and shape digital systems the way they want to for doing business. Interoperability between different software solutions or a free flow of information can no longer be guaranteed. Different specialized applications from different vendors cannot be combined in a modular way, nor can the software solutions be adapted to one's own needs and business cases. Within this set-up, innovation is then reduced to simply implementing updates for services that someone else has developed. Ultimately, this also has a negative financial impact on the company's own business model: as soon as a successful company uses a cloud service and its own business depends significantly on the use of this service, the service provider could charge an amount for the use of the cloud service that practically consumes the company's entire profit margin or causes it to shrink to a minimum. This in turn could result in the company no longer having sufficient financial resources to invest in research and development to drive innovation itself.
- **Data security**: Since cloud operators can fully control their customers' access to their own data, they also have the ability to restrict or block this access altogether. At the same time, hyperscalers have full access to their customers' data in their current cloud architectures. Against the backdrop of international industrial espionage, this poses a significant threat to economy and industry. In order to appease customers and to ensure them that no data will be leaked in the background or accessed by unauthorized third parties, the major cloud providers make extensive security promises which, however, do not offer any legal protection for a European company when push comes to shove. Considering a statute like the US CLOUD Act there is also always the risk that hyperscalers will have to allow their governments to access the data or to hand over this data due to legal regulations (The United States Department of Justice 2023). A shutdown of critical cloud infrastructures due to political pressure is also conceivable. In this case, central software applications would abruptly become unavailable. Such a case almost occurred in Venezuela in 2019, when the company Adobe was ordered by the US Trump administration to shut down its services in the South American country for political reasons. The service was prevented from being shut down across the country at the last moment. (Tremmel and Heise Online 2023). In light of this example, the dependence on individual providers represents an unacceptable risk for the European economy and industry.
- **Physical dependency and resilience**: Today IT services are run less and less "on premise", i.e., on servers or in data centers on the grounds of the respective company that is using these services. Instead, the software solutions run in the central cloud of a large provider, i.e., "somewhere", often abroad in non-European

data centers, sometimes the location unknown. But even these data centers are vulnerable to physical failures or threats. If, for any reason, the respective servers fail, break down, or lose connectivity, it becomes a significant problem for the company whose data is stored there and whose software applications are running there. From one moment to the next, services, data, communication and critical infrastructure of a company are lost. From a European country there is then no possibility of controlling or mitigating this crisis (Website Computerwoche 2023). Dependence on individual providers therefore has a negative impact on the resilience of economy and industry—in other words, on the ability of a system to deal with external disruptions in such a way that it nevertheless remains functional.

In light of rapidly changing technology and new business models, there are a number of additional challenges for companies beyond the risks mentioned above that result from their dependence on individual large providers.

For example, the use of AI is playing an increasingly important role for companies in order to be able to participate in business at all. AI is already used in numerous commercial areas today, from product logistics to communication and customer support. This importance of AI for value creation processes will increase even more in the future. In order to profitably leverage the use of AI for its own business case, a company must have extensive data sets at its disposal with which the AI can be trained and continuously improved, otherwise it has to purchase this data.

However, the large hyperscalers already have a considerable head start, as they have access to huge amounts of data and can thus precede in developing AI applications and bringing them to market. European companies currently have little opportunity to catch up. This illustrates once again how important it is for European economy and industry to be able to access their own data in a secure and self-determined way and to also decide for themselves who else can have access to this data. Projects such as Catena-X or Manufacturing-X show that the solution for German companies lies in cooperating with each other in order to develop a shared open infrastructure where they can share and make data available to each other so that the collective dependence on large hyperscalers does not increase even further.

There is already an awareness among the major German industry associations of the aforementioned problems and dangers. This is shown for example by the paper "Sustainably strengthening Europe's digital sovereignty" by the "Federation of German Industries" (BDI), (Position paper of the BDI 2023) published in 2020, or the white paper on Manufacturing-X (Whitepaper of the VDMA 2023) by the "Mechanical and Plant Engineering Association" (VDMA), published in September 2022.

The paper published in 2021 by the "German Academy of Science and Engineering" (acatech) "Digital sovereignty—status quo and fields of action" also highlights risks and possible solutions (Paper by Acatech 2023). These and other initiatives show that German economy and industry are ready to take on the existing challenges and to work together to find solutions to existing dependencies.

4 Industrialization and Digitalization: History Doesn't Repeat Itself, It Rhymes

The digitalization we are experiencing today is in many ways comparable to the industrial revolution of the eighteenth and nineteenth centuries. Back then, industry and society were confronted with dramatic changes and entirely new technologies that had far-reaching consequences for politics and the economy within a very short time frame as well. These new technologies significantly improved the lives of many people. They created the basis for prosperity and for a modern, open, and equitable society.

At the same time, this technology led to significant negative side effects, such as dependencies on monopoly-like structures, technology that had adverse health effects and was harmful to the climate, and ruthless exploitation of people's labor and the environment. These negative aspects posed challenges and problems that had to be addressed, yet not all have been conclusively resolved to this day.

The discourse on these challenges, however, prompted early on in almost all societies a decisive development for the infrastructures underlying industrialization, i.e., the road network, railroads and waterways, communications and energy networks: The consensus was that infrastructures should be designed so they could be used safely, they should be operated on predictable terms, and work reliably and continuously for everyone—for the government, economy, and civil society—without, for example, an individual competitor being able to capitalize on this. Previously there had been efforts by individual players to impose proprietary standards, for example in rail networks, where this resulted in different track widths, or in electricity in different networks running on different voltages. These developments restricted competition and favored a few market players above everyone else.

Deciding to use standardized infrastructures open to everyone and the enforcement of this decision was the essential basis for people all over the world to be able to communicate with each other. Both the economy and different nations benefited from the fact that customers and suppliers could reliably exchange goods and information with each other and that freedom of movement could develop at all. Thanks to open infrastructures, a diverse and innovative economic ecosystem emerged.

Nowadays in the age of digitalization, digital networks and digital platforms are taking over the central importance that classic transport and communications networks used to have. Therefore, the question arises anew today as to whether we will manage to make the crucial (digital) infrastructures accessible and open to all market participants and for the benefit of all, or whether we will let the growing development of monopolies happen unhindered. With increasing digitalization our prosperity, security and freedom depend on the secure availability of reliable digital infrastructures.

So, similar to the era of the industrial revolution, we must decide how we want to shape transformation processes for the benefit of all in the face of industry-wide upheavals. Digital sovereignty and open source software are undeniably crucial factors for this.

5 Digital Sovereignty and Open Source Software

The term "digital sovereignty" has had a steep career. It is on everyone's lips these days, has arrived in political discourse, and is used as a leitmotif throughout the German Federal Government's digital strategy. However, the term is interpreted and used quite differently, depending on who is using it. Digital sovereignty is understood in this article to mean the independent and self-determined using, shaping and controlling of digital technologies by the state, the economy and by individuals. The IT Planning Council, a political steering committee of the German Federal Government and the state governments, defines the term similarly, namely as "the capabilities and opportunities of individuals and institutions to exercise their role(s) in the digital world in an independent, self-determined and secure manner." (Key point paper of the IT-Planungsrat 2023) This understanding of "digital sovereignty" thus encompasses more than just data sovereignty (with regard to informational self-determination) or technical sovereignty (with regard to digital infrastructures, e.g., in a cloud stack).

The use of open source software forms a central basis for digital sovereignty. This is because, together with open standards, open source licenses ensure that the software systems that are being used can be independently verified, designed and substituted. Thus, individuals and institutions can use these digital technologies independently, self-determined, and securely.

Open source software is software whose license enables all people to understand the corresponding software (to take a look at the source code), to use it freely and without restriction, to modify it and also to redistribute it in modified form. Proprietary software, on the other hand, severely restricts the possibilities of use and reuse as well as adaptation of the source code by third parties by way of the corresponding licenses.

Existing dependencies on proprietary providers prevent companies, for example, from shaping and controlling their own IT architecture and switching between different providers—and thus stand in the way of the digital sovereignty of economy and industry. Open source software, on the other hand, shows a way out of this dependency.

The use of open source software offers various advantages that benefit companies in particular. For example open source software in combination with open standards makes interoperability possible, i.e., users of open applications can combine different solutions and switch between different providers; vendor lock-ins are thus a thing of the past. The source code can be viewed and modified openly, can therefore be adapted and designed as required, and can also be checked for security gaps and vulnerabilities. This also prevents third parties from influencing the IT systems, e.g., by controlling gateways and interfaces.

In times of geopolitical turmoil and uncertainty, resilience in particular proves to be a significant advantage of open source software. Because even if political or economic structures suddenly change and a central provider or its software fails or is shut down due to political reasons, there are alternatives that can instantly be used, because existing solutions can be operated independently or by other providers. Resilience is particularly important for companies against the backdrop of globalization and the interconnectedness of economic sectors and supply chains. Because if one part of the economy collapses, this immediately affects various other sectors as well. Switching between different software providers thus provides companies with the flexibility in turbulent times and also puts them in a much better negotiating position as opposed to cloud providers, reducing their susceptibility to pressure and extortion.

Innovative power and competitiveness of companies are supported and strengthened by open source software and consequently benefit the economy as a whole. The study "The impact of open source software and hardware on technological independence, competitiveness and innovation in the EU economy", commissioned by the European Commission, proves a significant impact of open source software on the competitiveness of European companies, on economic growth, on start-ups and SMEs and on technological independence (Open Source Study of the EU Commission 2023). Last but not least, the use of open source software has a positive effect on the mitigation of climate change, for example by enabling a much more efficient use of resources (OSB Alliance Website 2023).

However, in order for open source software to fully develop its potential, it must be used more widely and be applied consistently. This is the only way to create a market with healthy competition that benefits everyone. To achieve this, technical and political framework conditions are needed.

6 Technical Standards for Openness and Interoperability

For an interoperable digital ecosystem to emerge, where every company can find its niche, make innovative contributions, and network with others at any time, open standards and interfaces with open-source reference implementations are essential. Open standards are, so to speak, the crucial norms that enable the integration of a wide variety of solutions and applications across vendors. Standards in this case ensure the opening up to competition.

In order for published and documented open standards to actually form an interoperable ecosystem with various mutually compatible solutions in practice, open-source reference implementations of the technical aspects of a standard must always be developed for all defined open standards. Only in this way can software providers know exactly how their solutions must be programmed to fit the corresponding standard.

One example of this is the Sovereign Cloud Stack (SCS) (Website 2023). This is an open cloud stack funded by the German Federal Ministry of Economics and Climate Protection and developed under the umbrella of the Open Source Business Alliance. Organizations and companies can develop their own cloud technologies based on this free cloud standard and, above all, implement it themselves, retaining control over their own data and infrastructures. SCS integrates existing, well-tried components such as OpenStack and Kubernetes and makes them easily accessible.

Today, the Sovereign Cloud Stack is already being used productively by various cloud service providers and deployed in various private clouds. It is also an essential part of the reference architecture of GAIA-X, a project to build a high-performance, competitive, secure, and trusted data infrastructure for Europe.

7 A Political Framework to Strengthen Digital Sovereignty

The German government is increasingly focusing on using open source software to strengthen the digital sovereignty of the public administration and the economy. For example, the German Federal Government's 2021 coalition agreement states, "[…] we will secure digital sovereignty through the right to interoperability and portability, and by relying on open standards, open source, and European ecosystems" (Coalition treaty of the German Government 2023). In the German government's digital strategy, digital sovereignty is even elevated to a leitmotif, and here, too, the importance of open source software for strengthening digital sovereignty is made abundantly clear: "Germany's technological and digital sovereignty is the leitmotif of the German government's digital and innovation policy and significantly contributes to the overarching goal of Europe's strategic sovereignty. Technological and digital sovereignty are necessary to strengthen agency and to reduce dependencies. These, in turn, are conditions for competitiveness, innovation and resilience. In this respect, to achieve technological and digital sovereignty, we aim at […] consistently promoting open source approaches […]" (Website for the German Digital Strategy 2023).

The German government has initiated a number of projects to implement the goals listed in the coalition agreement and in the digital strategy. Some of these focus on the use of open source software in public administration. These include the Center for Digital Sovereignty (Website 2023a), the Sovereign Workplace for Public Administration (Workplace and for the Public Administration 2023) and the OpenCoDE Repository for Public Administration (OpenCoDE 2023). Other initiatives such as the Sovereign Tech Fund, which promotes the further development and improvement of critical open-source basic infrastructures, benefit the public administration as well as the economy, academia, and civil society (Sovereign Tech Fund Website 2023). Ventures supported by the German government, such as Catena-X or Manufacturing-X, are connected to the European GAIA-X project. These initiatives are intended to establish an infrastructure for a European digital ecosystem through cooperation between the participating companies where they can securely use and share data—as an alternative to the large hyperscalers on which German companies are still highly dependent today. Open source software plays a central role in these projects as well, so that an open infrastructure is created that can be used by a wide variety of companies in economy and industry, which can develop their various business models and use cases based on it. Many other projects in the German states and municipalities as well as on the European level aim in the same direction.

The aforementioned projects of the German government address the desire of economy and industry for greater political involvement in order to create alternatives to the existing dependencies and to enable functioning competition on the market once again. For example, the Federation of German Industries' paper "Using Open Source Software Strategically" from February 2023 states that without the use of open source solutions, "Germany as an industrial location can hardly be globally competitive in terms of technology." The German government should therefore "create a legal framework" in order to fully make use of the opportunities offered by open source software (Website 2023b).

Policymakers must therefore create frameworks and conditions and provide support in setting standards so that competition is strengthened, equal access to digital infrastructures is enabled, and all participants in an industry can benefit from jointly developed innovations. However, technical standards and political framework conditions alone are not enough to achieve this goal. Ultimately, there is also a need for players from the business community who have the ability and enthusiasm to shape digitalization sustainably with the help of open source software for the benefit of everyone.

8 Cooperation Instead of Walling-Off: A Paradigm Shift with a History

In April 2023, at an event hosted by the German Federal Ministry of Economics and Climate Protection on Manufacturing-X at the Hannover Messe, representatives of large German companies repeatedly emphasized that in the past they used to be successful primarily by walling-off and by guarding their trade secrets. Today, however, and in the age of the data economy where value is created by processing data, cooperation is the order of the day. Only together can we build open, reliable and secure digital infrastructures that are a real alternative to the big hyperscalers. Individual German companies could never manage to do this alone, but this challenge can be met by working together. These data infrastructures would have to be based on open source software, since insight into the source code would create trust in the corresponding data spaces, and this trust would be the necessary prerequisite for the cooperation of the diverse companies. At first glance, this seems like a remarkable paradigm shift.

Yet the approach of companies cooperating "for a greater common good" to overcome particular challenges and to achieve something from which everyone can benefit is not new at all. Nearly everyone has at one time or another driven a screw into a wall or bolted some piece of furniture together while assembling it. When you buy screws, all you need to know is the size and style, and you have a number of suppliers available that you can choose from, but they all meet the same standards and can be used interchangeably. Until the DIN-standard was introduced by the "German Institute for Standardization" at the beginning of the twentieth century, things were different. At that time, each machine manufacturer produced their own screws and their own components according to their own standards, which made the industrial production of goods made up of components from different manufacturers practically impossible. The DIN-standard created the prerequisite for interoperability, which also marked the starting point of an unprecedented success story. While manufacturers were initially skeptical about this regulation at the time, it soon became apparent that it was enabling completely new leaps in value creation and innovation for the entire industry.

Another example for the revolutionary change in economy and industry brought about by standardization is the invention of the freight container. Today, it is impossible to imagine life without the container and it is seen as a symbol of globalization. In the early 1950s, things looked different; goods were loaded individually, on pallets, packed in barrels, bales or sacks. Then Malcolm McLean, an American freight forwarder, invented the 20-foot standard container and advocated for this revolutionary idea against the resistance of shipping companies and transport firms. Logistics became enormously easier, and at the same time completely new possibilities opened up for business to develop, produce and send goods anywhere in the world. Many innovations that came about as a result of the simplified exchange of goods would probably not have been conceivable without the standardized freight container (Randelhoff and Zukunft Mobilität 2023).

In the software sector, we are at a similar crossroads today. It makes no sense for each company to come up with its own solutions that are incompatible with those of others when it would be more advantageous for everyone to develop and use common standards on the basis of which their own business ideas can then be capitalized on. Common foundations can accelerate or enable individual innovations in the market. In the open-source industry, this realization that all participants in the ecosystem benefit from sharing knowledge with one another to develop a better common foundation has been a matter of course for a long time. This cooperative approach has been spreading to other industries for some years now:

- Companies like RedHat, SUSE or Ubuntu, whose business models are built on different Linux distributions, compete with each other, but they also exchange information with each other and make sure that security updates are available to everyone, not just their own customers.
- The Sovereign Cloud Stack provides similar examples: When a security vulnerability was discovered in early 2023, various companies using SCS gave each other insights into their operating environments in order to be able to jointly close the gap as quickly as possible—a process that was hardly imaginable until then.
- STACKIT is a German sovereign cloud solution based on open source that was developed by the Schwarz Group for the retail industry. In 2022, the Schwarz Group opened the cloud to other companies as well as the public administration so that others could also use this alternative to hyperscalers (Website Supermarkt Inside 2023).
- In the automotive industry, a cross-manufacturer expert circle was appointed by the German Minister of Economic Affairs and Climate Actions, Habeck,

with the task of developing recommendations for action for politics, business and society in order to address the medium- and long-term challenges posed by the structural change driven by electrification, digitalization, networking and automation. In doing so, the circle also focused on the potential of open source software for strengthening Europe as an automotive location (Expertenkreis and Automobilwirtschaft 2023).

- Starling-X is an open source Internet of Things (IoT) cloud platform for edge computing optimized for high performance and low latency applications (Website 2023c). The technology is used by various telecommunication companies as well as industrial companies specialized in IoT or automation (Robinson 2023). Thus, different market players are also using the same open-source cloud stack for a wide variety of business models.
- ADAMOS is a vendor-independent IoT platform for mechanical engineering companies that offers among other things industrial apps (Adamos Website 2023). Different mechanical and plant engineering manufacturers have joined forces to enable and facilitate the implementation of digital business models for companies in the industry. In this way, the participating companies support each other in the digital transformation (Automationspraxis and (konradin Industrie) 2023). This offers advantages especially for SMEs when they are entering digitalized production for the first time.
- With the help of data analysis of machine data, the AXOOM platform supports the optimization of manufacturing processes in manufacturing, i.e., it develops smart factory solutions (Website Plattform Industrie 4.0 of the BMWK and BMBF 2023).
- Even in the agricultural sector, open-source data platforms are now being developed so that all participants in the value chain can have access to basic data. This avoids information asymmetries between companies and facilitates public sector-led monitoring. It also enables a joint early response to yield losses due to climate change, mismanagement or other reasons ("Landwirtschaft, Digitalisierung und digitale Daten" 2023).

9   Every Transformation is Hard

In some parts of "traditional" economy and industry, the shift to a collaborative approach is still difficult. As the examples of the DIN-norm and the standardized freight container show, sudden changes and drastic paradigm shifts are always accompanied by resistance and fears on the part of the companies involved.

For a successful digital transformation in light of the fourth industrial revolution the mindset of many people still needs to change. The use of open source software should not only be driven by fear of existing and increasing dependencies, but also by the knowledge of possibilities and opportunities. Extrinsic motivation must therefore become intrinsic motivation.

In the various aforementioned initiatives from the economy and politics for more digital sovereignty and open source, it is also important that buzzwords are filled with meaning and life and that the concepts developed also have depth. Often, the players still lack a deeper understanding of the functionalities and properties of open source software. It is also not enough if only the managing directors are convinced that the use of open source software is favorable. For a successful transformation, middle management must also be on board and pulling in the same direction. For a successful digital transformation in the industry with the help of open source, every individual is needed.

This requires also trust in open-source technologies being always open, always available and driven forward in the long term. To achieve this trust and drive these technologies forward to become the new open and foundational infrastructure of digitalization in more areas of the economy, further investments and safeguards will be needed. This can for example be achieved through a foundation which would invest in open-source technologies, helps them to build sustainable open business models and at the same time makes sure (through share holder agreements or other means) that the technologies themselves and their governance models remain open.

Today we are already in the midst of the fourth industrial revolution, with entirely new technologies conquering the market at ever shorter intervals. To ensure that we do not just passively let these changes happen to us and end up being the losers of the transformation, economy and industry must realize the importance of open source software and work together with policymakers so that we can successfully set the course for healthy competition and a digitalized ecosystem in which everyone can be innovative.

References

Adamos Website: https://www.adamos.com/en/, last accessed 2023/06/28.

BDI Website: https://bdi.eu/media/publikationen#/publikation/news/open-source-software-strate gisch-nutzen, last accessed 2023/06/28.

Coalition treaty of the German Government, S.2, https://www.bundesregierung.de/breg-de/aktuel les/koalitionsvertrag-2021-1990800, last accessed 2023/06/28.

Dan Robinson / The Register: https://www.theregister.com/2022/02/03/starlingx_60_takes_openst ack_to/, last accessed 2023/06/28.

Jana Zscheischler, Reiner Brunsch, Hans W. Griepentrog, Christine Tölle-Nolting, Sebastian Rogga, Gert Berger, Bernard Lehmann, Tanja Strobel-Unbehaun, Christian Reichel, Steffi Ober, Roland W.Scholz, "Landwirtschaft, Digitalisierung und digitale Daten", http://www.didat.eu/files/pdf/ vernehm/WBK04/WB_Kapitel_VR04_Landwirtschaft.pdf, last accessed 2023/06/28.

Key point paper of the IT-Planungsrat: „Stärkung der Digitalen Souveränität der Öffentlichen Verwaltung", S.1. https://www.it-planungsrat.de/fileadmin/beschluesse/2020/Beschluss2020-19_Entscheidungsniederschrift_Umlaufverfahren_Eckpunktepapier.pdf, last accessed 2023/06/28.

Martin Randelhoff / Zukunft Mobilität: https://www.zukunft-mobilitaet.net/9121/vergangen heit-verkehrsgeschichte/malcolm-mclean-containerschiff-erfinder-container-teu/, last accessed 2023/06/28.

OpenCoDE: https://opencode.de/de, last accessed 2023/06/28.

Open Source Study of the EU Commission: https://digital-strategy.ec.europa.eu/en/library/study-about-impact-open-source-software-and-hardware-technological-independence-competitiven ess-and, last accessed 2023/06/28.

OSB Alliance Website: https://osb-alliance.de/verbands-news/nachhaltige-digitalisierung-ist-nur-mit-open-source-software-moeglich, last accessed 2023/06/28.

Position paper of the BDI: „Europas digitale Souveränität nachhaltig stärken", https://bdi.eu/media/publikationen#/publikation/news/europas-digitale-souveraenitaet-nachhaltig-staerken, last accessed 2023/06/28.

Paper by Acatech: „Digitale Souveränität—Status Quo und Handlungsfelder". https://www.acatech.de/publikation/digitale-souveraenitaet-status-quo-und-handlungsfelder/, last accessed 2023/06/28.

SCS Website: https://scs.community/de/

Starling-X Website: https://www.starlingx.io/, last accessed 2023/06/28.

Sovereign Tech Fund Website: https://sovereigntechfund.de/de/, last accessed 2023/06/28.

Sovereign Workplace for the Public Administration: https://www.cio.bund.de/Webs/CIO/DE/digitale-loesungen/digitale-souveraenitaet/souveraener-arbeitsplatz/souverarner-arbeitsplatz-node.html, last accessed 2023/06/28.

Sylvester Tremmel / Heise Online: „Adobe drohte die Abschaltung seiner Creative Cloud in Venezuela": https://www.heise.de/news/Adobe-schaltet-seine-Creative-Cloud-in-Venezuela-doch-nicht-ab-4563862.html, last accessed 2023/06/28.

The United States Department of Justice: „Cloud Act Resources": https://www.justice.gov/criminal-oia/cloud-act-resources, last accessed 2023/06/28.

Website Supermarkt Inside: https://www.supermarkt-inside.de/schwarz-gruppe-oeffnet-deutsche-cloud-stackit/

Website for the German Digital Strategy: https://digitalstrategie-deutschland.de/static/fcf23bbf9736d543d02b79ccad34b729/Digitalstrategie_Aktualisierung_25.04.2023.pdf, last accessed 2023/06/28.

Website Expertenkreis Automobilwirtschaft: https://expertenkreis-automobilwirtschaft.de/media/pages/home/8653794fe6-1686745132/expertenkreis-transformation-der-automobilwirtschaft_kurzpapier_open-source-software.pdf, last accessed 2023/06/28.

Website Computerwoche: https://www.computerwoche.de/a/microsoft-fehler-legte-tausende-unt ernehmen-lahm,3553820, last accessed 2023/06/28.

Website Automationspraxis (konradin Industrie): https://automationspraxis.industrie.de/news/ada mos-waechst-mit-neuen-partnern-und-angeboten/, last accessed 2023/06/28.

Website Plattform Industrie 4.0 of the BMWK and BMBF: https://www.plattform-i40.de/IP/Red aktion/DE/Anwendungsbeispiele/491-axoom/beitrag-axoom.html, last accessed 2023/06/28.

Whitepaper of the VDMA: „Manufacturing X". https://vdma.org/documents/34570/55087429/VDMA-Whitepaper%20Manufacturing-X.pdf/7e799522-d86d-5004-32de-4388ee891a8c, last accessed 2023/06/28.

ZenDiS Website: https://zendis.de/, last accessed 2023/06/28.

Peter Ganten is CEO of Univention GmbH, an internationally active manufacturer of an open source platform for identity management, application integration and end-user portals, which he founded in 2002. He is also Chairman of the Board of the 'Open Source Business Alliance—Federal Association for Digital Sovereignty'. He is an advocate of a new generation of information technology that can be co-designed and controlled by developers, operators and user organizations in the spirit of digital sovereignty. Peter Ganten is also active in various other bodies as a member, advisor and expert, such as 'APELL—the European umbrella organization of open source industry associations'.

Miriam Seyffarth is Head of Public Affairs and Political Communications at the Open Source Business Alliance. In this position she communicates and represents the interests of the German Open Source Industry that is organized in the trade association. From 2016 to 2020, she worked as a political advisor to Member of Parliament Tabea Rößner on issues of digital infrastructure and digital consumer protection on the federal level. She has been active in NGOs in Berlin on digitization issues since 2010.

Nils Kuhlmann is the Assistant to the CEO of Univention GmbH Peter Ganten. In this position, he supports the CEO in all matters. Previously he has worked as a staff member of three Members of Parliament.

Future Challenges of Data-Driven Problem-Solving in Producing Companies in Context of Digital Sovereignty and Lessons Learned from Electronics Industry

René Wöstmann⑩, Lukas Schulte⑩, Florian Meierhofer, Gunter Beitinger, and Jochen Deuse⑩

Abstract Increasingly complex products and production technology, as well as customer requirements, pose major challenges for the manufacturing industry. Data-based techniques offer new possibilities for problem-solving but also bring new demands on operational IT architectures as well as competences and procedures. The article addresses future challenges of data-based problem-solving and highlights new roles and competence profiles in the context of changing IT and organizational structures by highlighting experiences from electronics manufacturing.

Keywords Problem-solving · Quality management · Machine learning · Competences · Roles

R. Wöstmann (✉) · L. Schulte · J. Deuse
Institute of Production Systems, TU Dortmund University, Leonhard-Euler-Straße 5, 44227 Dortmund, Germany
e-mail: rene.woestmann@tu-dortmund.de

L. Schulte
e-mail: lukas2.schulte@tu-dortmund.de

J. Deuse
e-mail: jochen.deuse@tu-dortmund.de; Jochen.Deuse@uts.edu.au

R. Wöstmann
RIF Engineering & Consulting GmbH, Joseph-Von-Fraunhofer Str. 20, 44227 Dortmund, Germany

L. Schulte · J. Deuse
RIF Institute for Research and Transfer e.V., Joseph-Von-Fraunhofer Str. 20, 44227 Dortmund, Germany

F. Meierhofer · G. Beitinger
Digital Industries, Siemens AG, Werner-Von-Siemens-Straße 50, 92224 Amberg, Germany

J. Deuse
Centre for Advanced Manufacturing, University of Technology Sydney, Sydney, Australia

U. Schmuntzsch et al. (eds.), *New Digital Work II*,
https://doi.org/10.1007/978-3-031-69994-8_3

33

1 Introduction

The ongoing digitalization has caused profound changes in many areas of the economy. In particular, in electronics manufacturing, digitalization has led to an increasing number of customer inquiries being processed through digital channels. Electronic products are typically manufactured using two methods: Surface Mount Technology (SMT) and Through Hole Technology (THT). SMT has established itself as the preferred manufacturing method, encompassing around 90% of consumer products (Brindley 1990).

Despite increasing automation and digitalization, quality assurance in electronics manufacturing remains a central aspect. Customer demands for product quality continue to be high, requiring a large number of inspection points and robust processes. A particular challenge in the production of electronic products is the high number of interactions that must be considered (Schulte et al. 2020a).

To meet these challenges, data-driven approaches and new information technology (IT) architectures are required that are particularly designed to handle large, heterogeneous data sets. These new approaches and architectures must combine frameworks for data integration, analysis, and deployment to enable effective problem-solving.

For manufacturing companies, it is crucial to remain digitally sovereign and to develop the necessary competencies and roles to address these challenges (Hartmann 2021). This paper provides an overview of data-driven problem-solving in the context of digitalization (Chap. "Successful Digital Transformation in Economy and Industry Requires Open Source") and discusses new roles and job profiles in problem-solving projects (in this Chapter). Finally, a good practice example from electronics manufacturing illustrates new possibilities and lessons learned from industrial practice (Chap. "The European Data Act and its Impact on Corporate Digital Sovereignty").

2 Data-Driven Problem-Solving in Terms of Digitalization

The main drivers of modern industrial value creation and productivity were based on work organization paradigms such as scientific management, increasing automation up to mass production, and the elimination of waste, often summarized under the term lean management (Womack et al. 1990). In order to solve problems, e.g. in product quality or the underlying production processes, the first procedures for systematic quality inspection were introduced at the beginning of the twentieth century, and tools for systematic quality improvement based on statistical procedures were introduced in the middle of the twentieth century (Jones 2014). New challenges arise from trends such as product and process flexibilization due to increasing and diversifying customer requirements as well as volatile markets. Products and manufacturing processes are also becoming more specialized and increasingly complex. At the same time, with increasing IT systems, higher computing and storage capacities at lower costs as well as increasing connectivity, the amount of data on products and their production processes as well as their usability is growing steadily (Abramovici et al. 2016). The application of conventional problem-solving methods from the fields of lean management (VDI 2012) and descriptive statistics (e.g. Six Sigma) is increasingly reaching its limits (Antony et al. 2019). Artificial intelligence (AI) and machine learning (ML) promise the identification and representation of complex multivariate correlations between problems and causes, and the first individual implementations in the manufacturing industry confirm the general potential. The demand for activity exists less in the research on new algorithms and procedures of ML, which are available in large numbers, but rather in their application within real production and IT structures as well as production systems being socio-technical systems, with the inclusion of the company organization and employees. In order to realize the potentials of ML, many companies need supportive activities during implementation. The lack of an overview and of transparency concerning the necessary hardware and software suitable for their own required IT structures is the first obstacle to this. Design support for ML architectures is needed to realize the integration of the actuator and sensor level via the control level to the production management level and, based on this, to be able to test, train, apply, monitor, and update ML models. Likewise, new role concepts are necessary in order to be able to develop the competencies of employees in a targeted manner. In addition to technical issues such as secure data exchange or interoperable platforms and services to avoid lock-in effects, employee integration is a central pillar of digital sovereignty. Only if this succeeds, new data-driven solutions can be understood and accepted. In the following, a role concept is presented that addresses both process-related and technical fields of action, but especially emphasizes changing role profiles in data-driven problem-solving. This is followed in Chap. "The European Data Act and its Impact on Corporate Digital Sovereignty" by a discussion based on an implementation example in electronics manufacturing.

3 Demands and Challenges for People and Organization

The following section provides an overview of the changing requirements of data-driven problem-solving in terms of people and organization. Based on an introduction to architectures, processes, and roles in Sects. 3.1 and 3.2 details tasks in different phases of projects.

3.1 Architectures, Processes, and Roles

The changes and challenges of digitalization and Industry 4.0 can be considered fundamentally driven by IT, which brings new possibilities for data collection and processing. For data-driven problem-solving, this subsequently means that new architectures and IT structures need to be implemented for organizations and employees, always based on a mostly heterogeneous brownfield of existing IT systems. These architectures enable a holistic view of required and existing data sources in the company, as well as training of machine learning models and deployment and feedback into processes and applications. In the work of Wöstmann (2023) a reference architecture for ML in the manufacturing industry is presented, which aims to simplify the implementation of the corresponding layers for users (Wöstmann 2023). It consists of the levels Asset, IT Systems, Edge Devices, Network Infrastructure and Databases, as well as Machine Learning and Application.

In parallel, processes and routines for data-driven problem-solving need to be considered. Process models from quality management, such as the Plan-Do-Check-Act (PDCA) or Define-Measure-Analyze-Improve-Control (DMAIC) cycle, have gained wide practical use. The Cross-Industry Standard Process for Data Mining (CRISP-DM) has established itself as the standard process model for structuring ML projects in industry (Schröer et al. 2021). It consists of the phases Business Understanding, Data Understanding, Data Preparation, Modeling, Evaluation, and Deployment. Structural differences in the process models are relatively small, limited to aspects such as the deployment phase in CRISP-DM. However, differences in the underlying tools and corresponding competency requirements for the actors involved in ML-based problem-solving are more significant.

The actors involved can be described heterogeneously. Based on the InDas (Schulte et al. 2020b), DPDA (Wöstmann et al. 2021), and ML2KMU (Reckelkamm and Deuse 2021) projects, relevant actors in ML implementation projects in an industrial context were identified and summarized in an interdisciplinary role concept. A detailed presentation of the tasks and competency requirements for IT, domain experts, data scientists, management, and the orchestrator, also known as the citizen data scientist, is provided in Deuse et al. (2021). Figure 1 gives an overview of roles and actors in data-driven problem-solving.

Fig. 1 Roles and actors in data-driven problem-solving, own illustration based on (Deuse et al. 2021)

In the following section, the main tasks for each actor in the design of architectures and application scenarios for the phases of CRISP-DM are outlined.

3.2 Tasks

Business Understanding

The fundamental activities of Business Understanding address the creation of an understanding of products, processes, facilities, goals, and problem statements. At the asset level, problem statements are analyzed, and objectives and key performance indicators (KPI) are defined. Management is responsible for a strategic perspective on the establishment of a shared vision, the assembly of the project team, and the provision of resources (e.g., budget and time). Domain Experts play a crucial role in the current state analysis, goal derivation, and delimitation of problem statements. In this context, an analysis of IT systems is also carried out, involving the IT department, to address potential intervention possibilities of ML as well as their requirements for the solutions to be developed. At the edge device level, the use of edge devices for decentral data acquisition is also discussed, involving Domain Experts and IT, to determine whether it is an option or requirement to be considered or neglected. The role of the Data Scientist assists in translating general problem statements and goals into potential machine learning problems, in order to obtain an early assessment of the possibilities and limitations of the application of ML. The Citizen Data Scientist acts as a liaison and orchestrator of the actors and performs important functions of project management Fig. 2 provides an overview of typical activities and the respective actors involved in the Business Understanding phase.

Digital sovereignty in this phase is significantly influenced by the choice of competence providers. While the majority of activities can be carried out by internal staff, external cooperation with Data Scientists or orchestrators may be necessary. It is important to develop a strategy for building up competences internally at an early stage in order to avoid becoming dependent on external service providers in the long term and to be able to carry out decisions independently.

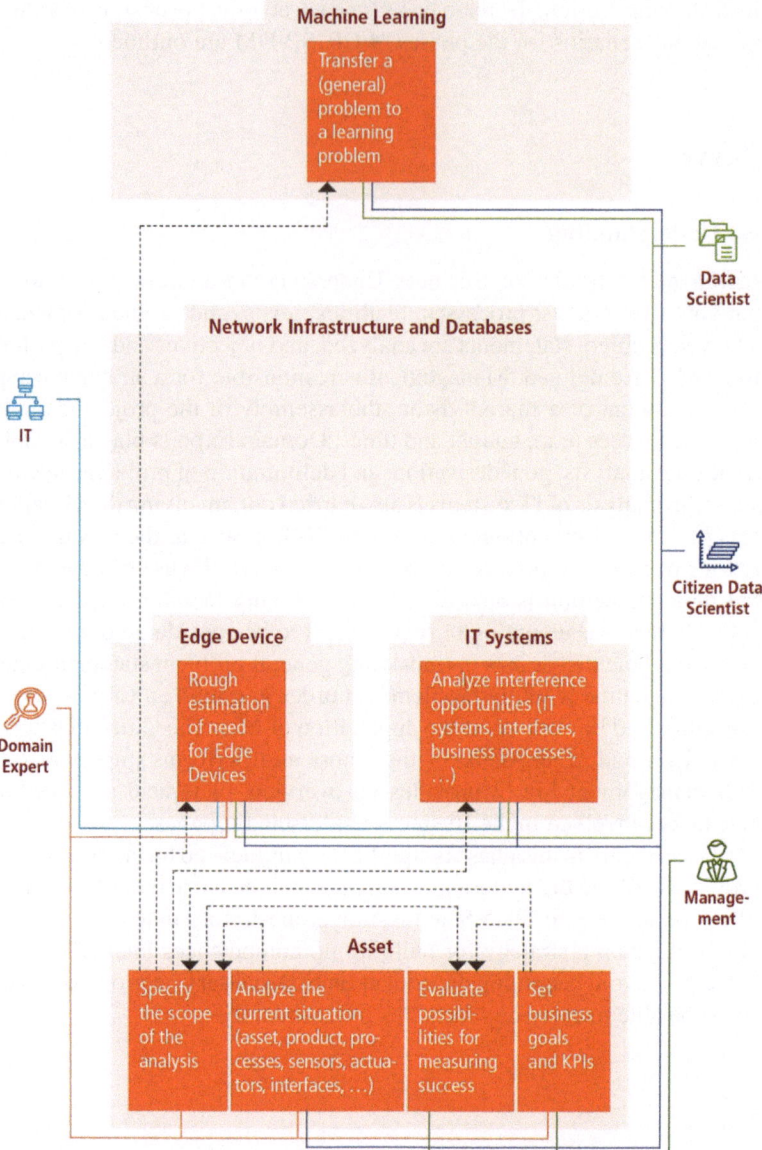

Fig. 2 Activities and tasks of the actors in business understanding, own illustration based on (Wöstmann 2023)

Data Understanding

The tasks of the Data Understanding phase consist of developing an understanding of the relevant data sources for (potential) influencing and target variables of an identified problem to be solved, as well as a more detailed evaluation of data quality and quantity. To do so, necessary information and underlying data on the product, process, and resources must first be specified. In this context, an analysis of existing IT systems and necessary database access is performed. Key actors are Domain Experts who know assets and IT systems from a user perspective, as well as IT departments who help implement data connections and ML environments. The use of edge devices and retrofitting can be a solution strategy if there is too little or no data available for a problem. To perform an initial exploratory assessment of data quality, a software solution for (exploratory) data analysis is also required. Therefore, it is recommended to implement an exploratory test environment, data connection management, and ML project management already in the Data Understanding phase.

Since working with data extracts is recommended, high computing power is not necessarily required. While Management may be involved in selecting platform solutions, the Citizen Data Scientist orchestrates essential activities such as assessing data quality. The Data Scientist assists with selecting and installing environments, as well as with exploratory analyses and visualizations, which are to be interpreted together with Domain Experts. Figure 3 provides an overview of key activities in this phase.

Digital sovereignty is crucial in the Data Understanding phase, as it involves identifying relevant data sources and assessing their quality and quantity. Ensuring data security and protecting data privacy is a key challenge, particularly when working with sensitive data. In addition, choosing the right ML platform, tools, and infrastructure can impact data sovereignty, as it determines the level of control over the data and the insights derived from it. Therefore, it is important to involve Domain Experts and IT departments in the selection process to ensure the chosen solution meets the organization's requirements and regulatory obligations. Additionally, it may be beneficial to consider open-source solutions to maintain greater control over the technology and data. Ultimately, ensuring digital sovereignty throughout the Data Understanding phase is critical for building trust in the organization's data-driven decision-making processes and protecting against potential risks and vulnerabilities.

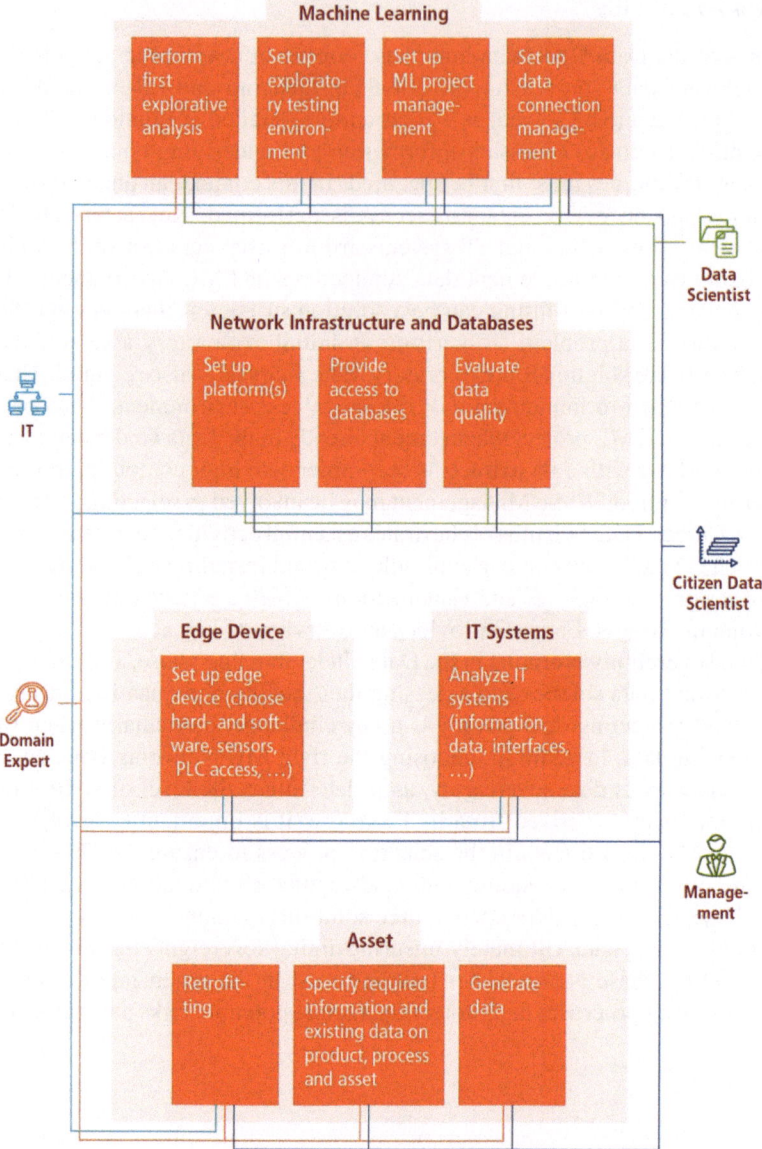

Fig. 3 Activities and tasks of the actors in data understanding, own illustration based on (Wöstmann 2023)

Data Preparation

The Data Preparation phase encompasses all relevant activities required for the preparation of data sets and connections for model training. The establishment of data connections begins at the asset level, where access and interfaces to machine and sensor data have to be prepared. Core tasks include the establishment of access to selected data sources, as identified in the Data Understanding phase, and the execution of data preparation steps at the network infrastructure and ML level. The creation of an exploratory test environment, suitable for building analytical processes, as well as appropriate ML project management for managing users, processes, versions, and environments, are essential prerequisites. Data Scientists and Citizen Data Scientists are responsible for merging heterogeneous data sources, cleaning data (such as removing incomplete values, handling outliers, estimating missing values, etc.), performing feature engineering, constructing and providing a suitable data set, and preprocessing processes for the training environment. Domain Experts are involved in the evaluation of these steps, and iterations, particularly with the Modeling and Evaluation phases, are standard practice and also included in the CRISP-DM process as the quality of pure data preparation cannot be reliably assessed without visualizations or the evaluation of (exploratory) model results and, therefore, closed analytical processes. Figure 4 provides an overview of the activities involved in the implementation of application scenarios in the data preparation phase.

Choosing the right infrastructure and platforms in Data Preparation is crucial for ensuring digital sovereignty. It can impact data security, privacy, and control over the data. By carefully selecting infrastructure and platforms, organizations can ensure that they maintain control over their data and are not reliant on external providers for critical data processing tasks. Initiatives such as Gaia-, Catena- or Manufacturing-X aim to counter lock-in effects on the large platforms of hyperscalers, but no concrete solution patterns are yet available. If operating an in-house computing center is not economically feasible, open data science frameworks can offer transparent solutions.

Modeling

In the Modeling phase, various modeling approaches have to be selected for the learning problems discussed in the Business Understanding phase, and they are transformed into tasks for training models using a large amount of data. Analysis processes have to be built and executed on the exploratory test environment, which was created in the data preparation phase. Therefore, a training environment has to be established to incorporate the testing and training data into the training process as extensively as possible to perform computational power-consuming tasks. The use of an ML project management system is also essential for managing experiments, processes, environments, and users. Data Scientists and Citizen Data Scientists perform the essential modeling tasks, with the support of IT departments in providing environments and data connections, while Domain Experts and Management are not directly involved. In the implementation process, this process is formally assigned to the subsequent evaluation phase, which is iterative in practice. Figure 5 provides an overview of the central activities in the modeling phase and the assignment of the respective actors.

A critical success factor for digital sovereignty in Modeling is the establishment of a training environment that allows for extensive incorporation of testing and training data into the training process. By having control over the training environment and

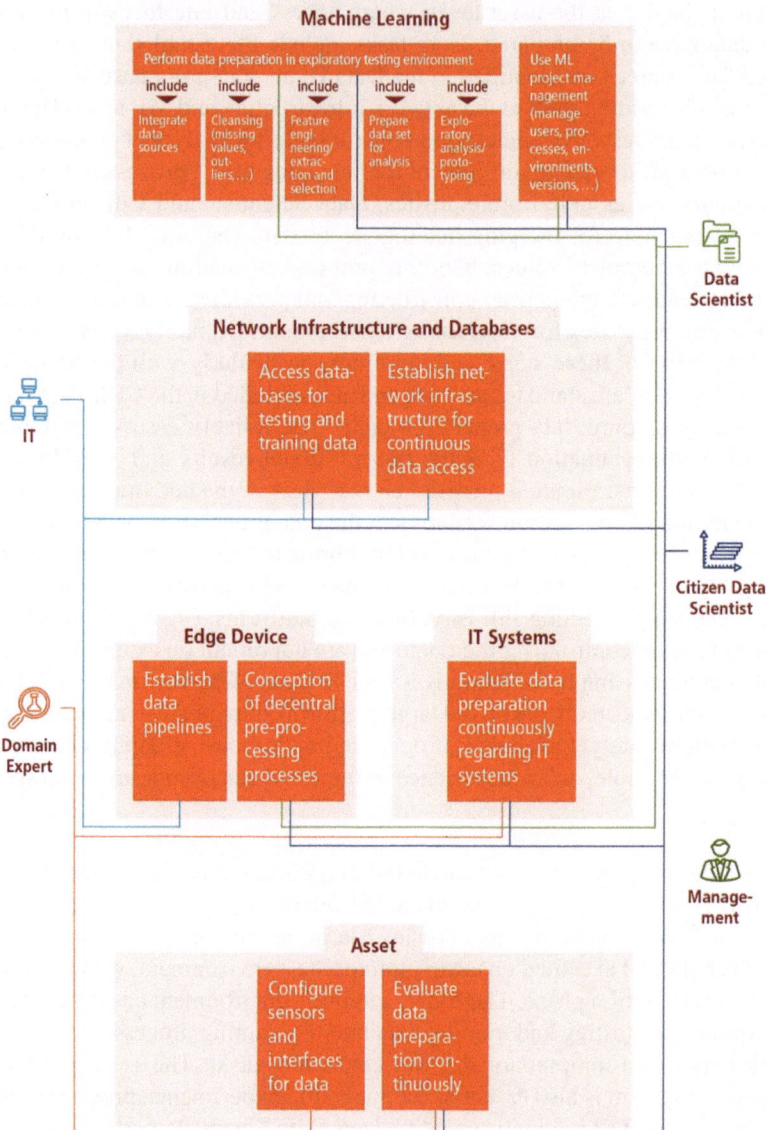

Fig. 4 Activities and tasks of the actors in data preparation, own illustration based on (Wöstmann 2023)

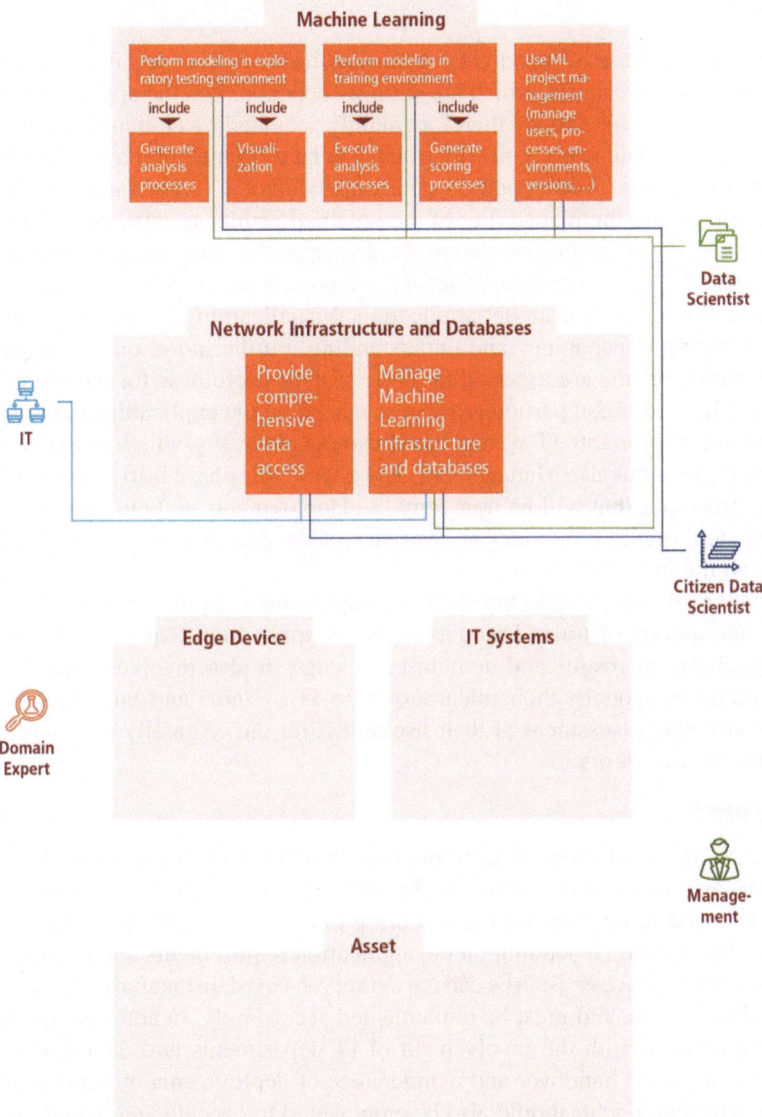

Fig. 5 Activities and tasks of the actors in modeling, own illustration based on (Wöstmann 2023)

data, organizations can maintain ownership and control over their models, enhancing their digital sovereignty.

Evaluation

The Evaluation phase is intended for assessing the trained models in terms of performance and quality criteria, which are strongly dependent on the application requirements and IT systems and are therefore mainly provided by Domain Experts and IT. Both the models and scoring processes have to be evaluated in terms of the application's requirements. Additionally, besides mathematical target variables, the statement contents and the plausibility of the obtained results are also evaluated, taking into account the domain knowledge. In this step, the analysis steps and possible deployment scenarios are also evaluated, which can result in new requirements for both data collection and model application. Visualization also plays an important role in creating transparency and understanding. Furthermore, on a strategic level, the evaluation results are assessed in terms of their usefulness for the originally set goals and KPIs. Here, a particular focus is placed on the applicability of the models and their integration into IT systems and business processes, which involves not only Domain Experts but also Management. The goal of this phase is to select models and scoring processes that will be transferred to Deployments and put into operation in the next step. Figure 6 provides an overview of the core activities of the Evaluation phase and the involved actors.

Achieving digital sovereignty in the evaluation phase requires assessing the performance and quality of trained models based on application requirements, including the plausibility of results and domain knowledge. It also involves considering the applicability of models, their integration into IT systems and business processes, and the strategic assessment of their usefulness for the originally set goals and key performance indicators.

Deployment

The aim of the final phase is to implement individual deployments and a deployment management environment. For this purpose, the selected models and analysis processes need to be converted into scoring processes and the type of deployment needs to be chosen. Depending on the application requirements and technical capabilities, scoring processes can be carried out server-based, in databases, in IT systems, or on edge devices, and must be implemented accordingly. In addition to setting up the environments with the involvement of IT departments and data scientists, the goal is a long-term handover and maintenance of deployments by users (e.g., business departments), who should also be empowered to execute and maintain scoring processes sustainably. This addresses the technical requirement of CI/CD (Continuous Integration/Continuous Delivery), which is prevalent in the context of MLOps (Machine Learning Operations). Depending on the strategy, these tasks can also be performed by machine and plant manufacturers or providers of database services, but the Citizen Data Scientist, as an internal service provider, also plays an important role in individual solution patterns. Therefore, deployment criteria also need to be defined for continuous evaluation. However, in addition to the technical level, the integration of users needs to be addressed in the deployment. Therefore, strategic tasks include the constitution and distribution of knowledge gained within the organization, the

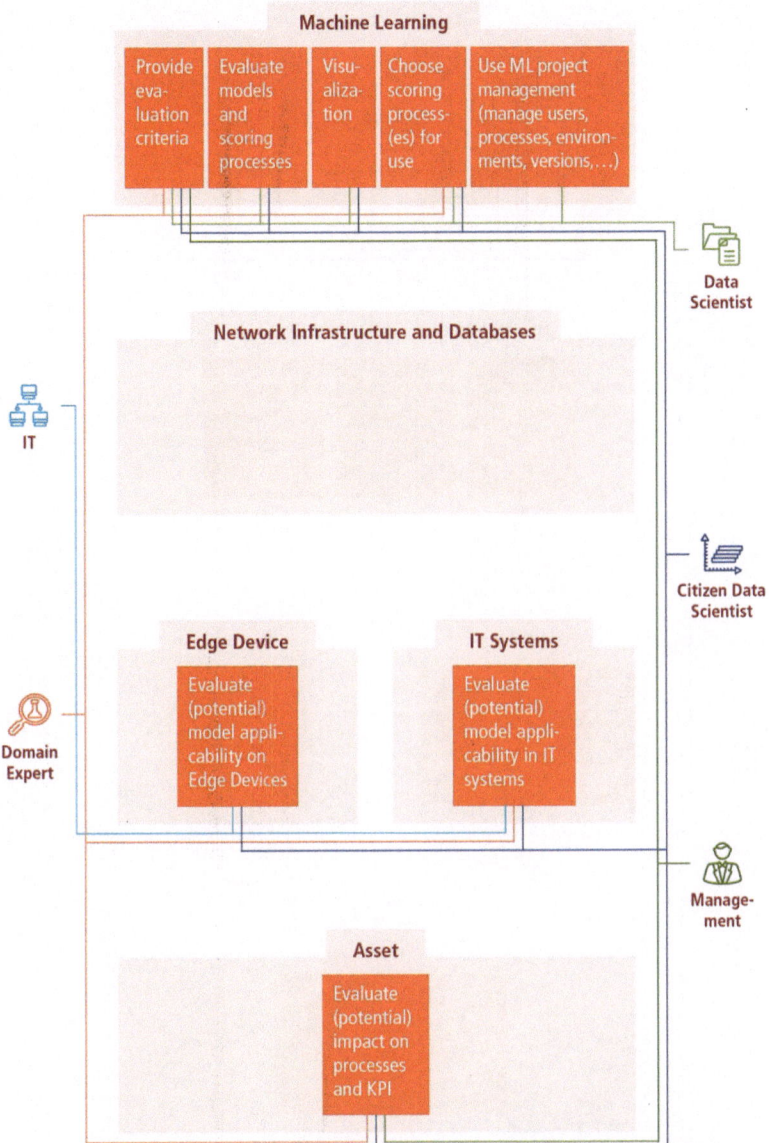

Fig. 6 Activities and tasks of the actors in evaluation, own illustration based on (Wöstmann 2023)

development of a strategy for deriving concrete actions based on model outputs, and their integration into the organizational structure and processes. Figure 7 provides an overview of the corresponding activities of the Deployment phase.

In the Deployment phase, digital sovereignty becomes crucial for organizations. This involves maintaining control over and ownership of scoring processes and

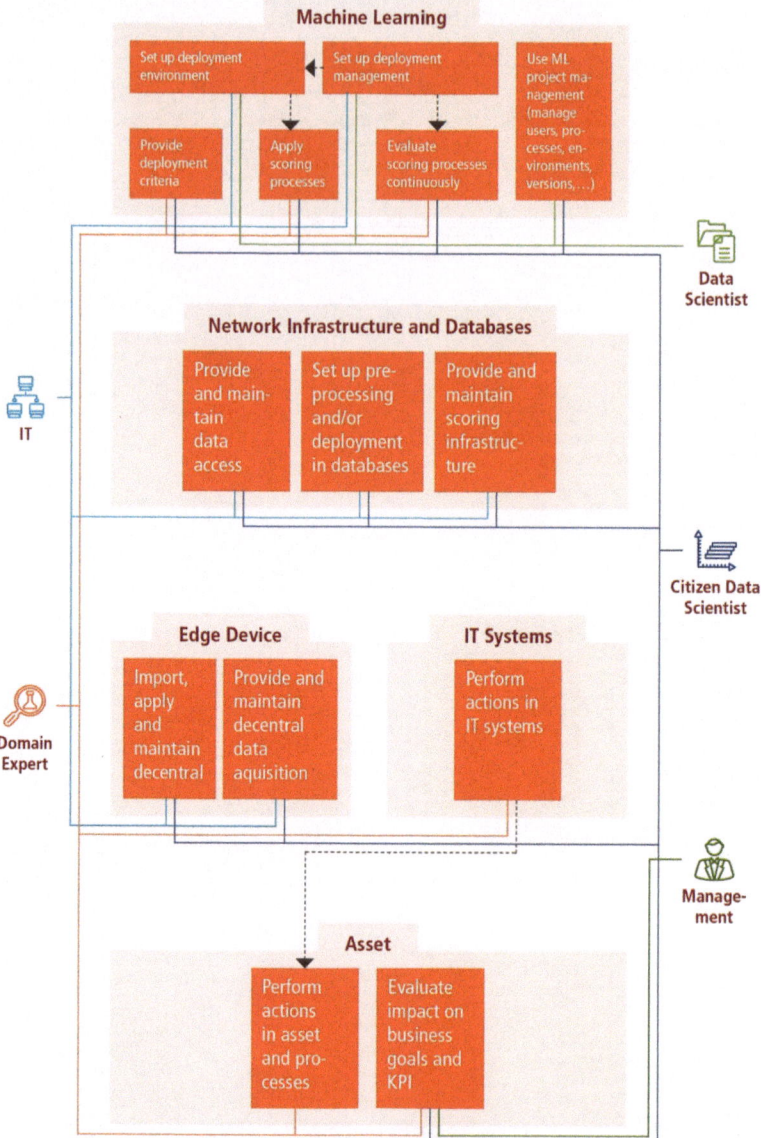

Fig. 7 Activities and tasks of the actors in deployment, own illustration based on (Wöstmann 2023)

choosing deployment types that align with technical capabilities and application requirements. Empowering users, such as Domain Experts in departments, to execute and maintain scoring processes sustains long-term usability. Internal competences also play an important long-term role in the interpretability and trustworthiness of deployments.

4 Implementation Example and Lessons Learned from Electronics Industry

In the following section, the challenges and suggested solutions are illustrated by an example from electronics manufacturing.

Business Understanding

Printed circuit boards (PCBs) serve as circuit carriers for the components to be mounted (e.g. capacitors, coils, and resistors). They provide the necessary structure and connection points to interconnect the components with the integrated circuits on the PCB (Brindley 1990). Surface-mount technology (SMT) is a common manufacturing technology used to securely connect these components to the PCB's integrated circuits through soldering (Risse 2012). To enhance productivity, multiple PCBs are usually combined into multi-boards, to streamline the manufacturing process and optimize production throughput. As presented in Fig. 8, an SMT production line consists of laser, solder paste printing, solder paste inspection (SPI), SMT assembly, soldering, and final automated optical inspection (AOI) processes. Due to long inspection times and high investment costs, an automated X-ray inspection (AXI) is not directly connected to the aforementioned process steps. SPI, AOI, and AXI represent the testing processes used and enable 100% monitoring of the manufactured products to ensure continuous compliance of product characteristics with customer requirements [Schmitt, Schulte, QU4LITY]. Due to the number and positioning of the testing processes, a deviation from quality can be identified before the finalization of the PCBs, following the Jidoka principle. This reduces scrap costs and ensures that products with high quality only are shipped to the customers (Schulte et al. 2020a). Additionally, the implemented testing processes help narrow down the sources of product defects, allowing for the correction of process-related compliance deviations.

In the case considered, the objective set by Management was to increase output at given production capacities. For this purpose, six relevant product variants were selected, each with two pages to be printed, for which the throughput time was to be reduced by an average of 0.5 s. At the same time, the prevailing average quality situation had to be maintained. In order to achieve these project goals, a project team consisting of two Domain Experts, two Managers, one IT employee, one external Data Scientist, and one external Citizen Data Scientist each was put together. Based on the described business objectives, the Domain Experts were able to define the resulting task: The solder paste printer could be identified as a bottleneck, so that the goal was to reduce the process time of the solder paste printer. Since a significant portion of final product quality errors (> 60%) can be attributed to the solder paste printer (Wilson et al. 2008), the objective was to find process parameters that would reduce processing time to the extent that it is no longer the bottleneck, while maintaining the quality rate at the SPI. Within this objective, the task of the external Citizen Data Scientist consisted in particular of analyzing the production processes to identify relevant data sources. In addition, possibilities for data transfer

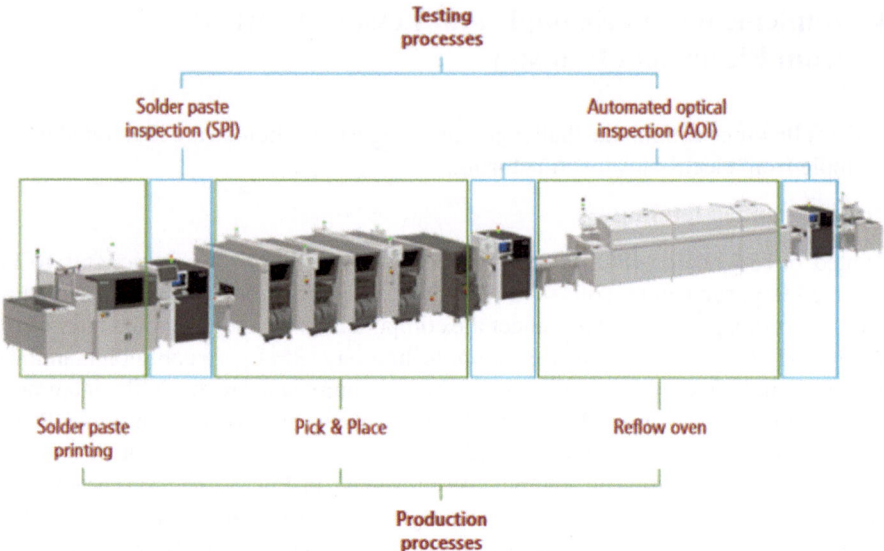

Fig. 8 Overview of an SMT production line, own illustration based on (Schulte et al. 2021)

from the respective plants were coordinated with the IT department. Since raw data could be derived natively from the plants within the framework of the project and the data analysis was to take place de-centrally, the use of edge devices was renounced. In further joint coordination with the external Data Scientist, the prediction of the print quality and process time could be translated as a regression task into the learning problem to be mastered, based on the set process parameters. The greatest challenge in this phase of the project was the transfer of knowledge about the specific process behavior of the solder paste printer to the external Citizen Data Scientist, who had to fulfill a coordinating function. This challenge was overcome by training the Citizen Data Scientist.

Data Understanding

Referencing Fig. 8, various testing systems capture the product quality at different levels of aggregation. For example, the SPI system captures the applied solder volume for each pad, while the AOI and AXI systems only provide information about defective components (e.g., misplaced components) but not about the specific connection point (pad) causing the issue. As a result, different data-driven control loops need to be established to adjust the production processes based on product quality. Within the use case, 241 components need to be positioned at various locations on six individual PCBs combined per printed multi-board. These components lead to 750 different pads whose quality needs to be controlled by the SPI. Since the solder paste printing process is influenced by a wide range of process parameters (e.g. squeegee speed, squeegee pressure) (Barajas et al. 2002), whereby the specific impact of these process parameters on quality characteristics and process time was unknown, it was necessary to investigate the effect of these parameters on the resulting quality and process efficiency. To overcome this challenge, process parameters with possible influences on the objective were recorded in an Ishikawa diagram in a structured manner in workshops led by the Citizen Data Scientist. Based on this diagram, relevant process parameters were selected together with the Domain Experts and test plans were drawn up. These experimental plans were used to provide systematically changed data in the subsequent process phases in order to identify causalities between process behavior and process parameters. In further preparation, the Citizen Data Scientist and Data Scientist specified the data quality of the natively extractable process data in order to assess the correct data collection of the process parameters and measurement characteristics identified in the workshops. Citizen Data Scientist, Data Scientist, and Domain Experts then used explorative analyses to work out the first common process backgrounds. In particular, the evaluation of the technical application limits of the solder paste used was a major uncertainty: Since the solder paste consists of flux and, in particular, solder balls, (Barajas et al. 2002) it had to be investigated whether the size of the solder balls would allow the apertures to be filled. The biggest challenge was to create a common understanding between Data Scientists, Citizen Data Scientists, and Domain Experts. For this purpose, graphical representations were used, which were combined with the usual vocabulary of the Domain Experts. In this context, scatterplots in particular enabled the holistic evaluation of the aperture widths and lengths of each pad for the Domain Expert: In the given application, the suitability of the solder paste for the apertures used had to be tested before the actual process improvement. Among other things, solder paste consists of metal powder, whereby the individual metal particles are formed into spheres. In the domain context, there is a minimum number of metal spheres that must be present in the aperture width and length. The use of scatterplots (cf. Fig. 9) enabled the domain expert to validate the suitability by means of a holistic representation of the metal spheres contained in the apertures.

Based on the expert knowledge of the domain expert, the technical applicability of the solder paste could then be confirmed. The Citizen Data Scientist's primary task was thus the transfer of the data gained knowledge to the Domain Expert.

Data Preparation and Modeling

As described in this chapter, the primary goal of Data Preparation is to prepare the data in order to enable the subsequent Modeling. For this purpose, the Data Scientist, under the guidance of the Citizen Data Scientist, prepared the raw data received, especially with regard to missing or faulty measured values. In addition, he developed a local database schema for the different data sources (process parameters of the solder paste printer and SPI measurement results) in order to make them usable for the subsequent modeling.

Together with the domain experts, the Citizen Data Scientist carried out various test plans to make the necessary data available for the data-driven process optimization. The Domain Experts in particular were able to narrow down an adequate test area that would not lead to premature wear of the printer in the long term, but still offered sufficient potential for improvement.

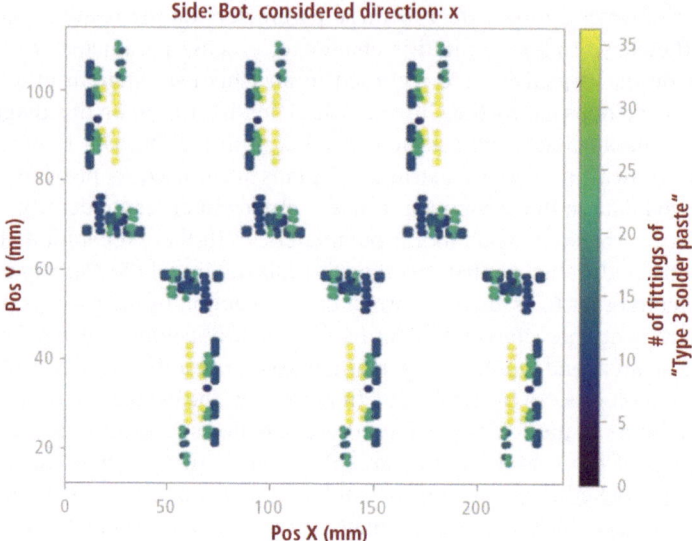

Fig. 9 Illustration of the filling of the widths and lengths of the apertures when using a type 3 solder paste (own illustration)

Based on the data collected in this way, the Data Scientist carried out exploratory analyses in order to record relevant characteristics for the learning task and the robust model training. In particular, the traversing speed was identified as the process parameter that has the most significant effect on process time and print quality. To verify the exploratively identified features, linear models were trained and evaluated with regard to the approximation quality between Data Scientists and Citizen Data Scientists.

Evaluation

The aim of this phase was, according to the previous presentation, the selection of trained models in preparation for the subsequent deployment. Based on Fig. 10 the Data Scientist determined the computational deviation of the linear model from reality for each pad and prepared it descriptively for the Domain Expert.

This representation enabled the Domain Expert to evaluate the causality of certain deviations. For example, the Domain Experts identified that fixed supports (so-called milling toolings) instead of pressurized supports enable less deflection of the PCBs and thus more uniform printing. In addition, the trained linear model, which shows the influence of the speed in particular on the volume filling and consequently the print quality, could be validated by the experience of the Domain Experts. Together with the Management, the potential influence of the process adaptation on the project goals set out at the beginning was then evaluated.

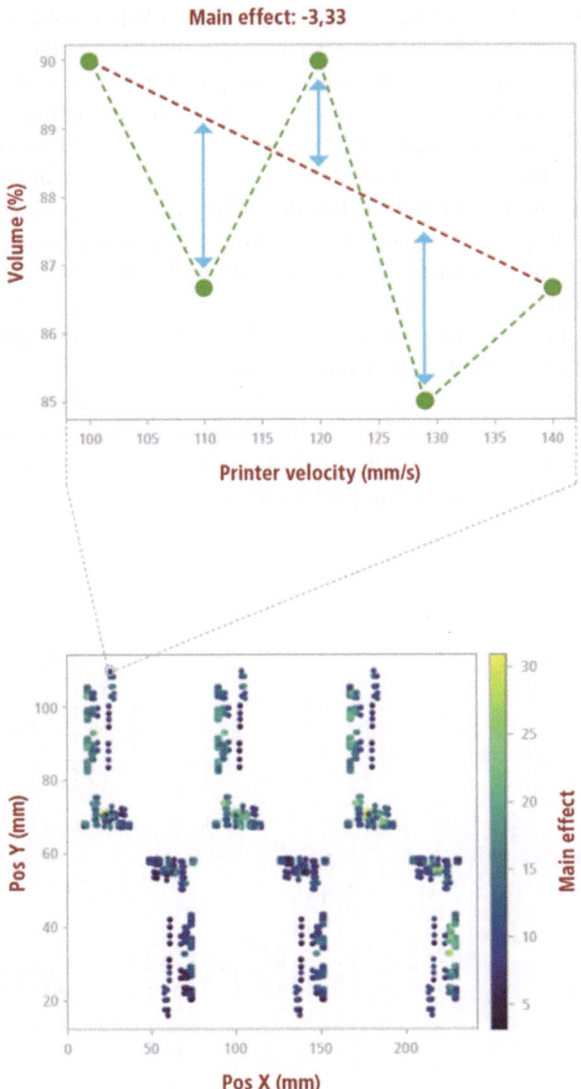

Fig. 10 Evaluation of the model performance per pad based on deviation: red shows the linear forecast and blue shows the real deviations from forecast. Cumulative deviations are recognized as main effect (own illustration)

Deployment

The aim of the project phase was to transfer the project results to the real SMT production line in order to achieve the identified potential business goals in reality. In the given use case, an instruction for action for the Domain Experts could be derived from the trained model, in particular by the Citizen Data Scientist. The integration into the real environment thus did not consist of the implementation of a trained model but of the derivation of the relevant process parameter settings and the adaptation of the parameters by the Domain Experts on the SMT production line under consideration. After transferring the derived process parameters to the production line, the project results were evaluated: At the end of the project, the average process time of the printer per product variant was reduced in the single-digit seconds range and the KPIs defined by the Management in the Business Understanding phase were exceeded. Figure 11 also shows that despite the reduced process time of the solder paste printer and thus reduced throughput time of the PCBs, the volume filling is within the predefined product specifications.

In addition, the structured collection and processing of the project results by the Citizen Data Scientist for the Domain Experts facilitated the training of further employees on the shop floor by the Domain Experts. This enabled the transfer of the data-based causal relationships beyond the analyzed SMT production line to other SMT production lines. Based on the project results, the Management was then able to conclusively evaluate the real impact on the company's goals.

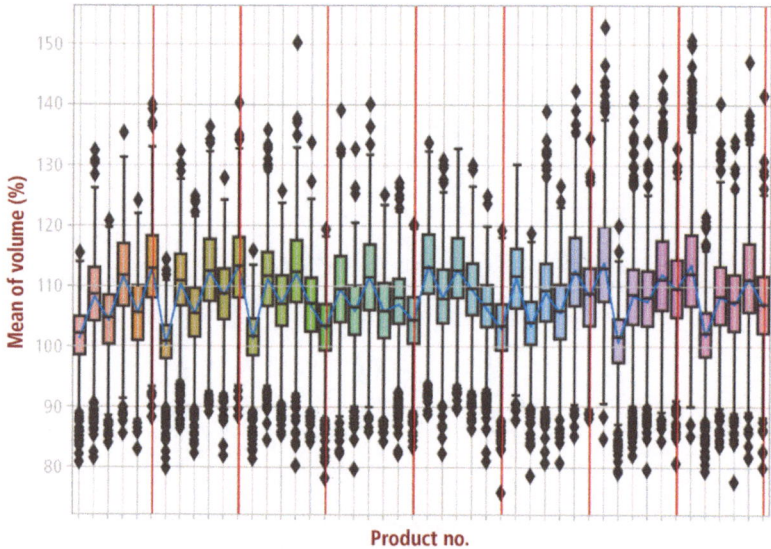

Fig. 11 Ongoing visualization of volume filling per product no. by boxplots. Red lines describe cleaning times while the blue line shows the mean value curve of the average volume filling of all pads per product no. (own illustration)

5 Conclusion and Outlook

The article highlights new roles and competence profiles in the context of data-driven problem-solving, which arise in the course of newly emerging requirements from applying ML as well as changing IT and organizational structures, illustrated through the example of electronics manufacturing.

In summary, it can be seen that descriptive methods are a good way to make interim project results available to the various stakeholders, while at the same time enabling joint solution finding. However, the role of the Citizen Data Scientist as an orchestrator corresponds to a key role: The Citizen Data Scientist ensures that the project is processed in accordance with the management's objectives and thus supports the management. As an orchestrator, the role of the Citizen Data Scientist requires not only a high level of process knowledge and in-depth knowledge of the methods and ML tools used but also a high degree of soft skills. In particular, a lack of process knowledge on the part of the Citizen Data Scientist can jeopardize the achievement of objectives, so this shortcoming must be addressed first. In the context of digital sovereignty, this shows that the Citizen Data Scientist, as the central project coordinator, must also have in-depth knowledge of internal and possibly confidential company processes. It follows that corresponding key positions must be trained and retained in the respective companies in order to be able to act with sovereignty in data-based projects in the long term.

During the implementation of the project, however, it became apparent that the constant availability and secure transfer of real data, especially for external data analysts, represent a particular challenge for the further integration of data-based process improvement projects. However, this service cannot be provided by the Citizen Data Scientist. Against this background, an expansion of the role model mentioned at the beginning can be discussed in the context of digital sovereignty with regard to an integration role of ML and IoT platforms within the company for the coordination of relevant data structures and integration options in such projects.

Acknowledgements The project "QUALITY" has received funding from the European Union's Horizon 2020 research and innovation programme under grant agreement No. 825030.

The "ML2KMU" project is part of the initiative "Digital sovereignty in the economy—using the example of technical-organisational systems for machine tool manufacturing" of the Institute for Innovation and Technology (iit) within VDI/VDE Innovation + Technology GmbH.

The project "InDaS" was funded by the Federal Ministry of Education and Research (BMBF), programme: "ICT 2020—Research for Innovation" (funding code 01S17063A), supervised by the DLR Projektträger.

The project "DaPro" was funded by the Federal Ministry of Economics and Climate Protection (BMWK) in the programme "Smart Data Economy" (funding code 01MT19004D), supervised by the DLR Projektträger.

References

Abramovici, M., Herzog, O. (Ed.): Engineering im Umfeld von Industrie 4.0. Einschätzungen und Handlungsbedarfe (acatech STUDIE). Herbert Utz Verlag, München (2016)

Antony, J., Sony, M., Dempsey, M., Brennan, A., Farrington, T., Cudney, E. A.: An evaluation into the limitations and emerging trends of Six Sigma: an empirical study. In: TQM **31** (2), 205–221 (2019)

Barajas, L.G., Kamen, E.W., Goldstein, A., Egerstedt, M., Small, B.: A closed-loop control algorithm for stencil printing. In *Proc. Surf. Mount Technol. Assoc. Int. Conf. (SMTA)* (pp. 51–58) (2002)

Brindley, K.: Newnes electronics assembly handbook, 1. Auflage. Burlington (1990)

Deuse, J, Wöstmann, R., Schulte, L., Panusch, T., Kimberger, J.: Transdisciplinary competence development for role models in data-driven value creation. The Citizen Data Scientist in the Centre of Industrial Data Science Teams. In: Sihn, W., Schlund, S. (Ed.): Competence development and learning assistance systems for the data-driven future, pp. 36–56. GITO Verlag, Berlin (2021)

Hartmann, E.A.: Digitale Souveränität in der Wirtschaft—Gegenstandsbereiche, Konzepte und Merkmale. In: Hartmann, E. A. (Hg.): Digitalisierung souverän gestalten. Innovative Impulse im Maschinenbau, pp. 31–43. Springer Vieweg, Berlin (2021)

Jones, E.C.: Quality Management for Organizations Using Lean Six Sigma Techniques. CRC Press, Taylor & Francis, Boca Raton (2014)

Reckelkamm, T., Deuse, J.: Kompetenzentwicklung für Maschinelles Lernen zur Konstituierung der digitalen Souveränität. In: Hartmann, E. A. (Hg.): Digitalisierung souverän gestalten. Innovative Impulse im Maschinenbau, pp. 31–43. Springer Vieweg, Berlin (2021)

Risse, A.: Fertigungsverfahren der Mechatronik, Feinwerk- und Präzisionsgerätetechnik, 1. Auflage, Wiesbaden (2012)

Schröoer, C., Kruse, F., Gómez, J.M.: A systematic literature review on applying CRISP-DM process model. In: Procedia Computer Science 181. pp. 526–534. Elsevir (2021)

Schulte, L., Schmitt, J., Meierhofer, F., Deuse, J.: Optimizing inspection process severity by machine learning under label uncertainty. In: Nunes, I. (eds) Advances in Human Factors and Systems Interaction. AHFE 2020. Advances in Intelligent Systems and Computing, vol 1207. Springer, Cham (2020)

Schulte, L., Killich, N., Deuse, J., Meierhofer, F.: Autonome Qualitätsprüfung 4.0, Reduzierung von Pseudofehlern in der Leiterplattenfertigung durch die Integration von Maschinellem Lernen. In: Industrie 4.0 Management **37** (6), 52–56 (2021)

Schulte, L., Schmitt, J., Stankiewicz, L., Deuse, J.: Industrial data science. Interdisciplinary Competence for Machine Learning in Industrial Production. In: Schüppstuhl, T., Tracht, K., Henrich, D. (Ed.): Annals of Scientific Society for Assembly, Handling and Industrial Robotics, pp. 161–171. Springer Vieweg, Berlin (2020)

VDI 2870 Blatt 1 - 2012–07 Ganzheitliche Produktionssysteme—Grundlagen, Einführung und Bewertung. Beuth Verlag, Berlin (2012)

Womack, J.P., Jones, D.T., Roos, D.: The machine that changed the world. Free Press, New York, London, Toronto, Sydney (1990)

Wöstmann, R.: Konzeption einer Referenzarchitektur für maschinelles Lernen in der Prozessindustrie und exemplarische Umsetzung in der Brauindustrie. Shaker Verlag, Düren (2023)

Wöstmann, R., Brünnhäußer, J., Büttner, G., Beckmann, F., Deuse, J., Stark, R.: Data Preparation for data analytics and artificial intelligence. In: ProductDataJournal **28** (1), 16–20 (2021)

Wilson, A.R., West, A.A., Velandia, D.S., Conway, P.P., Whalley, D.C., Quintero, L.H., Monfared, R.: Characterization of printed solder paste excess and bridge related defects. In *2008 2nd Electronics System-Integration Technology Conference* (pp. 1305–1310). IEEE (2008)

Dr.-Ing. René Wöstmann graduated in Industrial Engineering at TU Dortmund University, Germany. For more than nine years, he has worked in different roles as research assistant, consultant and head of the Digital Manufacturing research group at IPS as well as at the cooperation partner RIF. His main focus is on the introduction of Industry 4.0 and machine learning in numerous industrial digitalisation projects; his PhD work focused on machine learning architectures and required competences. Additionally, he works in international standardisation in terms of Digital Twins and Industry 4.0 standards. René Wöstmann is chief engineer and deputy head of the IPS.

Lukas Schulte, M.Sc. for more than five years has worked in different roles as research assistant, consultant and head of the Digital Manufacturing research group at IPS as well as at the cooperation partner RIF e. V. He is doing his doctorate in the area of objectifying industrial testing procedures to reduce quality costs. His work focuses on the design of zero-defect production through advanced data analytics.

Florian Meierhofer, B.Eng., has been working at Siemens since 2006 and has held various positions as Technologist, production engineer, IoT Expert, Head of Production Engineering PCBA technology and Production Manager. In his role as Head of Production Engineering PCBA, he was responsible for strategy and investments for PCBA production and the transformation of product-oriented PCBA production into a flexible technology-oriented PCBA cluster. In his current position as Production Manager, he is responsible for the PCBA Finalization production area in the Digital Lighthouse Factory EWA Amberg and is responsible for operational and production planning.

Dr.-Ing. Dipl.-Wirt.-Ing. Gunter Beitinger studied Industrial Engineering and Business Administration. After five years as a research collaborator at the University of Erlangen, he started his career with Siemens in 1999 and since then has held various responsibilities in countries like Mexico and the USA in different operational companies like SIEMENS VDO Automotive, Siemens Healthineers, Process and Drives (PD) and now Digital Industry (DI). He worked as a Management Consultant, as Director of Industrial Engineering and Plant Manager for a Motor Plant in Mexico. His current role is Senior-Vice-President Manufacturing & Head of Manufacturing Excellence at Digital Industries.

Univ.-Prof. Dr.-Ing. Jochen Deuse graduated in Mechanical Engineering (Dipl.-Ing.) at the University of Dortmund, Germany. He received his Doctor of Engineering (Dr.-Ing.) from RWTH Aachen University, Germany, while working as a research fellow at the Laboratory for Machine Tools and Production Engineering (WZL). During his industry career, Jochen Deuse held senior management positions in the Bosch Group in Germany and Australia before joining TU Dortmund University, Germany, as Professor and Head of the Institute of Production Systems (IPS). Jochen Deuse holds a dual appointment with TU Dortmund and University of Technology Sydney where he is Professor for Advanced Manufacturing/Industry 4.0 and directs the Centre for Advanced Manufacturing.

The European Data Act and Its Impact on Corporate Digital Sovereignty

Sebastian Straub[ID]

Abstract In the digital age, data is the foundation of the digital economy. Its availability is not only convenient, it is essential as a driver of innovation, business improvement, and economic growth. Recognising its transformative power, the European Union has initiated a series of legislative measures to improve data accessibility and ensure seamless data sharing across sectors to realise the overarching goal of a single market for data. At the very centre of this strategy is the European Data Act. The aim of this important piece of legislation is clear: the improvement of data availability across the board. By guaranteeing that data generated by connected products is easily accessible, the Act aims to create an environment in which consumers, businesses, and public authorities can benefit from a richer and more diverse data pool. However, this vision is not without its complexities. Many companies, especially those involved in the development of new technologies, are worried. They fear that the push for greater data accessibility could endanger their trade secrets, the basis of their competitive advantage. The European Union has recognised these concerns and is seeking to guarantee fairness between actors along the value chain. As a result, specific rules have been implemented to safeguard the interests of businesses. The key question is whether these provisions are sufficient to address manufacturers' concerns about preserving their digital sovereignty while not losing sight of the overarching goals of a functioning data economy. This article explains the specifics of the Data Act, identifies the relevant legal provisions, addresses the challenges it poses for businesses and shows the opportunities for stakeholders along the data value chain.

Keywords Data act · Data access · Digital sovereignty · Trade secrets · Data economy · Legal framework

S. Straub (✉)
Institute for Innovation and Technology (iit), Steinplatz 1, 10623 Berlin, Germany
e-mail: straub@iit-berlin.de

© The Author(s) 2025
U. Schmuntzsch et al. (eds.), *New Digital Work II*,
https://doi.org/10.1007/978-3-031-69994-8_4

1   Introduction

In the rapidly evolving digital landscape, data has become a critical element for businesses, driving innovation, growth, and competitive advantage. In this ongoing transformation from 'industrial capitalism' towards 'informational capitalism' (Schweitzer and Metzger 2023, p. 337), the question of who truly 'owns' and controls this data has come to the forefront of business and regulatory discussions. As part of its data strategy, the EU has launched a series of measures to regulate and streamline the data economy, ensuring that it remains fair, transparent and beneficial for all stakeholders (European Commission 2020, p. 13). So far, data regulation has been primarily related to personal data within the framework of the General Data Protection Regulation (GDPR). In order to achieve the goal of a single market, the EU has introduced further legislation that explicitly includes access to and use of non-personal data. Part of these measures is the adoption of the Data Governance Act (European Union 2022), which provides rules for the provision of data intermediary services. In addition to the DGA, the Data Act establishes a cross-sectoral governance framework for data sharing, addressing the following questions:

- Who, apart from the original manufacturer of the product, has the right to access the data?
- Under what conditions can this access be granted?
- On what basis are these rights and conditions determined?

These issues are not just theoretical but have practical implications for businesses, especially in the context of product manufacturing and the vast amounts of data these products generate. Making the question of data access a cornerstone of this legislation, the EU is not only shaping the way how data is shared in the future. It also answers the question of who data should be allocated to and who should benefit from the economic added value. This legislation represents a paradigm shift in how data is perceived and accessed, because it breaks the old paradigm that only those who have the technical sovereignty over data are allowed to determine how it is used (Paal and Fenik 2023, p. 253).[1] In light of these regulatory developments, the role of digital sovereignty for companies is becoming increasingly important. Companies are moving beyond their traditional role as passive data providers to a more proactive role in the data market. This requires a strategic shift that enables companies to use their data sovereignty as a competitive advantage.

---

[1] At the same time, the long-discussed approach of data ownership was finally rejected (Hennemann and Steinrötter 2022, p. 1481).

2 The Data Act: An Overview

The Data Act establishes a comprehensive governance framework for data sharing, addressing the challenges of accessing and using the vast amounts of data generated by machines and products. The underlying premise is that a significant amount of this data remains untapped due to limited access and rights (Recital 2 DA). By introducing the Data Act, the EU seeks to break down these barriers by strengthening the digital sovereignty of companies and consumers who own and use connected products. At the same time, manufacturers of these very products may experience a loss of digital sovereignty by being forced to share data with other entities. Recognising this, the EU is seeking to balance the interests of manufacturers, users, and third parties to ensure that the vast amounts of data generated is shared under fair, transparent, and non-discriminatory conditions.

3 Ensuring Data Accessibility

In order to comprehend the impact of the Data Act,[2] it is first necessary to consider the scope in Article 1(2). The Regulation primarily applies to **manufacturers** who place products on the EU market. These products, often embedded with sophisticated technology, generate and transmit data during their operation (e.g. connected machines, vehicles, household appliances or electronic consumer goods). In addition to manufacturers, the Data Act will also apply to providers of related services that are essential to the functioning of the product. The Act guarantees that users (individuals or companies) can access the data generated by the product or associated service and utilize it, including sharing it with third parties of their choice. Such third parties typically operate in commercial or professional capacities, ranging from service providers to aftermarket entities that provide services such as product maintenance (Straub 2022, p. 27). To guarantee users' right to data accessibility, manufacturers must design their products in such a way that the generated data is easily and securely accessible. This obligation under Article 3(1) DA states the principle of '**accessibility by design**' (Steinrötter 2023, p. 216) and highlights the importance of product design in the data economy. In order to enable users to exercise their right to access data, producers are obliged to provide a certain level of transparency. Before concluding a contract for the purchase, rent or lease of a product or a related service, a broad set of information must be provided to the user (Article 3(2) DA). This includes information about the nature and scope of the data, how the user can access the data or how he may request that the data be shared with a third party.

Example of 'accessibility by design' under Article 3(1) DA: Precision Machinery Ltd. is a hypothetical company specialising in the manufacture of industrial machinery. These machines are integrated with sensors and advanced software

---

[2] Hereinafter referred to as DA as amended on 15.11.2023 available at https://data.consilium.eur opa.eu/doc/document/PE-49-2023-INIT/en/pdf.

systems. This integration serves several purposes: measuring operational effi-
ciency, predicting maintenance needs, and understanding user interactions with
the machines. Gathering this data is important as it provides insight into machine
performance and potential areas for improvement. The machines are controlled via a
cloud-based analytics platform. This platform, which is integrated with the machine,
analyses the data collected by the machine's sensors to provide insights into oper-
ational efficiency and predictive maintenance. In accordance with Art. 3(1) DA,
companies such as Precision Machinery Ltd. are potential subjects of data access
requests. The scope of the Data Act extends beyond manufacturers to include related
services. Services such as the cloud-based analytics platforms are essential to the
operation of the machines and enhance their functionality, and therefore fall within
the scope of the regulations. Precision Machinery Ltd is thus required to adopt a
design approach that focuses on data accessibility. Accessibility by design, as set out
in Article 3(1) DA, requires Precision Machinery Ltd. and its cloud-based analytics
platform to ensure that the data generated by the machines or available on the
platform is easily and securely accessible (Fig. 1).

4 Empowering Users with Data Access Rights

The right of the user to access their data according to Article 4 DA is the crucial
cornerstone of the Data Act. It applies to all data generated by the use of the product.
This includes data intentionally generated by the user and data generated as a byproduct of user actions. It also includes data generated without any user interaction,
e.g. when the product is in standby mode or switched off (Recital 14a DA). It also
covers environmental data generated as a result of use (e.g. room temperature data).
In contrast, derived data, which is information generated by analysis of usage data
by the data owner, cannot be accessed (Wiebe 2023, p. 230).

The right to data access is directed towards the **data holder**, typically the manufacturer of the machine or the device, who, by virtue of their control over the product's
technical design, holds the key to the data. According to Article 4(1) DA the data
holder shall make the generated data available to the user without undue delay, free
of charge, easily, securely, in a structured, commonly used and machine-readable
format, continuously and in real-time. This shall be done on the basis of a simple
request by electronic means. To protect the users' right to access data any agreement
between the data holder and the user that narrows the access rights is inadmissible.

Example for the user's right to access under Article 3(1) DA: Consider Airline A,
a major airline operating a fleet of commercial aircraft. These aircraft are equipped
with numerous sensors and systems that collect a wide range of data—from engine
performance and fuel consumption to flight path efficiency and passenger comfort
indicators. Despite the amount of data generated, Airline A has historically faced
significant challenges in accessing this data. Aircraft manufacturers maintained tight
control over the data, providing limited insight or charging high fees for more detailed
access. This limitation was a problem for Airline A as they tried to optimise flight

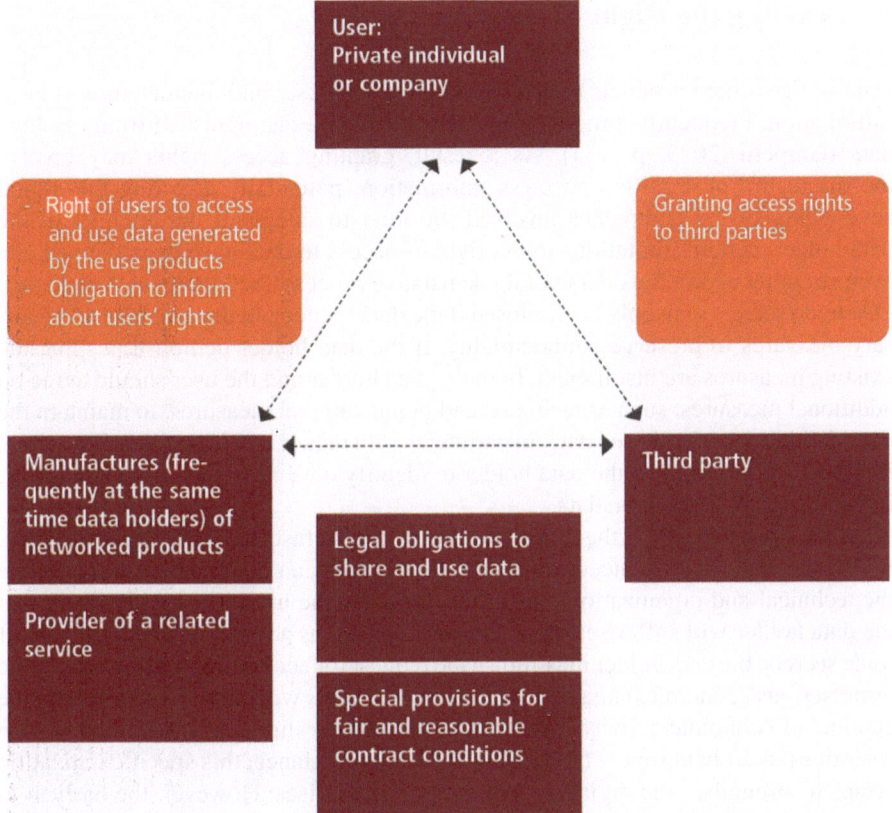

Fig. 1 Overview of the relationships and demands of the actors (own illustration)

operations, enhance passenger experience and improve maintenance schedules. They believed that a more detailed analysis of the data could lead to significant improvements. The introduction of the Data Act could change this dynamic. Aircraft manufacturers are now required to provide Airline A with easy, timely, and free access to all data generated by their aircraft. This includes not only basic flight data but also the detailed operational and environmental data collected by sensors. With Article 4 DA, Airline A can now integrate this data into their operational systems, analyse it in real-time and make data-driven decisions. The Data Act could therefore improve the digital sovereignty of companies that operate with connected vehicles. At the same time, the momentum is shifting away from manufacturers, who have traditionally had de facto control over the data.

5 Limiting the Right of Data Access

Data of networked products typically comprise both user and manufacturer-related information. Frequently, product usage data cannot be distinguished from machine data (Grapetin 2023, p. 174). As a result, granting access rights may involve the disclosure of sensitive business information, potentially affecting the digital sovereignty of the companies involved. In order to safeguard the interests of the data holder, certain limitations to the right of access to data have been defined and even strengthened in the course of the legislative process. According to Article 4(6) DA, trade secrets may only be disclosed if the data holder and the user take all necessary measures to preserve confidentiality. If the data holder demonstrates that the existing measures are insufficient, both the data holder and the user should agree on additional measures, such as technical and organizational measures, to maintain the confidentiality of the shared data, in particular with regard to third parties. Therefore, it is the responsibility of the data holder to identify data classified as a trade secret, including the relevant metadata.

In the final version of the Data Act, the right to refuse access is even extended. In exceptional circumstances, where the data holder can demonstrate that, despite the technical and organizational measures taken by the user, it is highly likely that the data holder will suffer serious economic damage as a result of the disclosure of trade secrets, the data holder may refuse the request for access (Article 4(8) DA). The term 'serious economic damage' refers to severe events with an adverse effect on the conduct of economic activity, which may in particular threaten its viability or pose a serious risk of bankruptcy (Recital 31 DA). At first glance, this specific regulation seems to strengthen the digital sovereignty of enterprises. However, the barriers to invoking this exemption should not be underestimated. The data holder is obliged to give detailed reasons for its refusal on a case-by-case basis, in writing and without undue delay. The justification provided needs to be based on objective elements demonstrating the exceptional circumstances and the risk of serious harm. In addition, the exceptional circumstances require the identification of specific risks of serious damage expected from a particular disclosure and the reasons why the measures taken to protect the requested data are not sufficient. Therefore, factors such as the enforceability of trade secrets protection in the third country where the user chooses to process the data, the nature and level of confidentiality of the data requested, the uniqueness and novelty of the product, the concrete factors why the damage would be very likely or very serious, should be taken into account (Recital 31 DA). In conclusion, the denial of the users' access right will in practice be very challenging for companies. A detailed argument suggesting that a specific case endangers the economic survival of the data holder or brings them close to bankruptcy will often not be feasible, especially when the actual risk lies in the aggregation capabilities of the users and data recipients (Grapetin 2023, p. 177).

In addition to that, Article 4(10) DA contains a further restriction regarding the use of the provided data. According to this, the user may not use the data for the development of a product that competes with the product from which the data originated, nor may he pass the data on to another third party for this purpose. Whether a product competes with the product from which the data originates depends on whether the two products compete in the same product market (Recital 31 DA). The prohibition is intended to prevent free-riding on the data owner's reputation and innovation efforts, which would be contrary to the principles of EU competition law (Lorenzen 2022, p. 262).

Exemplary case in which data access is refused: Consider this fictional example of Manufacturer R, an industrial robotics company. They produce robots equipped with state-of-the-art sensors and AI capabilities that are used in various manufacturing processes. These robots collect vast amounts of data, including operational efficiencies, machine learning insights, and unique manufacturing techniques. Car manufacturer C uses these robots on its production lines. They are interested in accessing the data generated by the robots in order to optimise their manufacturing processes. However, R is concerned about sharing this data. They fear that revealing detailed operational data and machine learning insights could expose their trade secrets and lead to serious economic harm, such as losing their competitive edge or even risking bankruptcy. R relies on Article 4(8) DA. They argue that, despite the technical and organisational measures that C could take, the disclosure of such sensitive data poses a high risk of serious economic harm. They provide a detailed written justification explaining the unique nature of their AI algorithms, the confidentiality of the data and the specific risks they believe would arise from disclosure. C has challenged this refusal, resulting in a detailed review of the case. The review will take into account factors such as the enforceability of trade secret protection in the regions where C operates, the uniqueness of the robotics technology, and the specific reasons why R believes disclosure would be likely to cause serious harm. This scenario shows the complexity of applying Article 4(8) DA. While at first glance it appears to strengthen the digital sovereignty of companies, the strict requirements for justifying a refusal of access to data make it difficult to exercise this right. Companies must provide detailed and case-specific reasons to demonstrate the likelihood of serious economic harm.

6 Transference of Data to Third Parties

At the user's request, the data holder must also make the data generated by the use of a product available to **third parties** (Article 5 DA). In this case, the same requirements apply as for making the data available to the user under Article 4(1) DA. In particular, the data must be provided immediately, securely, in a structured, commonly used and machine-readable format, continuously and in real-time. Although the user is not required to pay any fees, the third party accessing the data may be charged for obtaining it. The conditions and compensations under which the data is made available are regulated in Articles 8 and 9 DA.

Example: For instance, consider Company R, a commercial real estate company that owns and manages several office buildings. These buildings are equipped with

*a range of smart devices and systems, including smart HVAC (heating, ventila-
tion and air conditioning) systems, energy-efficient lighting and advanced security cameras. These systems, manufactured by Company C, collect extensive data
on energy consumption, temperature preferences, occupancy patterns and security
events. Traditionally, C. had exclusive access to this data. R could use the systems
for their basic functions but had limited access to the detailed data. However, under
Article 5 DA, the situation for R could improve noticeably. R wants to work with
analysis company A, a third-party energy management and building optimisation
company. To do this effectively, they need detailed data from their smart building
systems. With Article 5 DA, R can request C to transfer the collected data directly to
A. This data transfer must be immediate, secure, in a structured, commonly used and
machine-readable format. There should be no charge to R as the user for this data
access, although C may charge A a compensation for the data. Equipped with data
from R, A can now analyse patterns to suggest optimisations for energy consumption,
improve the efficiency of HVAC systems and enhance overall building security. As a
result, the case shows how Art. 5 DA could potentially empower companies like R by
shifting digital sovereignty to their advantage. They gain control over the data generated from their own properties, allowing them to make more informed decisions and
collaborate more effectively with third-party service providers. On the other hand,
the digital sovereignty of C is diminished, as it is no longer authorised to use the
data exclusively. At the same time, its interests are protected as the access rights for
third parties are limited and data access can be denied in certain situations.*

7   Limiting the Right to Transfer Data to Third Parties

In order to protect the interests of the data holder, Article 5 DA provides restrictions
referring to the right of the user to share data with third parties, which will in many
cases be competitors of the data holder. By limiting the transfer of data to third
parties, Article 5 helps to ensure that companies retain control over their sensitive
data, thereby strengthening their digital sovereignty. An important exception applies
to the transfer of data to **large platform services**. These are not permitted data
recipients if they have been designated as gatekeepers within the meaning of the
Digital Markets Act[3] due to their market position. The data holder can therefore
refuse to provide data to these services. Similar to the user's right of access to data,
Article 5 DA also contains provisions for the protection of **trade secrets**. Trade
secrets shall only be disclosed to third parties to the extent strictly necessary to fulfil
the purpose agreed between the user and the third party. Once the third party has
obtained access to the data, it may only process it for the purposes agreed with
the user, without interference from the data holder (Recital 34 DA). The purpose

[3] Regulation (EU) 2022/1925 of the European Parliament and of the Council of 14 September 2022
on contestable and fair markets in the digital sector and amending Directives (EU) 2019/1937 and
(EU) 2020/1828 (Digital Markets Act).

of the data processing, as defined in the agreement between the user and the third party, is therefore decisive. Thus, the right to process the data received is limited to the agreed purpose of the contract. If the contractual basis ceases to exist, the third party is obliged to delete the data (Article 6(1) DA). In addition, the third party must implement all technical and organizational measures agreed upon with the data holder to ensure that business secrets remain confidential. If the data holder demonstrates that these measures are inadequate, they and the third party must concur on further protective actions. In particular, it is the responsibility of the data holder to identify data that is classified as a trade secret, including the relevant metadata.

Moreover, a new paragraph 8a has been introduced in Article 5 DA, which establishes a further exception for refusing access to data to third parties. In rare cases, the data holder might face severe harm from the exposure of trade secrets. This may happen even if the third party has taken all the necessary technical and organizational measures. If the data holder believes this to be the case, they can deny the access request. However, they must provide a well-supported explanation in writing promptly. If the data holder decides not to share the data based on this Article, they must inform the national authority.

Example of exceptional circumstances under Article 5(11) DA: In the agriculture sector, consider Company A, which has developed an advanced connected tractor that collects unique data on soil and crop conditions. Company B, on behalf of the owner of the tractor, requests access to this data for analysis. A, fearing replication of its proprietary technology and potential market loss, refuses the request. The company's concerns are based on the risk of serious economic harm if its trade secrets embedded in the data are disclosed. A provides a detailed justification to B, highlighting the uniqueness of the data and the confidentiality concerns, particularly given B's international reach. Consequently, A notifies the competent authority under Article 37 DA.

In the same context, Article 6 DA sets out a list of obligations for the third party receiving the data at the user's request. The provision initially focuses on a strict purpose limitation regarding the use of the received data. This means that the use of the data received is strictly linked to the purpose of the contract the third party and the user agreed upon. In order to protect the interests of the data owner, the third party is also prohibited from using the data for the development of competing products or passing it on to large platform services (gatekeepers in the sense of the Digital Markets Act).

8 Obligations for Data Holders When Sharing Data

Contractual disparities are recognized as a major barrier to data sharing (Paal and Fenik 2023, p. 258). Recognizing this, Chapter III of the Data Act sets out the obligations and conditions under which data holders are required to make data available to data recipients.

When granting data access to a data recipient, the contractual terms must comply with the **FRAND** principles (Article 8(1) DA). This means that the terms of the contract must be fair, reasonable, non-discriminatory and transparent. To match theses modalities the European Commission will provide a non-binding model of contractual terms (Article 41 DA). These standardised terms shall help companies negotiate fairer data-sharing agreements with companies that have much more bargaining power (Recital 62 DA). It is envisaged that an autonomous expert group focusing on B2B data sharing and cloud contracts will assist the Commission in this task. Data holders and data recipients are expected to mutually agree on the terms of data sharing. However, any contractual terms relating to access to data, its use or any liability arising from its breach will not be binding if they do not comply with the conditions of Article 13 DA or if they modify the rights of the user as set out in Chapter II.

Importantly, data holders are prohibited from favouring certain data recipients, including their associated businesses (Article 8(3) DA). If a recipient feels they have been discriminated against, the onus is on the data holder to prove otherwise. Exclusivity in data sharing is only allowed if explicitly requested by a user. In addition, both parties are only required to provide information that is necessary to verify compliance with the agreed terms or their legal obligations.

For example, consider a scenario where an individual owns a vehicle equipped with advanced data collection capabilities, such as a modern electric car. The owner, interested in contributing to electric vehicle research, requests the manufacturer to provide access to the vehicle's data for a specific research institute specializing in electric mobility studies. In compliance with the Data Act, the manufacturer then establishes a data access agreement directly with the research institute. According to Article 8(3) DA, the manufacturer must ensure that the terms of this agreement are not discriminatory compared to similar agreements with other research institutes or partners. If the research institute believes that the terms of the agreement are less favourable than those offered to comparable entities, the manufacturer must promptly demonstrate that the terms are fair and equitable.

Furthermore, Article 9 DA states that any **compensation** agreed between a data holder and a data recipient for making data available shall be reasonable and may include a margin. However, for micro, small or medium-sized enterprises, the compensation should only cover the direct costs associated with the request for data sharing. In addition, data holders are required to provide recipients with clear information on the calculation of the remuneration, so as to ensure that the recipient can confirm that the provisions of Article 9 DA are complied with.

For example, in a data-sharing agreement involving three entities, Company A, which owns environmental sensors, collects data about air quality. Company B uses these sensors and wishes to analyse the data for environmental optimisation purposes. To facilitate this, B asks Company A to grant access to the data to Company C, an SME with expertise in data analysis. The compensation negotiated between A and C primarily covers the direct costs of data preparation and transmission, in line with Article 9(4) DA. Company A ensures transparency by providing C with a detailed breakdown of these costs, thereby complying with the Article 9 DA principles of fair and non-discriminatory compensation.

Furthermore, rules for dispute resolution are also set out in Article 10 DA. Both data holders and data recipients have the right to refer disagreements on the terms of data sharing to certified dispute resolution bodies. These bodies, certified by the Member States, must be impartial, have the necessary expertise and be easily accessible. However, this does not infringe upon the parties' rights to seek legal remedies in national courts.

Further provisions to ensure the digital sovereignty of businesses are also included in Article 11 DA. The regulation focuses on technical safeguards and provisions against unauthorised use or disclosure of data, allowing data holders to implement technical safeguards to prevent improper access and to ensure compliance with agreed data-sharing terms. However, these measures should not interfere with the user's ability to effectively share data with third parties or violate the legal rights of third parties. If a data recipient obtains data by fraudulent means, provides false information, exploits technical vulnerabilities, uses the data for unauthorized purposes, or discloses the data without the consent of the data holder, the recipient must immediately destroy the data and any copies. In addition, they must cease making, selling or using any products, derivative data or services derived from such data, and destroy any infringing goods. However, these measures are not required if the use of the data has not caused significant damage to the data owner, or if taking such measures would be disproportionate to the data owner's interests.

9 Unfair Contractual Terms Unilaterally Imposed on Another Enterprise

Usually, contractual rules should recognize the importance of contractual freedom in business-to-business interactions. Thus, not every contractual term should be subject to an unfairness assessment, but only those terms that are unilaterally imposed (Recital 59 DA). Therefore, Article 13 DA focuses on contractual terms that are unilaterally imposed on other enterprises. It refers to situations where one party imposes a particular contractual term and the other party is not in a position to modify that term despite efforts to negotiate (so-called 'take-it-or-leave-it' situations). According to Article 13(1) DA, a unilaterally imposed contractual term concerning access to and use of data, or liability and remedies for breach or termination of data-related obligations, shall not be binding on the other party if it is unfair. The party introducing a contractual term shall bear the burden of proving that the term has not been unilaterally imposed (Specht-Riemenschneider 2022, p. 822). If the other party tries to negotiate the term and fails, this is sufficient to assume that the term was unilaterally imposed (Art. 13 (6) DA). According to Article 13(3) DA, a contractual term is unfair if it is intended to protect the party who introduced it from liability, in particular if the party acted intentionally or with serious negligence. In addition, Article 13(4) DA contains a list of contract terms which are presumed to be unfair. In this context, it is considered unfair if it allows one party to access and misuse the other's data, prevents the other party from using or obtaining its own data, or allows the contract to be terminated at very short notice without consideration for the other party's situation.

Scenario for an unfair term within a data licence agreement: Consider a high-tech CNC milling machine produced by Manufacturer A, which specialises in machine tools. Company B, a medium-sized precision manufacturer, uses this CNC machine to produce complex metal components. The machine is equipped with sensors and software that collect detailed data about the manufacturing process, including operating times, tool wear, energy consumption and machine efficiency. Company B wants to analyse this data to improve efficiency and contracts Company C, which specialises in data analysis for the manufacturing industry, to perform the analysis. However, there is a clause in the data licence agreement between A and C that exempts A from liability for any loss or corruption of the data, even if caused by intentional or grossly negligent acts. This clause could be considered unfair under Article 13(3) a) DA as it unfairly limits A's liability.

10   Data Sharing Obligations Towards Public Sector Bodies

Chapter V of the Data Act defines the relationship between companies and public authorities (B2G). The provisions affect the digital sovereignty of companies by requiring data holders to make data available to public sector bodies or Union institutions (i.e. national, regional and local authorities, bodies and institutions governed by public law of the Member States). However, the request is only admissible if an exceptional need for the requested data is demonstrated. Such a need arises in three scenarios: responding to a public emergency, preventing or recovering from such an emergency, or fulfilling a specific public interest task in the absence of data (Art. 15(1) DA). The request for data must comply with the criteria set out in Article 17 DA, in particular by demonstrating the exceptional necessity. To protect companies' digital assets, trade secrets must be disclosed to public bodies only to the extent necessary for the purpose of the request. Public bodies are also prohibited from using the data for purposes incompatible with the original request. In addition, they are obliged to take appropriate measures to maintain the confidentiality of such trade secrets, such as technical and organisational measures (Art. 19(2) DA). If the request is necessary to respond to an emergency, data access must be granted free of charge (Article 20(1) DA). In cases where the request is necessary to prevent or recover from such an emergency or to fulfil a specific public interest, the data holder may claim compensation. However, the compensation is limited to the technical and organisational costs plus a reasonable margin incurred in fulfilling the request (Article 20(2) DA). Public authorities may transfer the data they have collected to individuals or organisations for scientific research or analysis, provided that this is in line with the original purpose of the data request. The beneficiaries must, though, be non-profit-making or act in the public interest as recognised by Union or national law.

From a broader viewpoint, the provisions of Chapter V of the Data Act seek to balance the interests of data holders, public bodies and the wider societal good. At first glance, the stipulations seem very broad and overreaching. However, from a practical perspective the strict criteria under which data access must be granted are likely to make state access to data a rare occurrence, reserved for situations such as health emergencies, natural disasters or cybersecurity incidents (Recital 57 DA).

11 Privileges for Small and Medium-Sized Enterprises

To reduce the regulatory burden on smaller companies and those that have recently achieved medium-sized status or have newly launched products, the Data Act grants privileges to certain stakeholders. Article 7 DA states that the data-sharing obligations do not apply to data generated from the use of products or related services provided by companies that are classified as **micro or small enterprises**.[4] This exemption applies as long as these micro or small enterprises do not have any partners or affiliates. In addition, the same exemption from data-sharing obligations applies to data generated by products or related services provided by medium-sized enterprises. However, this exemption only relates to medium-sized enterprises that have met the criteria for this category for less than one year or to products that have been placed on the market by a medium-sized enterprise for less than one year.

Furthermore, small and micro-enterprises are only obliged to share data with public sector bodies where the data is necessary to respond to a public emergency. Unlike larger companies, they are also allowed to claim compensation for providing the data.

The proposed provisions are designed to help preserve the digital sovereignty of SMEs. By reducing the data-sharing obligations, these companies retain control over the data generated by their products and services. Such control is important for SMEs to protect their proprietary information, especially from potential competitors or larger companies. It also allows them to manage their resources more effectively by avoiding the burden of complex data-sharing requirements that may be more manageable for larger companies. In summary, the regulations for the protection of SMEs appear to be favourable, as competitive disadvantages primarily affect SMEs and the burden of complying with the obligations of the Data Act would affect them disproportionately.

However, the adequacy of these privileges in protecting SMEs within the broader digital economy is a matter of debate. While the exemptions provide immediate relief, they may not fully address the challenges faced by many SMEs that rely on collaboration with other businesses. In particular, collaboration with partners along

---

[4] According to the Recommendation 2003/361/EC, Micro enterprises: up to 9 employees and up to €2 million turnover p.a., Small enterprises: up to 49 employees and up to €10 million turnover p.a.

the value chain could pose a risk of losing the privileges granted to SMEs by the Data Act.

12 Seamless Cloud-Switching

Chapter VI of the Data Act emphasises the importance of facilitating switching between cloud providers, which is recognised as a key requirement for open innovation in the European data economy (Recital 76 DA). This is achieved by addressing commercial, technical, contractual, and organisational barriers, ensuring that enterprises can make decisions based on their immediate needs without undue hindrance. Providers, under these regulations, must guarantee portability, ensure functional equivalence post-switch, and accommodate a termination by the customer within a maximum of 30 days (Article 24 DA). Article 25 DA outlines the financial aspects and proposes that fees for requested changes should either be waived or limited to marginal costs for specific transactions. This provision has a practical impact on the decision-making process in companies, as it reduces the overall effort involved in switching providers. Moreover, it promotes competition between cloud providers.

In addition, Article 26 DA sheds light on the technical nuances of switching. It mandates that providers offering infrastructural data processing services ensure functional equivalence for customers switching to a similar service from a different provider. For services beyond pure infrastructure, providers are required to offer open interfaces freely and ensure compatibility with open interoperability specifications or European standards. If such standards are absent, providers must export all data in a commonly used, machine-readable format upon customer request. Furthermore, Art. 26(3) DA emphasises open interfaces and compatibility with open interoperability specifications or European interoperability standards, which fosters an environment where businesses can seamlessly integrate services from different providers. Chapter VII (Article 27 DA) complements this by introducing regulations on third-country access and transfers concerning non-personal data, drawing parallels with the provisions of Article 44 ff. GDPR for data with personal reference (Hennemann and Steinrotter 2022, p. 1485).

The provisions in Chapter VI of the Data Act have the potential to enhance the digital sovereignty of companies by giving them more control over their data. In terms of digital strategy and infrastructure decisions, the provisions allow for greater autonomy and flexibility. The ability for businesses to switch cloud providers without being locked into a particular technology or platform is a key aspect of these regulations. This makes it easier for organisations to make decisions based on their specific needs, rather than solely on their existing cloud infrastructure.

13 Regulatory Oversight and Penal Measures

Oversight and execution of the stipulated provisions laid down in Art. 33 DA are delegated to the respective national jurisdictions. To this end, each Member State designates one or more competent authorities whose task, among other things, is to provide information about the content and obligations of the regulation. Additionally, these authorities are also responsible for dealing with complaints about possible infringements and are empowered to impose sanctions, which may include penalties and fines. The punitive spectrum includes fines that can escalate to EUR 20 million or an equivalent of 4% of the aggregate global turnover from the preceding fiscal year.

14 Future Outlook and Conclusion

The Data Act has the potential not only to reshape the existing data economy but also to stimulate better availability and use of data. By creating data access rights for users, third parties and public bodies, it challenges the de facto dominance of manufacturers over the data generated by their products. The concern of many enterprises is that data including sensitive corporate information could be revealed. Those trade secrets are normally shielded, particularly from rival entities. In response to these concerns, the Data Act includes measures that aim to protect the confidentiality of trade secrets. However, these contractual and technical measures do not always seem to be sufficient. At the same time, a complete refusal of a data access request is only allowed if the disclosure of trade secrets could cause fundamental economic harm. Furthermore, much justification needs to be provided to invoke these exemptions. The numerous requirements of the Data Act, especially with regard to the technical and contractual design of data sharing, bring new challenges for companies' 'digital compliance' (Gergen and Daum 2022, p. 521). Therefore, companies must effectively implement data management standards and define clear organisational responsibilities in order to reduce liability and fine risks.

On a more positive note, smaller companies and those that have recently achieved SME status are privileged and face fewer regulatory burdens. In addition, the access rights for enterprises with significant market power within the meaning of the Digital Markets Act have been restricted. However, this should not hide the fact that the overall regulatory burdens and thus also the legal uncertainty will increase, particularly in the light of the not-insignificant fines.

Apart from these risks, the Data Act includes certain provisions that can enhance the digital sovereignty of businesses. Specifically, it streamlines the process of transitioning between cloud providers by removing several barriers, such as commercial, technical, contractual, and organisational obstacles, that previously hindered the switching process. This allows businesses to align with their evolving needs, minimises the chances of vendor lock-in and promotes digital interoperability, allowing businesses to utilise a range of cloud services without compatibility issues.

While the aim of the Data Act is to democratise the data landscape, the legal implications carry real-world consequences for businesses, particularly for manufacturers of connected products. The strengthening of user rights may therefore be accompanied by greater legal uncertainty which potentially could decrease the intended benefits for the data economy.

References

European Commission: A European strategy for data, COM.: 66 final. (2020). https://eur-lex.eur opa.eu/legal-content/EN/TXT/?uri=CELEX%3A52020DC0066

European Union: Regulation (EU) 2022/868 of the European Parliament and of the Council of 30 May 2022 on European data governance and amending Regulation (EU) 2018/1724 (Data Governance Act) (2022)

Gergen, P., Daum, A.: Data economy und digital compliance—Der geplante data act und seine Implikationen für die unternehmens-compliance. RDi **12**, 514–521 (2022)

Grapetin, S.: Datenzugangsansprüche und Geschäftsgeheimnisse der Hersteller im Lichte des data act. Rdi **4**, 173–182 (2023)

Hennemann, M., Steinrötter, B.: Data Act – Fundament des neuen EU-Datenwirtschaftsrechts? NJW **21**, 1481–1686 (2022)

Lorenzen, B.: Geschäftsgeheimnisschutz und data act. ZGE **3**, 250–567 (2022)

Paal, B., Fenik, M.: Access to data in the data act proposal. ZfDR **3**, 249–262 (2023)

Schweitzer, H., Metzger, A.: Data access under the draft data act, competition law and the DMA: Opening the data treasures for competition and innovation? GRUR Int. **4**, 337–356 (2023)

Specht-Riemenschneider, L.: Der Entwurf des data act—Eine Analyse der vorgesehenen datenzugangsansprüche im Verhältnis B2B, B2C und B2G, MMR, 809–826 (2022)

Steinrötter, B.: Verhältnis von Data Act und DS-GVO—Zugleich ein Beitrag zur Konkurrenzlehre im rahmen der EU-digitalgesetzgebung. GRUR **4**, 216–226 (2023)

Straub, S.: Die aktuelle EU-Gesetzgebung im Bereich Digitalisierung und Datenwirtschaft— Mögliche Auswirkungen für Forschungs—und Entwicklungsprojekte. Institut für Innovation und Technik (iit), Berlin (2022)

Wiebe, A.: The data act proposal—Access rights at the Intersection with database rights and trade secret protection. GRUR **4**, 227–238 (2023)

Sebastian Straub works as a fully qualified lawyer at the Institute for Innovation and Technology (iit). He advises research and development projects on legal issues in the field of AI regulation and data economy. He previously worked as an attorney specialising in IT law and as a scientific advisor in a research association.

Concepts for Data Sovereignty in Digital Value Chains: Data Cockpits—Data Usage Control—Data Trustees

Denis Feth, Christian Jung, and Andreas Eitel

Abstract Digital value chains require the exchange of data. This data is always sensitive in one way or another—whether due to data protection, trade secret protection, or the very individual protection needs of data providers and data consumers along the value chain. It is therefore important to preserve the data sovereignty of all parties involved in the value creation. This requires three solution blocks, which we address in this article: (1) offering user-friendly ways to express data sovereignty needs and understanding their consequences; (2) devising technologies to enforce these needs; and (3) providing infrastructure in the form of digital platforms and their special form, data trustees, to enable a trustworthy data exchange between all involved parties.

Keywords Data sovereignty · Data cockpits · Data usage control · Data trustees

1 Self-Determined Data Sharing and Economy

In today's interconnected world, data has emerged as one of the most valuable and transformative assets, revolutionizing industries, economies, and societies. This creates added value, innovations, and new (occasionally disruptive) business models. The phrase "data as commodity" underlines the idea that data emerged as a product and is no longer merely a byproduct of digital business. Instead, data (including trade secrets and personal data) is a valuable resource with immense potential to drive innovation, decision-making, and competitiveness.

However, unlike material goods, data can easily be digitally replicated and shared. Thus, data providers should have the greatest possible insight and control over the processing and sharing of their data. This concept is referred to as "data sovereignty".

D. Feth (✉) · C. Jung · A. Eitel
Fraunhofer IESE, Fraunhofer-Platz 1, 67663 Kaiserslautern, Germany
e-mail: denis.feth@iese.fraunhofer.de

A. Eitel
e-mail: andreas.eitel@iese.fraunhofer.de

© The Author(s) 2025
U. Schmuntzsch et al. (eds.), *New Digital Work II*,
https://doi.org/10.1007/978-3-031-69994-8_5

To make data sovereignty work effectively, efficiently, and satisfactorily, several aspects need to be considered. More specifically, this includes legal regulations, the market situation, technologies used, and the participation of data providers, all of which need to be aligned with each other.

In this article, we address the latter two aspects by tackling the following three questions:

1. Which tools are required so that data providers can inform themselves and express their needs—and how do these tools need to be designed?
2. Which background technologies are needed so that data controllers can implement the data providers' individual data sovereignty needs?
3. How can trustworthy infrastructures be established so that data providers and data consumers can interconnect to exchange and use data in a trustworthy and secure manner?

To answer these questions, we first introduce some basic terminology in Sect. 2 to create a common understanding. In Sect. 3, we discuss requirements, structures, and designs of data cockpits (Question 1). These data cockpits need to be linked to a technical security system that enforces the sovereignty needs of the users' data. Enforcement technologies and their interplay with data cockpits (Question 2) are therefore addressed in Sect. 4. In Sect. 5, we discuss how data providers and data consumers can find each other and how data trustees can mediate sensitive data between them while safeguarding the interests of both parties (Question 3). We conclude our work in Sect. 6.

2 Background and Terminology

For many of the concepts we address, there are no universally accepted definitions. Therefore, we want to clarify some terms within the scope of this article to avoid misunderstandings. The terminology relates to our understanding of digital ecosystems and platforms (the term *platform* is a perfect example of a highly overloaded term), data sovereignty, data providers, and data consumers. We further address the concept of data trustees in Sect. 5.

2.1 Digital Ecosystems and Digital Platforms

Digital value is increasingly being generated within so-called *digital ecosystems*. According to MYDATA Control Technologies (2023), "a digital ecosystem is a socio-technical system connecting multiple, independent asset providers and consumers for their mutual benefit. [...] A *digital platform* is a software system that forms the technical core of a digital ecosystem, is directly used by providers and consumers via Application Programming Interfaces (APIs) or User Interfaces (UIs)—such as a digital marketplace—and facilitates the matching of a provider and a consumer in relation to an asset within a digital ecosystem service."

To make this definition a bit more tangible, the following examples describe the digital ecosystem service for two famous digital ecosystems:

- Airbnb Lodging offers the brokerage of private accommodation from private providers for travelers.
- Uber Ride offers the brokerage of transportation services from private drivers to passengers.

Digital ecosystems offer a wide range of opportunities for their participants. These include the acquisition of new customers as well as the exploitation of new business areas, and the initiation of innovations in their own industry. Economies of scale and network effects are a central component of digital ecosystems. In all of this, data plays a major role. For example, the asset providers and the platform provider generally process personal data to provide the asset. There are even various examples where the traded asset itself is data, for example:

- Caruso[1] offers the brokerage of vehicle data from manufacturers to the automotive aftermarket.
- Aviatar[2] offers the brokerage of maintenance data from aircraft parts manufacturers for airlines.
- GovData[3] offers the brokerage of administrative data from public institutions to citizens and companies.

In this context, there is an increasing demand both by legislation (in the context of the GDPR) and by the users (i.e., primarily data providers and data consumers) themselves that users be granted certain information and co-determination rights regarding the use of "their" data. This kind of informed self-determination is also referred to as *data sovereignty*.

2.2 Data Sovereignty

For the term "data sovereignty", although heavily researched and widely used, there is no broadly accepted definition. However, this is the common understanding we derived from literature (Ethikrat 2017; Goldacker 2017; Gesellschaft für Informatik 2020; Feth and Polst; Smart Data Forum 2019) and use throughout this article: Data sovereignty describes a subarea of digital sovereignty. It aims to give data providers the greatest possible control over and insight into the use of their data. This includes the ability of data subjects to exercise their right to informational self-determination and to protect their privacy. This understanding is summarized in the following Fig. 1.

---

[1] https://www.caruso-dataplace.com/.

[2] https://www.aviatar.com/.

[3] https://www.govdata.de/.

Fig. 1 Data sovereignty (own illustration)

2.3 Data Provider

Data providers are natural or legal persons who make their data available in digital ecosystems. If the asset of a digital ecosystem is data, data providers are a special form of asset providers. A distinction must be made between (a) the data subjects or companies holding certain rights regarding the data (e.g., patients), and (b) the entity that technically manages and provides access to the data (e.g., hospitals). Of course, data providers expect to benefit from data sharing. In addition to monetary incentives, there are also a lot of non-monetary incentives, such as convenience or contribution to the public good. On the other hand, data providers also need to consider the risks associated with data sharing. Risks may include the loss of intellectual property and data protection issues. This is where data sovereignty comes into play. Many data providers have well-founded reservations about sharing data and will be willing to do so only if they keep control over their data.

2.4 Data Consumer

Data consumers are natural or legal persons that process data stemming from data providers. On the one hand, this means that data providers must consider the data providers' data sovereignty needs and implement corresponding measures to fulfill legal requirements. For example, a corresponding consent, contract, or other legal basis is required for processing in compliance with the law.

On the other hand, data consumers have demands with respect to data quality. These include, for example, the general availability of the data, its timeliness and accuracy, and requirements of a structural nature (e.g., standardized data formats). Furthermore, the data cost needs to be considered, including the price charged by the data providers, possible brokerage fees of the data trustee, and expenses for making the data usable (e.g., converting data formats or harmonizing data). Thus, there is a tradeoff between data sovereignty needs of data providers and data quality needs of data consumers that need to be balanced.

3 Data Cockpits: User-Friendly Interfaces for Transparency and Self-Determination

Data sovereignty in digital ecosystems is particularly challenging because they consist of a highly dynamic and hard-to-understand network of participants that have a (commercial) interest in the data. Thus, the following requirements with respect to data sovereignty are particularly challenging in this context:

- Data providers must be able to *understand, interpret, and verify* with reasonable effort how their data is used and shared.
- Data providers must be given the opportunity to *control the processing* of their data.
- Data providers must be able to understand the *impact* of certain decisions on their legitimate interests (e.g., by giving consent).
- Data providers must be *free and uninfluenced* in their decisions.

These requirements directly relate to the user experience (UX) and usability of data sovereignty measures. In very general terms, the interaction between data sovereignty and UX can be summarized as follows: On the one hand, lack of data sovereignty can have a negative impact on important UX aspects such as satisfaction or trust. Therefore, data sovereignty can be a prerequisite for good UX. On the other hand, data sovereignty will only be achieved if the measures are also implemented in a usable way. Otherwise, users will not use them (at least not correctly), and thus they will be indirectly deprived of their sovereignty. For example, choosing incorrect settings can even have the exact opposite effect of what data providers actually wanted to achieve.

For data providers to be able to make use of their rights and effectively exercise data sovereignty, they must first be given the opportunity to inform themselves appropriately and formulate their needs. Even if some of the required options already exist, they are often not very user-friendly in practice. Privacy statements, for example, are designed for legal experts, not for end users. In the case of configuration options, attempts are often made to influence the user inappropriately ("nudging"), and the consequences of an action are rarely communicated clearly.

In our projects, we research so-called data cockpits. A data cockpit offers a central point for users regarding the control of their data. Due to the aforementioned complexity of digital ecosystems, as well as due to the comprehensive legal framework, such cockpits comprise several interlinked functions, including:

- general information related to data protection,

- data access and export,
- overviews of data consumers, services, and permissions,
- data transaction and usage logs,
- data protection settings and consent management,
- data correction and deletion, and
- reporting of data protection issues or incidents.

Especially in distributed systems, there are several implications to fulfill the responsibility of such a data cockpit: First, the users should have a central point for configuring and understanding all data usage. In digital value chains, there should be a single responsible entity for all settings related to the use and dissemination of users' data across all services and participants. Second, all services and participants must provide information about data usage, so that the data cockpit can present them to the user (e.g., visualization of data usage). Third, data sovereignty and data protection settings should apply to all services and participants in the digital value chain. It is recommended that such settings are handled centrally. In digital ecosystems, this central entity is usually provided technically by the digital platform and legally by the platform operator.

Designing data cockpits in a comprehensible, intuitive, and effective way is a major challenge. A user-friendly data cockpit must support even inexperienced and lay people in accessing and exercising their rights appropriately. However, our experience shows that the design of such an interface and the determination of an appropriate granularity for the presentation strongly depend on the domain and on the users. In most cases, users are IT laypersons and therefore prefer a setting option that is as simple as possible and understandable for them. We also experienced that abstracting from the concrete data and specifying the usage rights when using the service is more effective than specifying the usage restrictions without context.

The following screenshots give an impression of how a data cockpit might look in a digital ecosystem that brokers health data between patients and medical personnel (e.g., doctors, nursing services, and health insurance companies). For example, Fig. 2 shows an overview of "services" (in the sense of health support services) used by the patient. Figure 3 shows a similar, but data-focused perspective. Taking a closer look, different data usage permissions apply for each service, related to different data categories (e.g., core data, blood pressure, blood sugar, or footsteps), as shown in Fig. 4.

Depending on the concrete patient and service, these usage permissions can become quite complex and go far beyond simple "yes or no" decisions. For example, patients may want to grant access only for a certain time or duration, allow usage only for concrete purposes, or they may want to mask or filter data before release. You can see such usage permissions in Fig. 5. In the selected option, the patient grants access to his blood sugar level data only in averaged form—meaning raw data must be smoothed before being passed on.

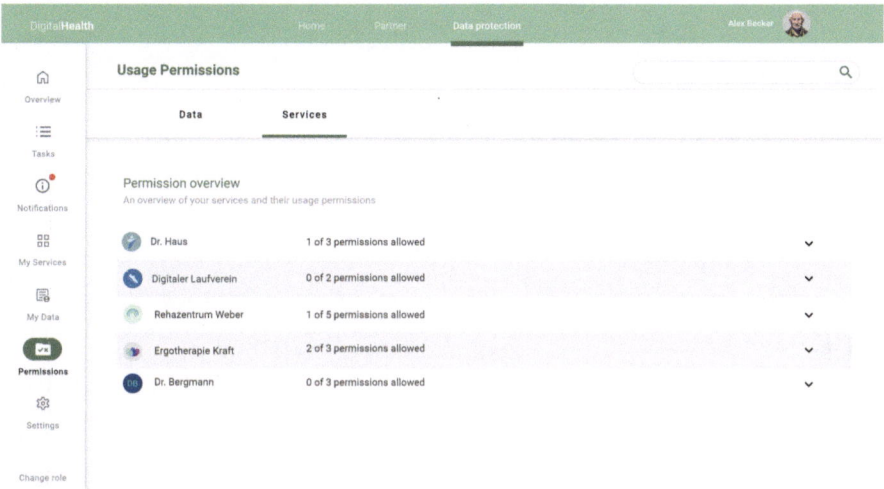

Fig. 2 Data cockpits in a health ecosystem: Service overview (Data subject perspective, own illustration)

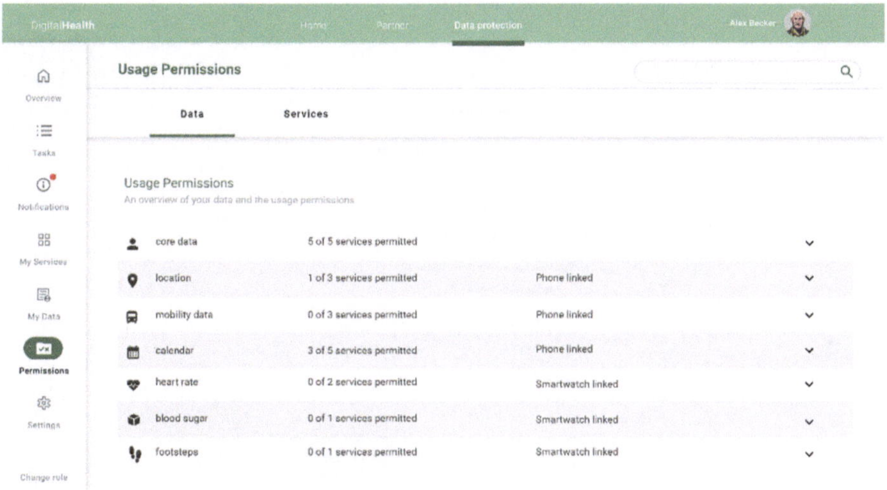

Fig. 3 Data cockpits in a health ecosystem: Data overview (Data subject perspective, own illustration)

As this would go beyond the scope of this article, we cannot go into more detail regarding the functions described above. However, we want to clarify that data cockpits also provide functions and views for the entities that consume data. For example, Fig. 6 shows one way for medical providers to request data from their patients.

The examples shown underline the fact that data cockpits must be strongly aligned with the specific digital ecosystem. Therefore, it is not possible to develop a universal

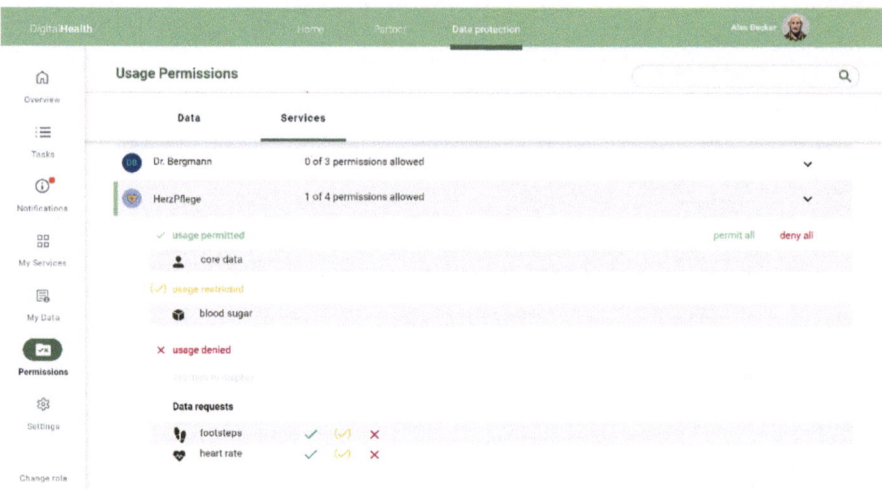

Fig. 4 Data cockpits in a health ecosystem: Permissions overview (Data provider perspective, own illustration)

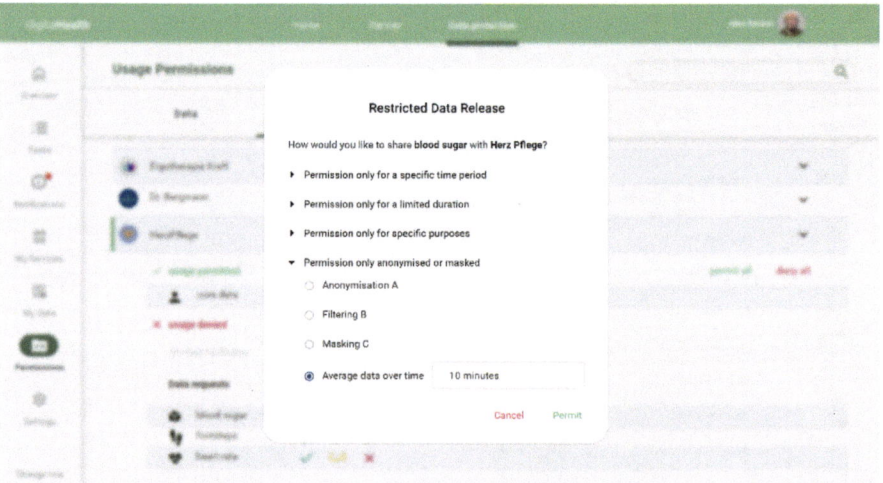

Fig. 5 Data cockpits in a health ecosystem: Granting usage permissions (Data provider perspective, own illustration)

data cockpit. Instead, individual development is necessary. For this purpose, we provide a framework consisting of user needs and requirements, legal constraints and interpretations, UI mockups and interaction patterns (best practices and worst practices) as well as architecture blueprints and required technical interfaces (APIs).

We have not yet addressed the question of how the settings from data cockpits can be implemented technically. The examples given in Fig. 5 show how complex

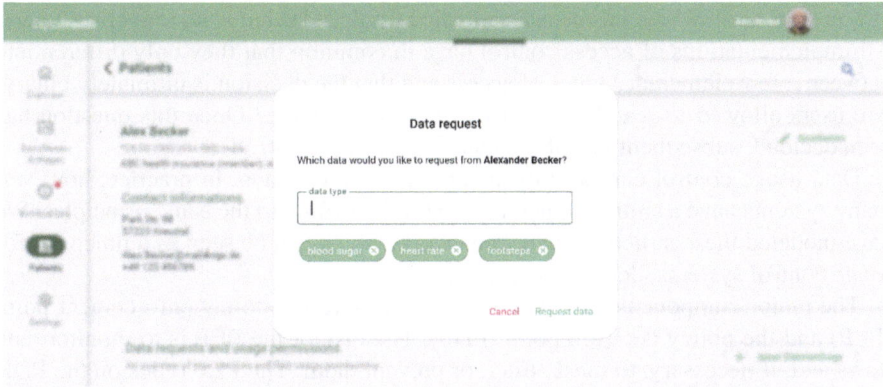

Fig. 6 Data cockpits in a health ecosystem: Requesting data (Data consumer perspective, own illustration)

these settings can become and illustrate that standard solutions also reach their limits here. In the next section, we will therefore focus on the topic of data usage control as a technical means of enforcement.

4 Data Usage Control: Technical Enforcement of Data Sovereignty Needs

Data usage control refers to the ability of individuals or organizations to dictate how their data is used (i.e., accessed, processed, and utilized) by others. It encompasses a range of mechanisms and practices that empower data providers and data owners to retain control over their data even when sharing them with third parties (i.e., data consumers).

There are several related concepts such as digital rights management (Frattolillo 2017; Lin et al. 2005; Ma 2017; Liu et al. 2003; Stamp 2003; Elshazly et al. 2017), which addresses the management of usage of digital content, or data leakage prevention (Alneyadi et al. 2016; Katz et al. 2014; Wu et al. 2011; Stamati-Koromina et al. 2012), which proactively prevents unintended data usage. For the sake of completeness, we should also mention access control (Sandhu 1993; Sandhu and Samarati 1994; Kalam et al. 2003; Goyal et al. 2006), usually granting or preventing access to data. Usage control, in contrast, additionally controls future data uses (Jung et al. 2014; Kelbert and Pretschner 2012; Sandhu and Park 2003). It thus combines access control rules (provisions) with future data usage rules (obligations).

Data usage control is a technical building block for implementing data sovereignty. It enables data providers to freely determine how their data is used. To this end, fine-grained conditions (so-called policies) for data use are specified and implemented using corresponding control mechanisms. The conditions extend the concept of access control, which has been established for decades but is no longer sufficient. All implementations of access control have in common that they only differentiate between a few elementary types of access and that the decision is ultimately binary: Are users allowed to read, write, or execute the resource? Once this question has been decided, subsequent use of the data is no longer controlled.

Data usage control can be implemented in various ways. In practice, however, many systems have a similar structure or work according to the same principles. We have modeled these structures in a framework, which can be seen as a blueprint for usage control systems, depicted in Fig. 7.

The major components for data usage control are the policy enforcement point (PEP) and the policy decision point (PDP). The task of the PEP is to monitor data flows and, if necessary, to mask, filter, or prevent them. The PEP relies on the PDP, which interprets the usage restrictions set by the data providers or other stakeholders and makes a decision for the specific case. Apparently, the degrees of freedom in the usage restrictions are determined by the technical capabilities of the PEP. Based on this basic scheme, there is a whole range of possible expansion points:

- Actions can be performed in addition to data masking and filtering, such as logging data usage or deleting data sets. Therefore, we need a component, called policy execution point (PXP), to execute the actions demanded by the PDP.
- To make a decision, the PDP may need additional information that cannot be provided by the PEP directly. For example, the PEP may be able to tell the username of the data consumer but does not know its role. Therefore, we need a component, called policy information point (PIP), to supply such information, e.g., from a directory service.
- The policy management point (PMP) manages the policy lifecycle. In a minimal setup, the PMP stores policies in a database (policy retrieval point, PRP) and deploys or revokes policies at the PDP. However, policy management can also become more complex including version histories, conflict detection, and so on.
- Eventually, policies need to be specified in a format the PDP can technically interpret. They are the machine-readable pendant of what was set by the users in data cockpits, as described earlier (e.g., what does it technically mean to smoothen data). As these policies can become very complex, special policy editors, so-called policy administration points (PAP), are used (Rudolph et al. 2019a, 2019b).

The interaction between these components, sequence diagrams, and further information about policy formalization and specification can be found in the position paper "Usage Control in the International Data Spaces" (Steinbuss et al. 2021). Moreover, Fraunhofer IESE developed the MYDATA Control Technologies that implement essential components of the IND2UCE Framework. Further information about these technologies, the MYDATA policy language, and details about their use can be found on the developers' webpage (MYDATA Control Technologies 2023).

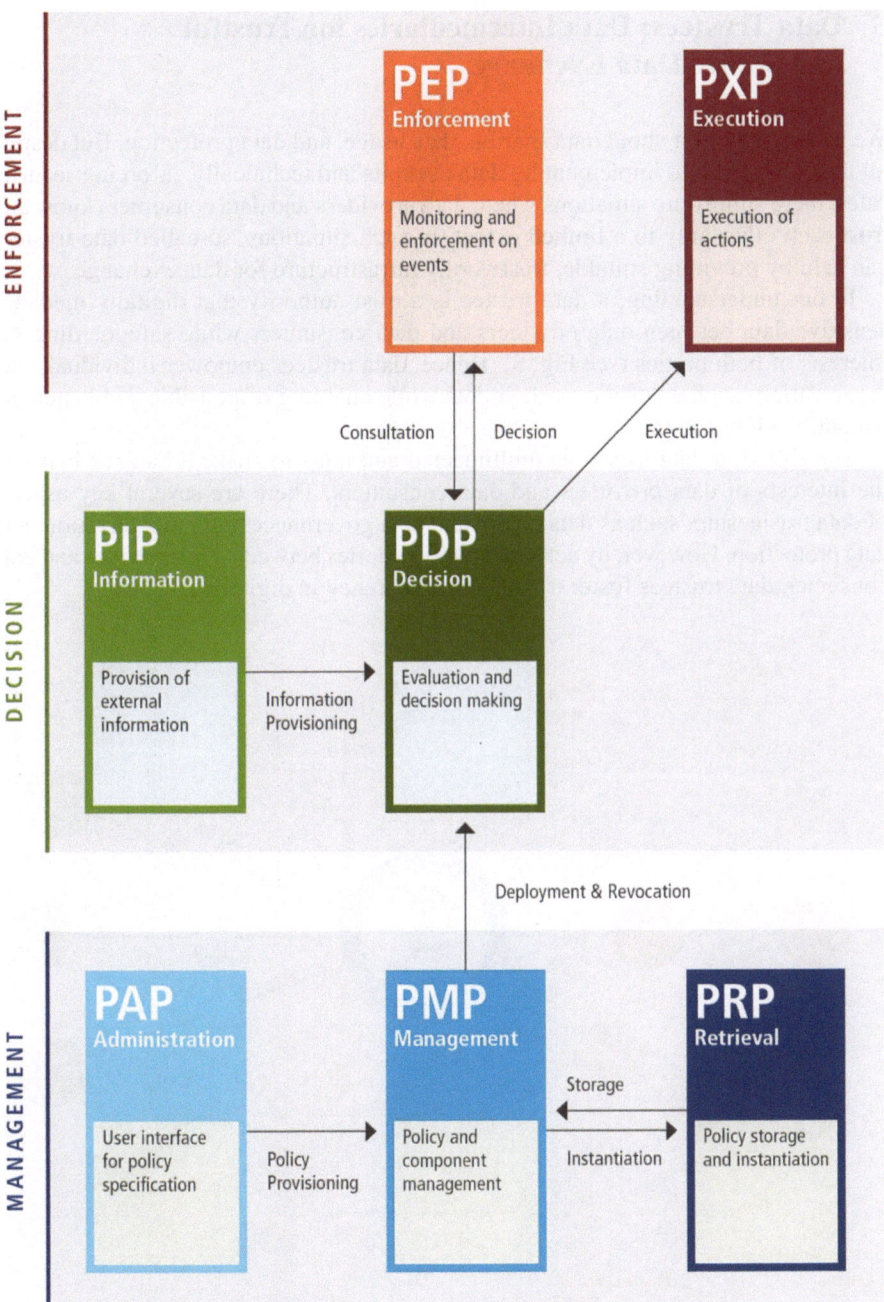

Fig. 7 Usage control framework "IND2UCE" by Fraunhofer IESE (own illustration)

5 Data Trustees: Data Intermediaries for Trustful and Secure Data Exchange

We have talked a lot about data sharing, data usage, and data protection. But despite all the effort put into implementing data cockpits and technically enforcing security rules: there simply are situations where data providers and data consumers know and trust each other only to a limited extent. In such situations, so-called data trustees can help by providing suitable, trustworthy infrastructure for data exchange.

In our understanding, a data trustee is a trust authority that digitally mediates sensitive data between data providers and data consumers while safeguarding the interests of both parties (see Fig. 8). Hence, data trustees empower individuals and organizations with control over their data while enabling responsible and beneficial data utilization.

The role of a data trustee is multifaceted and aims to strike a balance between the interests of data providers and data consumers. There are several key aspects of data trusteeship, such as data custodian, data governance, data monetization, and data protection. However, by acting as intermediaries between data provider and data consumer, data trustees foster trust and transparency in digital value chains.

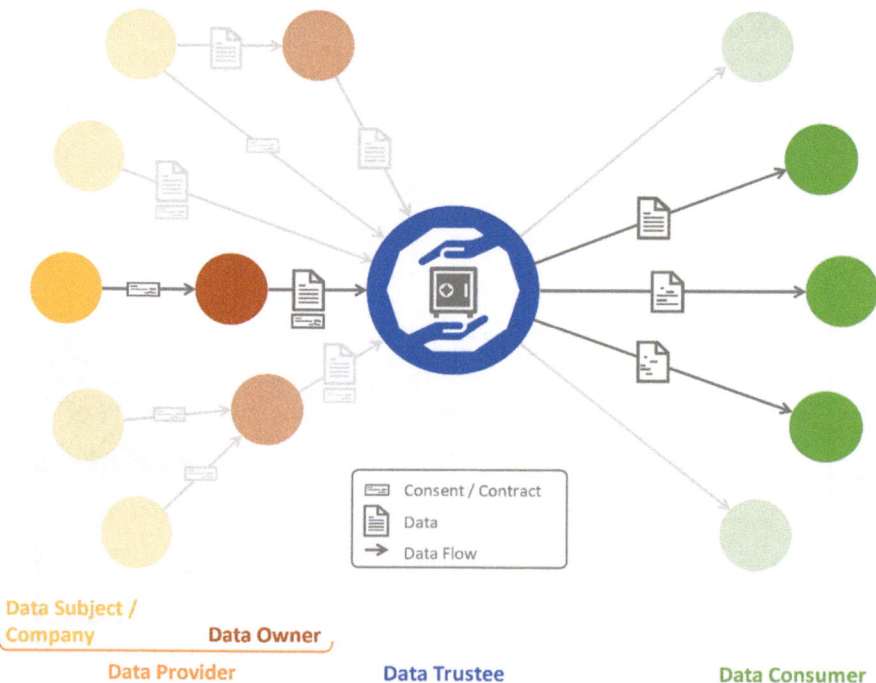

Fig. 8 Different roles in data trustee models

Therefore, data trustees can be found in almost all domains, such as mobility, energy, finance, agriculture, logistics, smart cities, and medicine. Accordingly, data trustees differ from one another in their implementation. We will take a closer look at some questions that are of particular interest to us.

1. **What are the core objectives of the data trustee?** Differences start with the core objectives of the specific data trustee model (e.g., promoting digital sovereignty, data sovereignty, economic exploitation, innovation, science and research, fair competition, or compliance with legal requirements).
2. **Who is involved?** Data trustees address different target groups regarding both data providers and data consumers (e.g., companies, citizens, administration, or research). In the concrete data trustee model, these groups differ in their data sovereignty needs, sometimes significantly.
3. **How can the data trustee's business models become sustainable?** In the context of value chains, how a data trustee is financed is a relevant question. Who pays for the data and its transmission? Here, too, a wide variety of options are available, ranging from fixed basic amounts to transaction fees, cross-financing, shareholder participation, and government subsidies. This topic should not be underestimated in the context of data sovereignty, as it is directly related to the issue of trust. Indeed, data trustees are mostly expected to be independent and neutral. Transparency about transactions and cash flows within a data cockpit can positively influence the acceptance of the data trustee.
4. **How is data stored, processed, and shared?** Where the data is stored (e.g., centrally at the data trustee or only temporarily at the intermediary) and which "additional services" the data trustee offers (e.g., data conversion, data verification, data enrichment) also have a major influence on how data providers and data consumers accept the trustee model. In most cases, the data provider himself decides on a concrete data release. However, there are also cases in which the data trustee is given a certain amount of leeway and, within this scope, brokers the data itself in the interests of the data provider. Both cases are equally challenging when thinking about data cockpits and technical enforcement by means of data usage control. In the first case, data providers may have to process many requests and handle each request individually. In the second case, the consequences of transferring responsibility to the data trustee must be clear. In addition, the data provider retains his data subject rights in this case as well, so the cockpit must continue to enable transparency and intervention.

Notwithstanding all these differences, the data trustee is ideally suited to act as a point of contact and implementer of data sovereignty due to its central position. Meeting this responsibility is in the fundamental interest of the data trustee. However, there are various things to consider, also due to the fairly new legal situation (e.g., the Data Governance Act). Therefore, we are working on a cross-industry guide to help new data trustees develop their model. In our guide, we go into detail about the legal situation and provide corresponding interpretation support. We have also compiled an extensive collection of requirements, including a catalog of security and data protection requirements based on the BSI's "IT-Grundschutzkompendium"

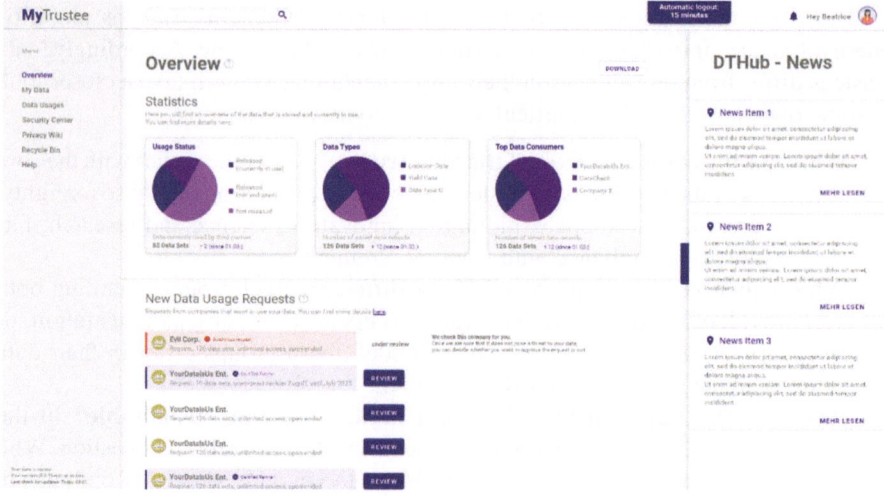

Fig. 9 Data trustee cockpits: Dashboard with usage statistics and requests (own illustration)

and a security testing guide based on the OWASP Web Security Testing Guide. Moreover, we provide various software building blocks that data fiduciaries can use as a blueprint or solution—including our data usage control solution, MYDATA.

In principle, many of the concepts of data cockpits can be directly applied to data trustees as well. For example, on the left-hand side, the dashboard for a data trustee in Fig. 9 gives the data provider access to his data, usages by data providers, security settings, and privacy-relevant documentation. In the middle, it provides usage statistics and overviews about pending usage requests. On the right-hand side, the user is informed about updates from the data trustee.

However, in contrast to many "normal" ecosystems (e.g., eBay, Airbnb, or Uber), two things are particularly important: First, data is not just a means to an end (e.g., meta-data), but the traded good itself. Second, mutual trust plays a special role. This is also reflected in the user interface of the data trustees. Thus, "live views" are particularly relevant. For example, Fig. 10 shows an overview of provided data and offers transparency of data access.

In addition, the reputation of data providers plays a crucial role in decision-making, which is why corresponding schemas, views, and processes must be implemented. In the lower part of Fig. 9, data requests are directly labeled according to the reputation of the data consumer (e.g., "certified partner" or "suspicious request") in the concrete data trustee model. In Fig. 11, you can also see more details about a data usage request that has been classified as suspicious.

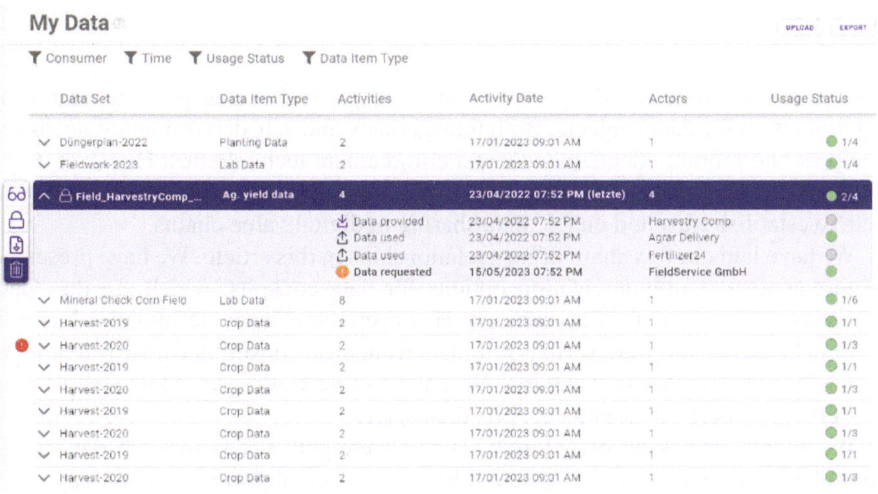

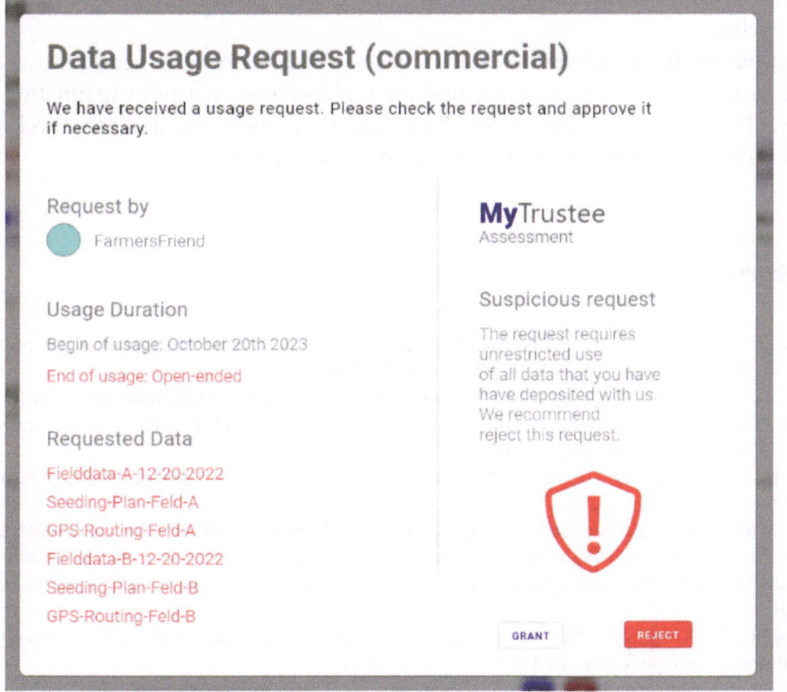

Fig. 10 Data trustee cockpits: Data overview and access log (own illustration)

Fig. 11 Data trustee cockpits: Data usage requests (own illustration)

6 Conclusion

Data economy requires data sovereignty—in particular transparency and self-determination for data subjects. Both transparency and self-determination needs are complex and require adequate technical enforcement technologies. However, technical enforcement alone does not help if there is a lack of trust. Data trustees can help to establish trust and enable data sharing in digital value chains.

We have laid out this chain of interrelationships in this article. We have presented requirements and implementation options for data cockpits, which are the "data sovereignty front end" for data subjects. However, there is no one-fits-all solution in data cockpits, and there are various pitfalls—from legal constraints to the (intentional or unintentional) use of dark patterns. We are therefore working on a kit to help digital platform providers build data cockpits themselves.

We address the issue of enforcing data sovereignty needs through data usage control. Usage control extends the traditional approach of protecting data only when it is accessed, thus enabling data to be protected along the entire value chain.

Finally, we expect data trustees to play an increasingly important role in the future. As trusted data intermediaries, they are in the ideal position to enable trustworthy data exchange.

Ultimately, these and various other aspects, which we have not been able to address in this article due its limited scope, must interact for data sovereignty to function effectively, efficiently, and satisfactorily. Therefore, much more research and development work on this complex interaction will be needed in the future.

References

Alneyadi, S., Sithirasenan, E., Muthukkumarasamy, V.: A survey on data leakage prevention systems. J. Netw. Comput. Appl. **62**, 137–152 (2016)

Elshazly, A.R., Nasr, M.E., Fouad, M.M., Abdel-Samie, F.S.: High payload multi-channel dual audio watermarking algorithm based on discrete wavelet transform and singular value decomposition. Int. J. Speech Technol. **20**(4), 951–958 (2017)

Ethikrat, D.: Big data und gesundheit—Datensouveränität als informationelle freiheitsgestaltung, (2017)

Feth, D., Polst, S.: Benutzerfreundliche umsetzung von datensouveränität in digitalen ökosystemen. Online: https://www.iese.fraunhofer.de/content/dam/iese/dokumente/leistungen/benutzerfreundliche_umsetzung_von_datensouveraenitaet_in_digitalen_oekosystemen-fraunhofer_iese.pdf

Frattolillo, F.: A digital rights management system based on cloud. Telkomnika (Telecommun. Comput. Electron. Control.) **15**(2), 671–677 (2017)

Gesellschaft für Informatik: Schlüsselaspekte digitaler Souveränität. Arbeitspapier der Gesellschaft für Informatik e.V, Berlin (2020)

Goldacker, G.: Digitale Souveränität. Kompetenzzentrum Öffentliche Informationstechnologie (ÖFIT), Berlin (2017)

Goyal, V., Pandey, O., Sahai, A., Waters, B.: Attribute-based encryption for fine-grained access control of encrypted data. In Juels, A., Wright, R., Di Capitani Vimercati, S. de (eds.), Proceedings of the 13th ACM conference on Computer and communications security, Alexandria, Virginia, USA, ACM, New York, NY, p. 89 (2006)

Jung, C., Eitel, A., Schwarz, R.: Enhancing cloud security with context-aware usage control policies. In: Plödereder, E., Grunske, L., Schneider, E., Ull, D. (eds.), 44. Jahrestagung der gesellschaft für informatik: Big data—Komplexität meistern, Stuttgart, 22.09.2014–26.09.2014, GI, pp. 211–222 (2014)

Kalam, A.A.E., Baida, R.E., Balbiani, P., Benferhat, S., Cuppens, F., Deswarte, Y., Miege, A., Saurel, C., Trouessin, G.: Organization based access control. In: Proceedings, POLICY 2003: IEEE 4th international workshop on policies for distributed systems and networks 4-6 June, 2003, Lake Como, Italy, Lake Como, Italy, 4-6 June 2003, IEEE, Piscataway, N.J., pp. 120–131 (2003)

Katz, G., Elovici, Y., Shapira, B.: CoBAn. A context-based model for data leakage prevention. Inf. Sci. **262**, 137–158 (2014)

Kelbert, F., Pretschner, A.: Towards a policy enforcement infrastructure for distributed usage control. In: Atluri, V. (ed.), Proceedings of the 17th ACM symposium on access control models and technologies, Newark, New Jersey, USA, ACM, New York, NY, pp. 119–122 (2012)

Koch, M., Krohmer, D., Naab, M., Rost, D., Trapp, M.: A matter of definition: Criteria for digital ecosystems. Digital Business **2**(2), 100027. https://doi.org/10.1016/j.digbus.2022.100027. https://www.sciencedirect.com/science/article/pii/S2666954422000072

Lin, E.T., Eskicioglu, A.M., Lagendijk, R.L., Delp, E.J.: Advances in digital video content protection. Proc. IEEE **93**(1), 171–182 (2005)

Liu, Q., Safavi-Naini, R., Sheppard, N.P.: Digital rights management for content distribution. In: Proceedings of the Australasian information security workshop conference on ACSW Frontiers 2003—Volume 21, Australian Computer Society, Inc, Darlinghurst, Australia, Australia, pp. 49–58 (2003)

Ma, Z.: Digital rights management. Model, technology and application. China Communications **14**(6), 156–167 (2017)

MYDATA Control Technologies.: MYDATA control technologies—developer documentation. Version 4.2.1 (last updated 2023–08–10), Fraunhofer IESE, (2023)

Rudolph, M., Polst, S., Feth, D.: Usable specification of security and privacy demands: matching user types to specification paradigms. Mensch Und Comput. (MuC) Work., (2019a)

Rudolph, M., Polst, S., Doerr, J.: Enabling users to specify correct privacy requirements. In: Requirements Engineering: Foundation for Software Quality, REFSQ, pp. 39–54 (2019b)

Sandhu, R.S.: Lattice-based access control models. Computer **26**(11), 9–19 (1993)

Sandhu, R.S., Samarati, P.: Access control. Principle and practice. IEEE Commun. Mag. **32**(9), 40–48 (1994)

Sandhu, R., Park, J.: "Usage control. A vision for next generation access control. In: Gorodetsky, V., Popyack, L., Skormin, V. (eds.), Computer network security: second international workshop on mathematical methods, models, and architectures for computer network security, MMM-ACNS 2003, St. Petersburg, Russia, September 21–23, 2003. Proceedings, Lecture Notes in Computer Science, vol. 2776, Springer, Berlin, Heidelberg, pp. 17–31 (2003)

Smart Data Forum.: Datensouveränität. (2019). Online: https://www.digitale-technologien.de/DT/Redaktion/DE/Downloads/Smart-Data-Forum/wissen-datensouveraenitaet.pdf

Stamati-Koromina, V., Ilioudis, C., Overill, R., Georgiadis, C.K., Stamatis, D.: Insider threats in corporate environments. A case study for data leakage prevention. ACM Int Conf Proceeding Ser

Stamp, M.: Digital rights management. The technology behind the hype. J. Electron. Commerce Res. **4**(3), 102–112 (2003)

Steinbuss, S., et al.: Usage control in the international data spaces. Int. Data Spaces Assoc. (2021). https://doi.org/10.5281/zenodo.5675884

Wu, J., Zhou, J., Ma, J., Mei, S., Ren, J.: An active data leakage prevention model for insider threat. In: Proceedings—2011 international symposium on intelligence Information Processing and Trusted Computing, IPTC (2011)

Denis Feth is head of the Security Engineering department at Fraunhofer IESE. The department develops and offers methods and tools to ensure secure data exchange and trading that meet the individual protection needs of data providers and data users. Mr. Feth's research focusses on usable security and privacy as well as data trust models.

Dr. Christian Jung was head of the Security Engineering department at Fraunhofer IESE until 2023. His research focuses on context-aware security mechanisms and data usage control for digital ecosystems. In his doctoral thesis, Mr. Jung addressed the topic of context-aware security for mobile devices.

Andreas Eitel is Expert 'Cyber Security' in the Security Engineering department at Fraunhofer Fraunhofer Institute for Experimental Software Engineering (IESE). He leads and conducts IT audits for the Fraunhofer-Gesellschaft and advises on IT secu-rity. Mr. Eitel's research focusses on network security.

Challenges for Scenario-Based Foresight and Potential for Digital Technologies: Insights from Practice

Patrick Ködding, Mathis Jahn, Christian Koldewey, and Roman Dumitrescu

Abstract Scenario-based foresight helps companies to systematically deal with future environmental developments and successful strategic business planning. However, the established management tool is used less frequently in today's fast-paced and complex world. Emerging digital technologies, such as generative artificial intelligence, can increase the efficiency of scenario-based foresight and the quality of the results. In the literature, there are already analyses of use cases for digital technologies. However, comprehensive current insights from practice are lacking in this regard and regarding the current status quo of scenario-based foresight in companies. For this purpose, we conducted an explorative interview study with eight scenario experts of industrial companies and consultancies. As a result, we obtained seven key messages that help address the research gap we raised. We describe the status quo of scenario-based foresight in companies and the challenges they face. Additionally, we analyze the current use of digital technologies in scenario projects and the further potential of digital technologies. Finally, we derive insights into how users and technologies should interact and which tasks they should perform in practice. By gathering practical insights through interviews, we contribute to research on user-centered design and the integration of digital technologies in scenario-based foresight.

Keywords Strategic foresight · Corporate foresight · Scenario planning · Scenario technique · Digital technologies · Digital sovereignty · Interview study

P. Ködding (✉) · M. Jahn · C. Koldewey · R. Dumitrescu
Heinz Nixdorf Institut, Universität Paderborn, 33102 Paderborn, Germany
e-mail: Patrick.Koedding@hni.upb.de

M. Jahn
e-mail: jahnm@hni.upb.de

C. Koldewey
e-mail: Christian.Koldewey@hni.upb.de

R. Dumitrescu
e-mail: Roman.Dumitrescu@hni.upb.de

R. Dumitrescu
Fraunhofer-Institut für Entwurfstechnik Mechatronik, 33102 Paderborn, Germany

1 Introduction

Scenario-based foresight, i.e., scenario technique and scenario planning, faces a dilemma: on the one hand, scenario-based foresight is a well-established tool that companies can use, especially in uncertain times, to support making meaningful decisions in complex environments and systematically develop pictures of the future. A systematic and clear picture of the future like this is a prerequisite for successful, strategic business planning (Gausemeier et al. 2019). On the other hand, we are now in such a rapidly changing world, characterized by strong volatility, uncertainty, complexity, and ambiguity, that some scientists even speak of a global poly-crisis (Homer-Dixon et al. 2021). According to a McKinsey study, companies tend more often to short-termism instead of long-term strategic planning (Barton et al. 2018).

Reinforced by the already existing traditional challenges of scenario-based foresight, such as the high time expenditure for a scenario project or the high methodological complexity, the use of scenario-based foresight has consequently declined significantly in recent years. Between 20 and 30 years ago, around 40% of companies used methods and tools from the field of scenario and contingency planning. During the financial crisis in 2008, this figure even exceeded 60%. Today, this figure drops to only about 19%, although satisfaction with the results remains constantly high. Those are the findings of a survey of more than 1,000 top managers conducted by the management consultancy Bain & Company (Bain and Company, Inc. 2018; Rigby and Bilodeau 2018; Rigby et al. 2023).

The rapid developments in the field of digital technologies[1] and, in particular, applications of generative artificial intelligence (AI) are creating opportunities to cope with these challenges. Digital technologies are capable of collecting data, evaluating data, generating knowledge from data, and supporting decisions. Thus, in principle, they have the potential to increase the efficiency of scenario-based foresight while maintaining at least approximately the same quality.

In our previous work, we conducted a systematic literature review to identify and analyze existing use cases for digital technologies in scenario-based foresight (Ködding et al. 2023a). Confirming insight from Bauer et al., van Belkom, and Steinmüller, we showed that digital technologies can be used to analyze large amounts of data. In this way, scenario experts can be supported, especially in time-intensive research tasks, such as the identification of influence factors (Ködding et al. 2023a; Bauer et al. 2022; Steinmüller 2022; van Belkom 2020). Creative tasks such as projection development or tasks that require company-specific knowledge, such as the interpretation of scenarios, however, are performed by humans. In this regard, digital technologies only have a complementary role (van Belkom 2020; Ködding et al. 2023a).

However, the literature still provides little information about the current practice in companies. This means that there is only limited information in the field of scenario-based foresight about (1) which digital technologies are already being used

---

[1] For the definition of digital technologies, we refer to Lipsmeier et al. and Berger et al. (Lipsmeier et al. 2018; Berger et al. 2018).

by companies and to what extent, (2) what challenges arise in the interaction of digital technologies and humans, and (3) what further potential there is for digital technologies. In this context, at least initial work deals with best practices. Schühly et al. present an AI solution for monitoring scenarios (Schühly et al. 2020). Jungwirth and Haluza state that the capabilities of generative AI like ChatGPT to systematically develop consistent scenarios are still limited (Jungwirth and Haluza 2023). Spaniol and Rowland also state that AI-generated scenarios, i.e., scenarios developed independently by generative AI, are usually too generic and superficial. In contrast, AI-assisted scenario-based foresight, i.e., scenario-based foresight in which AI and humans work together, can improve the efficiency of scenario-based foresight and the evidence of the results. The quick provision of raw material for scenario development or identified white spots for influence factors are mentioned as examples (Spaniol and Rowland 2023). The investigations of Fischer et al., whether and how ChatGPT can support real scenario projects, support these findings. Accordingly, generative AI can be used in principle as an impulse generator or additional expert in the scenario process. However, humans must always evaluate the quality of AI-generated results. The use of AI applications and the interaction with these applications also result in completely new tasks for the scenario experts, e.g., sufficiently precise prompt engineering (Fischer et al. 2023). However, it has not yet been clarified how exactly the role of human scenario experts will change and how a digitally sovereign interaction between humans and technology should be designed (Steinmüller 2022; Bauer et al. 2022).

Against this background, we derive three research questions that we analyze through an explorative interview study:

- What is the status quo of scenario-based foresight in practice?
- To what extent are digital technologies used in scenario-based foresight, and what are the potentials of digital technologies?
- What should the interaction between humans and digital technologies look like in scenario-based foresight?

The structure of the paper is as follows: Sect. 2 presents the research design of the explorative interview study. Section 3 displays the results. Section 4 concludes the paper with a discussion.

2   Research Design

We conducted an explorative interview study with eight scenario experts from industrial companies and consultancies to answer the research questions. Specifically, our research design consists of two overarching phases: data collection and data analysis:

Data collection: For the data collection, we used a 4-step procedure following the suggestions of Eisenhardt (1989): preparing the interview guide, selecting interview partners, conducting the interviews and documenting the interviews. First, we developed the interview guide for our semi-structured interviews. Based on our research questions, we divided our interview guide into five sections: (1) status quo of scenario-based foresight in the companies, (2) use of tools and digital technologies for scenario-based foresight, (3) challenges, (4) potentials, and (5) digital sovereignty of scenario-based foresight practitioners. Subsequently, we present the questions of our interview guide:

(1) First, we wanted to explore the current status quo regarding the use of scenario-based foresight within industrial companies and consultancies. For this reason, we posed the following questions:

    1. For what purposes is scenario-based foresight used?
    2. How is the planning of scenario projects conducted?

(2) Second, we wanted to know which tools and digital technologies are used in practice for conducting scenario projects. Therefore, we asked the following questions:

    3. Which tools and digital technologies are used for scenario-based foresight?
    4. To what extent can the tools and digital technologies be tailored to the company's needs?

(3) Third, we wanted to identify the challenges that companies face during scenario projects in general and specifically regarding the use of digital technologies. For this purpose, we raised the following questions:

    5. What are the main challenges in applying scenario-based foresight?
    6. What are the biggest challenges in applying digital technologies to scenario-based foresight?

(4) Fourth, we wanted to explore the potential of digital technologies for scenario-based foresight by asking the subsequent questions:

    7. What would be the greatest added value to be created using digital technologies for scenario-based foresight?
    8. What new methodological steps are conceivable through the use of digital technologies?

(5) Finally, with the help of the subsequent questions, we wanted to find out which aspects of digital sovereignty matter the most regarding the use of digital technologies:

    9. Which tasks should humans still perform in the future?
    10. Which tasks should be performed by digital technologies, e.g., AI applications, in the future?

Furthermore, the interview guide included a few more basic questions for exploring possible answers in more detail, e.g., *what are the technical challenges when using digital technologies? What are the non-technical challenges?*

Second, we selected the interview partners. With our sample, we wanted to cover the different users of scenario-based foresight sufficiently. Therefore, we selected

ID	Position	Company type	Experience
I-0	Chief scientist scenario technique	Industrial company	24 years
I-1	Head of new market and portfolio strategy	Industrial company	9 years
I-2	Senior expert product innovation	Industrial company	10 years
I-3	Head of foresight and strategy	Consultancy	13 years
I-4	Head of trendscouting and innovation management	Industrial company	6 years
I-5	CEO	Consultancy/software dev.	9 years
I-6	Deputy director	Consultancy	24 years
I-7	Partner	Consultancy	25 years

Fig. 1 Overview of the interview partners

interviewees from four consultancies that conduct scenario projects with companies from many different branches and four industrial companies from drive technology with broad expertise in scenario-based foresight. According to the suggestions for case study research from Eisenhardt (1989), this number of interviews is considered sufficient for our research purpose. Furthermore, the heterogeneity of our interview partners and their companies, e.g., the differing company sizes, helped us gain insight into a wide range of scenario-based foresight users, increasing our results' validity. Figure 1 characterizes the interviewed scenario experts.

Third, we conducted the eight interviews digitally via Microsoft Teams. The interviews were conducted by two interviewers and lasted 45–60 min each. The first interview (I-0) served not only to gather insights into the issues raised in our research questions but also to review and improve the interview guide and create a final version.[2]

Fourth, for documentation, we transcribed the recorded interviews using the simple transcription rules of Dresing and Pehl (2011). We also anonymized all interviews to prevent inferences about the interview partners and their companies.

Data analysis: We conducted a summarizing qualitative content analysis according to Mayring (2019) to analyze the interviews. The procedure for data analysis consists of two phases: coding the interviews and deriving key messages. Figure 2 shows the steps of data analysis for each phase.

---

[2] Nevertheless, the guide was adapted again to the newly collected insights in the further course of the explorative interview study. This is in line with the suggestions from Eisenhardt for theory-building case research (Eisenhardt 1989). For example, the question, *"Who develops the scenarios?"* was replaced by the question *"How is the planning of scenario projects conducted?"* because the planning phase of scenario projects emerged as an important topic in the first interviews.

Fig. 2 Procedure for
analyzing the interviews

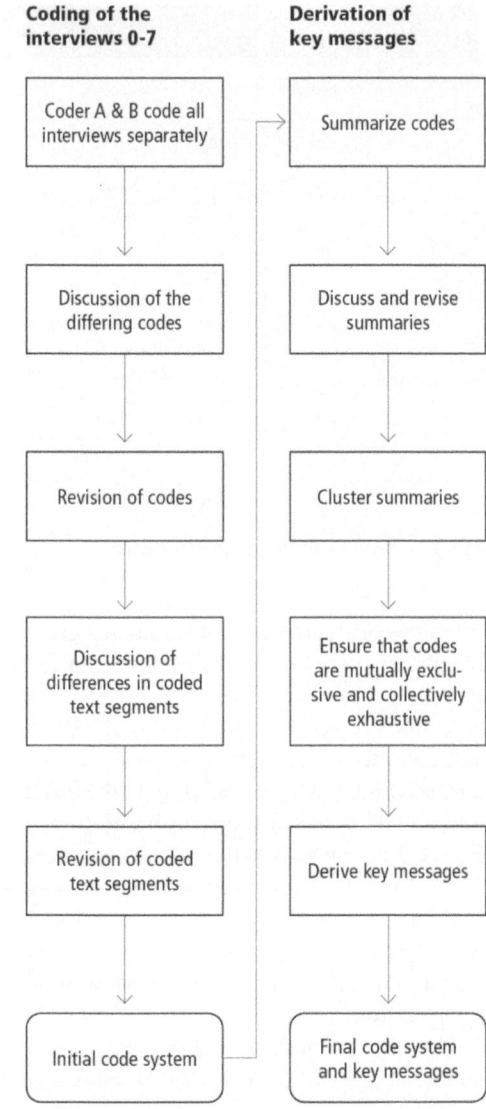

In the first phase, we (coders A and B) coded all interviews independently using the software Citavi. The coding was done inductively. Then, we discussed the differences between our two code systems and created one common code system. After revising the codes, we analyzed and discussed differences in the coded text segments. We achieved full inter-coder reliability by resolving all differences in the coded text segments. I.e., our initial common code system included aligned text segments for all interviews.

In the second phase, we derived key messages from the interviews. Key messages (or so-called themes) summarize something important about the data (in this case, interview transcripts) regarding the research questions. They represent some level of response pattern or meaning within the analyzed data set (in this case, the set of interview transcripts) (Braun and Clarke 2006). For deriving the key messages, we first summarized all the paraphrased and generalized text segments that belonged to the same code. Second, we discussed and revised our summaries. Afterward, we clustered the summaries. We made sure that all codes and text segments were mutually exclusive and collectively exhaustive across all clusters. I.e., each text segment was uniquely assigned to one code resulting in the final code system. Finally, we formulated key messages based on the clusters. As a result, we obtained seven key messages from the interviews.

3   Results

The seven key messages from the interviews serve to answer our research questions. In the following, we will first present our findings regarding the status quo of scenario-based foresight. Second, we display our insights regarding the use of digital technologies. Third, we show our findings concerning the digital sovereignty of users conducting scenario-based foresight. All sources are indicated with the respective IDs of our interview partners, e.g., [I-0].

3.1   Status Quo of Scenario-Based Foresight

The interviewees of companies interviewed use scenario-based foresight in particular for developing and reviewing strategies, but also for innovation purposes [I-0, I-1, I-2, I-3, I-4, I-5, I-7]. The basic methodical procedure for the development of scenarios is *"very strongly standardized in terms of the process"* [I-1]. That is, the methodological framework, i.e., the overarching phases, is fixed [I-0, I-7]. However, the exact design of the individual methodological steps is for each scenario project [I-5, I-6, I-7]. On the one hand, the objective and the content-related context are specific to each scenario project [I-3, I-4]. On the other hand, the individual steps can be methodologically simplified (e.g., influence factors can be evaluated via influence and relevance analysis, via a ranking according to importance and uncertainty or via dot sticking), the degree of involvement of external partners can be varied, different tools can be used, or different data sources to be analyzed can be used [I-5, I-6, I-7]. In addition, the application of the scenarios at the end of a project is carried out methodically individually, depending on the application purpose [I-7]. Consequently, *"there is not that one process for a scenario project"* [I-6], so different scenario projects *"also need different planning"* [I-5]. To simplify the planning of a scenario project, criteria catalogs are sometimes used, including the aspects mentioned above [I-5].

Key message 1: The basic methodological framework for scenario development is generally the same in scenario projects. However, the individual steps are often customized for each project. The application of scenarios is usually done individually, depending on the application's purpose.

In scenario projects, we can distinguish between large, methodologically comprehensive projects and smaller, methodologically rather simplified projects. In the industrial companies surveyed, smaller scenario projects tend to be carried out [I-1, I-2, I-4].

[The] overall process, [...] the methodical, clean and consistent, we use that, but rarely [I-1].

This is due to the challenges fundamentally associated with scenario projects, particularly large ones. First, conducting a scenario project requires a large amount of time [I-0, I-1, I-2, I-3, I-4, I-5]:

In our industrial practice, however, I find, we often do not have the time for a full scenario project. That is the biggest challenge: time needed for a really methodologically coherent, complete walk-through [I-1].

Examples of particularly time-consuming steps are the initial research and the associated evaluation of large amounts of data for deriving influence factors or the consistency evaluation of projections [I-0, I-4, I-5].

Furthermore, the lack of methodological expertise challenges companies [I-2, I-7]. For this reason, industrial companies often use the external support of consulting firms for large scenario projects [I-1]. From a methodological perspective, for example, the consistency assessment, in particular, requires a high level of expertise and experiential knowledge [I-7]. When it comes to planning scenario projects, the interviewees from all consulting companies highlight the clear definition of the objective, topic, and scope of a scenario project as challenging and critical for the success of such a project [I-3, I-5, I-6, I-7]. *"This clarification of content is [...] a very important point at the beginning"* [I-7].

Furthermore, once the scenarios have been developed, communicating them to uninvolved people proves to be challenging:

It is extremely difficult to fully explain to someone who was not involved in the project how these scenarios work and what the thinking behind them is [I-1].

Consequently, it takes much effort to describe and prepare scenarios in such a way as to create a common understanding [I-4, I-7].

Key message 2: Key challenges in conducting methodologically complete scenario projects are the high expenditure of time and insufficient methodological expertise. When planning scenario projects, it is particularly important and challenging to define the objective and scope of the project clearly. After scenario development, communicating the scenarios to people not involved in the process is a major challenge.

3.2 Digital Technologies

In small, methodologically less extensive scenario projects that industrial companies carry out themselves, digital technologies are hardly ever used [I-1, I-6]:

I don't need the software there [I-1].

Furthermore, it does not pay off for industrial companies to purchase special software for scenario-based foresight or develop elaborate scenario software independently [I-1, I-3]. Instead, freely available tools for scenario-based foresight [I-2], simple, proprietary Excel tools [I-4], or tools that serve another primary purpose are used. In particular, trend databases are mentioned as such tools. They provide input for collecting influence factors [I-1, I-3, I-4]. Trend databases are either used manually as knowledge bases or to evaluate defined data sources with the help of text mining.

In larger projects, industrial companies often work with external partners, such as consultancies. In addition to content input and methodological expertise, these companies usually also provide traditional professional scenario software. This is software in which *"you can simply map the project in it throughout; from the collection of influence factors to the influence matrix, key factors [...]"* [I-3]. In other words, this classic scenario software supports the scenario process *"from the start of the project to the future map"* [I-7].

However, consultancies, in particular, use tools beyond classic scenario software. That is, tools that can either provide input for the scenario process or support the preparation of scenarios and scenario transfer:

All the rest [of tools] that is on the outside can then be added on a case-by-case basis [I-7].

For example, AI tools are sometimes used to flesh out the scenarios textually or visually or to monitor the scenarios [I-5, I-7]. However, rather simple survey tools or tools for digital collaboration are also used [I-3, I-4, I-6, I-7].

Key message 3: In large, methodologically comprehensive scenario projects, where industrial companies often work with consultancies, scenario software is usually used to support the process from defining influence factors to scenario generation. In smaller scenario projects, industrial companies rarely use software solutions. In contrast, consulting companies sometimes also use AI tools in scenario projects and tend to develop software solutions themselves.

In the course of the scenario-based foresight, a wide range of potentials for digital technologies were also mentioned in the interviews. Accordingly, a digital, methodological assistant can ensure that the entire scenario process is *"methodically guided through once"* [I-4]. On the one hand, it would address a lack of methodological knowledge, and on the other hand, it enables project participants to focus on content-related discussions [I-0, I-2, I-4, I-5].

Digital technologies can also increase efficiency in particularly time-consuming tasks in scenario-based foresight, e.g., by analyzing large amounts of data:

So, the biggest added value right now is certainly the ability to collect really big data faster and more accurately [...]and analyze it accurately, systematically, and repeatedly [I-5].

Through this capability, digital technologies can support a collection of influence factors that is as complete as possible. In particular, suggesting influence factors that would otherwise not have been necessarily considered by companies but have an impact on the defined object of investigation, as well as identifying potential white spots, are seen as advantages [I-0, I-1, I-2, I-3, I-5, I-6, I-7].

Digital technologies could also improve the monitoring of scenarios, for example, through data-driven analytics and automatic notifications to decision-makers when fundamental elements of scenarios change:

So, you don't have to be proactive or active about it anymore, but you just hand off this monitoring of weak signals to a software system and it proactively informs you. [...] so it's a proactive notification, an automated, recurring scanning [I-5].

However, digital technologies can support or take over even more tasks of scenario-based foresight. They could support the consistency assessment of projections by an automatic pre-assessment or suggestions for assessing individual projection pairs [I-0, I-3, I-6]. Furthermore, they could simplify or automatically perform sensitivity analyses of scenarios [I-0, I-3, I-6]. Digital technology could also promote creativity in projection development through suggestions [I-3, I-7]. Another potential is seen in improving to flesh out scenarios textually or visually [I-6].

Key message 4: Potential for the use of digital technologies exists throughout the entire process of scenario development and monitoring of scenarios. The greatest potential of digital technologies is seen in analyzing large volumes of data to automatically propose a list of influence factors that is as complete as possible and in a digital assistance system to support the methodological steps.

Digital technologies may or may not change the methodological scenario process. "*There are, after all, two extremes*" [I-7]. On the one hand, digital technologies can specifically support individual steps of scenario-based foresight. In this case, no methodological changes are to be expected:

I'm not sure that digital technologies would necessarily change anything about the scenario technique now. Digital technologies are still just tools [I-2].

On the other hand, the use of digital technologies, especially AI applications, provides the opportunity to design a completely new process for scenario-based foresight:

You just have to see, when you think of these tools, [that you] probably can't think of the classic process. But rather, you probably have to completely rethink the scenario process [I-7].

Regardless of whether digital technologies and AI applications will support individual steps in the classic scenario process or change it completely, their use will result in new tasks for the user. One of these new tasks is prompt engineering, i.e., the description of tasks to be performed by AI applications. The user must "*formulate and describe everything in great detail so that a tool like this can continue to use that*" *automatically"* [I-7]. Otherwise, *"if you ask the wrong questions, you get the wrong answers"* [I-4].

Key message 5: Digital technologies that specifically support individual steps of scenario-based foresight will change the methodology little or not at all. However, if digital technologies, especially AI applications, can develop consistent scenarios independently in the future, the scenario process will have to be completely rethought and redesigned.

3.3 Digital Sovereignty

Digital technologies are able and allowed to support many tasks of scenario-based foresight or to perform them completely themselves, e.g., the analysis of large amounts of data [I-5]. However, digital technologies should not be complete black box systems [I-1, I-4, I-5]. Rather, they should ensure the secure handling of corporate data [I-5]. Digital technologies must also be comprehensible to the user and be able to transparently show why they arrive at which results:

> Transparency, what this digital technology actually does and how it arrives at its results. I think that is very important [I-1].

However, this does not mean that the procedures of digital technologies and their results must be fully explainable:

> [...] currently it is so that it is not at all possible here to ensure the explainability. So that is also not possible and also not necessary [I-5].

Key message 6: With regard to digital sovereignty, digital technologies need not and cannot be fully explainable. However, it should be comprehensible and transparent for the user at any time, for which reasons which results are obtained.

Despite the required transparency of digital technologies, human users should always check their results for plausibility [I-2], [I-7]:

> Of course, as a human being, you always have to critically examine and question the whole story. That will always remain because all the tools only reproduce what knowledge there is [I-7].

Digital technologies may take over individual tasks and provide suggestions for decisions. However, regarding divergence, a characteristic feature of socio-digital sovereignty related to the concepts of scope for action and degrees of freedom (Hartmann 2020; Hartmann 2021), humans should have extensive intervention capabilities throughout the whole process or be able to adapt results [I-1, I-3]. The actual decisions should also continue to be made sovereignly by humans:

> So, IT should propose it, and then we decide. In every step, ultimately, IT can support and make suggestions [I-1].

> But making the decision and then following through with the decision and implementing it should continue to be done by humans [I-5].

Moreover, humans should continue to lead the actual discussions about the future since these discussions represent a significant added value in scenario projects [I-0, I-1, I-2, I-6]: *"The discussion about the future must take place"* [I-0]. This includes, above all, the development of projections and the interpretation of scenarios for the respective application purpose: *"Of course, people still have to interpret themselves"* [I-5]. At the same time, the interviewees estimate digital technologies as still limited in performing the aforementioned tasks. On the one hand, they doubt whether digital technologies can creatively develop projections relying on a historical database [I-2, I-7]. On the other hand, they note that digital technologies do not have sufficient domain knowledge or company-relevant knowledge for interpretation [I-1, I-5].

Key message 7: Humans should critically question the results of digital technologies and check their plausibility. In particular, humans should continue to discuss the future in scenario-based foresight, i.e., in particular, develop projections, interpret and apply the scenarios, and make decisions.

4   Discussion

Research outcome: We examined three research questions in our explorative interview study with eight industry companies and consultancies. Our first research question addressed the status quo of scenario-based foresight. We discovered that the overarching methodical phases are standardized but the individual steps are customized to a respective project context of a company. This aligns with our previous work on a reference model for scenario-based foresight (Ködding et al. 2023b). Additionally, industry companies tend to conduct smaller scenario projects rather than methodological complete projects due to the well-known key challenges of scenario-based foresight: the high expenditure of time and the methodological complexity. In the relevant literature, however, the importance and the challenge of planning a scenario project and defining its objective and scope are not emphasized much.

The second research question addressed the status quo and potential of digital technologies in scenario-based foresight. Our findings suggest that the use of digital technologies is rather heterogeneous. In large scenario projects, scenario software is usually employed to support the whole process, whereas, in small scenario projects, hardly any software is used. When comparing industry companies to consultancies, it can be stated that consultancies use digital technologies more regularly and that they even use self-developed or existing AI applications. According to the interviewees, there is potential for digital technologies throughout the entire process of scenario development and monitoring of scenarios. Supporting the work of Bauer et al. Steinmüller and van Belkom, the interviewees assigned the biggest potential to the analysis of large amounts of data, e.g., for identifying influence factors (Bauer et al. 2022; Steinmüller 2022; van Belkom 2020). However, the interviewees also highlighted the potential for a digital methodological assistant. On top, they raised the question of whether or not the whole process of scenario-based foresight can or should be rethought as soon as digital technologies and especially AI applications can develop consistent scenarios independently.

The third and final research question aimed at digital sovereignty and the interaction between humans and technology. In this regard, we can conclude that digital technologies do not need to be fully explainable but rather provide results transparently. Furthermore, in line with the relevant literature, humans should always check the plausibility of such results. Also, humans should continue to conduct the steps that require company- and domain-specific knowledge or creativity.

Evaluation and limitations: We base the evaluation of our interview study on the evaluation criteria for case study research and qualitative content analysis by Yin and Mayring (2018, 2019):

Intersubjective traceability is ensured because our research process and the derivation of key messages are transparently described and substantiated in the interview transcripts. Therefore, in terms of *reliability*, other researchers should be able to repeat our analysis. Due to the purely qualitative data of our interview study, we could not meet the requirements for *triangulation*. Therefore, we considered all data identified as equally important. The next limitation is related to the date of the interviews. Since the first five interviews were conducted before the release of the generative AI tool ChatGPT, the first interviewees did not have access to the same information and tool landscape. However, since we treated all data equally important, we should have limited possible bias in the analysis. Moreover, the explorative interview study was only conducted with eight companies. We might have yielded further insights if we had conducted more interviews. Therefore, we cannot claim a statistical representation of our findings. However, due to the extensive experience of our interview partners, e.g., the interviewed consultants conduct scenario projects with companies from many different branches, and because we felt that we already reached the point of theoretical saturation, i.e., further interviews might not yield significantly more insights (Glaser and Strauss 2008), we are confident that we covered the most relevant aspects.

Implications for future research: The findings and the limitations of the paper lead to two recommendations for future research: (1) First, a procedure model should be developed that enables the individual design of the scenario process for scenario projects. For this, various criteria should be considered, such as the application purpose, the digital technologies to be used, the interaction between humans and technology, or the available resources. In this way, the existing methodological chaos (Martelli 2001; Millett 2003; Bradfield et al. 2005) can be reduced in practice. (2) Second, foresight practitioners should be provided with a structured overview of digital technologies in scenario-based foresight. To this end, the findings from practice and research must be consolidated and expanded. The identified digital technologies and their potential as revealed in the interview study can be expanded through a tool study. The existing use cases from the literature can be complemented by an exploratory analysis of potentials that adequately considers the manifold possibilities of emerging digital technologies, e.g., generative AI.

References

Andre, L., Michael, B., Daniel, R., Christian, K.: Framework for the identification and demand-orientated classification of digital technologies. In: 2018 IEEE International Conference on Technology Management, Operations and Decisions (ICTMOD), 2018 IEEE International Conference on Technology Management, Operations and Decisions (ICTMOD), Marrakech, Morocco. IEEE, pp 31–36 (2018)

Bain & Company, Inc. (Hrsg.).: Management tools. scenario and contingency planning. (2018). Available online at https://www.bain.com/insights/management-tools-scenario-and-con tingency-planning/ (accessed 18 July 2021)

Barton, D., Manyika, J., Koller, T., Robert Palter, T., Godsall, J., Zoffer, J.: Measuring the economic impact of short-termism. Available online at https://www.mckinsey.com/~/media/mckinsey/ featured%20insights/long%20term%20capitalism/where%20companies%20with%20a%20l ong%20term%20view%20outperform%20their%20peers/mgi-measuring-the-economic-imp act-of-short-termism.ashx (accessed 18 July 2021)

Bauer S, Kollosche I, Uhl A, de Melo G, Fritzsche K.: Die digitale Vermessung der zukunft. Welche rolle spielt künstliche intelligenz in foresight zur gestaltung von nachhaltigkeitstransformationen? [Digital surveying the future. what role does artificial intelligence play in foresight to shape sustainability transformations?] 9th ed. (2022). Available online at http://gerard.dem elo.org/papers/foresight-zukunft.pdf

Berger, S., Denner, M.S., Roeglinger, M.:The nature of digital technologies. Development of a multi-layer taxonomy. In: Proceedings of the 26th European Conference on Information Systems (ECIS), 26th European Conference on Information Systems, Portsmouth, United Kingdom, pp. 1–18 (2018)

Bradfield, R., Wright, G., Burt, G., Cairns, G., van der Heijden, K.: The origins and evolution of scenario techniques in long range business planning. Futures 37(8), 795–812 (2005). https:// doi.org/10.1016/j.futures.2005.01.003

Braun, V., Clarke, V.: Using thematic analysis in psychology. Qual. Res. Psychol. 3(2), 77–101 (2006). https://doi.org/10.1191/1478088706qp063oa

Dresing, T., Pehl, T. (eds.): Praxisbuch Transkription—interview, transkription & analyse. anleitungen und regelsysteme für qualitativ forschende. [Handbook of transcription—interview, transcription & analysis. Instructions and rule systems for qualitative researchers], 2nd edn. Dr. Dresing und Pehl GmbH, Marburg (2011)

Eisenhardt, K.M.: Building theories from case study research. Acad. Manag. Rev. 14(4), 532 (1989). https://doi.org/10.2307/258557

Fischer, D., Joachim, V., Tranaes, S., Jung, H.H.:KI in der Vorausschau. Kritische Evaluation der Anwendung von generativer KI am Beispiel von ChatGPT in der Szenario-Technik. [AI in foresight. Critical evaluation of the application of generative AI using the example of ChatGPT in scenario engineering.] (In press) In: Dumitrescu, R. (ed.), Vorausschau und Technologieplanung. 17. Symposium für Vorausschau und Technologieplanung, Berlin, Berlin, 14-15. September 2023. Paderborn, Heinz Nixdorf Institut, Universität Paderborn (2023)

Gausemeier, J., Dumitrescu, R., Pfänder, T., Steffen, D., Thielemann, F.: Innovationen für die Märkte von morgen. Strategische planung von produkten, dienstleistungen und geschäftsmod-ellen. [Innovations for tomorrow's markets. strategic planning of products, services, and business models]. München, Hanser (2019)

Glaser, B.G., Strauss, A. L.: Grounded theory. Strategien qualitativer Forschung. [Grounded theory. Strategies of qualitative research.] Bern, Huber (2008)

Hartmann, E.A.: Digitale souveränität in der Wirtschaft—Gegenstandsbereiche, Konzepte und Merkmale. [Digital Sovereignty in Business–Subject Areas, Concepts, and Characteristics]. In: Hartmann, E.A (ed.), Digitalisierung souverän gestalten. Berlin, Heidelberg, Springer Berlin Heidelberg, pp. 1–16 (2020)

Hartmann, E.A.: (2021). Digitale souveränität: soziotechnische bewertung und gestaltung von anwendungen algorithmischer systeme. [Digital Sovereignty: Socio-technical assessment and design of algorithmic systems applications]. In: Hartmann, E.A (ed.), Digitalisierung souverän gestalten II. Berlin, Heidelberg, Springer Berlin Heidelberg, pp. 1–13 (2021)

Homer-Dixon, T., Renn, O., Rockstrom, J., Donges, J.F., Janzwood, S.: A call for an international research program on the risk of a global polycrisis. SSRN Electron. J. (2021). https://doi.org/10.2139/ssrn.4058592

Jungwirth, D., Haluza, D.: AI-based scenario generation for future planning. An exploratory study using GPT-3. J. Curr. Trends Comput. Sci. Res. 2(2), 57–67 (2023). https://doi.org/10.33140/JCTCSR

Ködding, P., Ellermann, K., Koldewey, C., Dumitrescu, R.: Scenario-based foresight in the age of digitalization and artificial intelligence—identification and analysis of existing use cases. Procedia CIRP 119, 740–745 (2023a). https://doi.org/10.1016/j.procir.2023.01.015

Ködding, P., Koldewey, C., Dumitrescu, R.: A reference process model for scenario-based foresight. (In press) In: Proceedings of the XXXII ISPIM innovation conference. Innovation and circular economy, ISPIM innovation conference, Ljubljana, Slowenien (2023b)

Martelli, A.: Scenario building and scenario planning: state of the art and prospects of evolution. Futur. Res. q. 17(2), 57–74 (2001)

Mayring, P.: Qualitative inhaltsanalyse—abgrenzungen, spielarten, weiterentwicklungen. [Qualitative content analysis—Delimitations, varieties, further developments]. https://doi.org/10.17169/fqs-20.3.3343

Millett, S.: The future of scenarios: Challenges and opportunities. Strategy & Leadership 31(2), 16–24 (2003)

Rigby, D., Bilodeau, B.: Management tools & trends. (2018). Available online at https://www.bain.com/contentassets/caa40128a49c4f34800a76eae15828e3/bain_brief-management_tools_and_trends.pdf

Rigby, D., Bilodeau, B., Ronan, K.: Management tools & trends 2023. (2023). Available online at https://www.bain.com/insights/management-tools-and-trends-2023/ (accessed 8 Feb 2023)

Schühly, A., Becker, F., Klein, F.: Real time strategy: When strategic foresight meets artificial intelligence. Emerald Publishing Limited, (2020)

Spaniol, M.J., Rowland, N.J.: AI-assisted scenario generation for strategic planning. Futures & Foresight Science (2023). https://doi.org/10.1002/ffo2.148

Steinmüller, K.: Essay: Kann Künstliche Intelligenz Zukunftsforschung? Ein spekulativer Impuls. [Essay: Can Artificial Intelligence Do Future Research? A Speculative Impulse]. Zeitschrift für Zukunftsforschung, (2022)

van Belkom, R.: The impact of artificial intelligence on the activities of a futurist. World Futur. Rev. 12(2), 156–168 (2020). https://doi.org/10.1177/1946756719875720

Yin RK (2018) Case study research and applications. Design and methods. Los Angeles/London/New Dehli/Singapore/Washington DC/Melbourne, SAGE.

Patrick Ködding is a Ph.D. candidate and research associate at Heinz Nixdorf Institute of the University of Paderborn. He works at the chair of Advanced Systems Engineering. Its focus is on strategic planning and innovation management as well as on systems engineering for intelligent systems. Patrick Ködding has also been a member of the doctoral network 'Digital Sovereignty in Business'; with its focus on mechanical engineering of the future – a project of the Institute for Innovation and Technology (iit) funded by Dr. Johannes Heidenhain GmbH. His research topics are foresight, strategy development, and the use of artificial intelligence in product engineering.

Mathis Jahn is a master's student of mechanical engineering with a focus on mechatronic systems. He works as a research assistant at the chair of Advanced Systems Engineering at the Heinz Nixdorf Institute of the University of Paderborn. His research focus is data-driven product engineering.

Dr.-Ing. Christian Koldewey is chief engineer at the chair of Advanced Systems Engineering at Heinz Nixdorf Institute of the University of Paderborn. There, he is responsible for the business engineering department. He obtained his Ph.D. from Paderborn University under the supervision of Prof. Dr.-Ing. Jürgen Gausemeier. His current research focuses on the transformation towards smart services and data eco-systems in manufacturing as well as general topics in digital transformation and circular economy

Prof. Dr.-Ing. Roman Dumitrescu is Director at the Fraunhofer Institute for Mechatronics Design IEM and Head of the chair for Advanced Systems Engineering at Heinz Nixdorf Institute of the Paderborn University. His research focuses on the product engineering of intelligent technical systems. Prof. Dumitrescu is also the managing director of the technology network Intelligent Technical Systems OstWest-falenLippe ('it's OWL'). In this cluster, he is responsible for strategy, research, and development. He is a member of the research advisory board of the research association 3-D MID e. V. and head of the VDE/VDI technical committee 'Mechatronically Integrated Assemblies'.

Hybrid AI-Driven Advances in Prognostics and Health Management Within Manufacturing Environments

Christopher Braun and Marco F. Huber

Abstract Industry 4.0 will benefit significantly from the ongoing advancements in artificial intelligence, particularly with regard to predictive maintenance. By continuously monitoring and analysing real-time data of a system, proactive maintenance actions can be taken before any major issues arise. Incorporating prognostics and health management allows for assessing the health of a system and predicting its future state based on current operating conditions. However, a major challenge in health modelling within manufacturing environments is modelling systems that are experiencing trend-based degradation. The field of theory-guided data science offers potential solutions by integrating prior knowledge about the system directly into data-driven methods, providing a hybrid approach to effectively implement prognostics and diagnostics in the context of Industry 4.0.

Keywords Industry 4.0 · Hybrid AI · Predictive maintenance · Prognostics and health management · Theory-guided data science

1 Introduction

The exponential progress of Artificial Intelligence (AI) over the last decade has initiated a transformative revolution, profoundly impacting various industries and fundamentally reshaping human interactions with technology. The advances in AI can be mainly attributed to the confluence of factors including increased computational power, the availability of big data, and significant progress in algorithm development (Fink et al. 2020). The ever-increasing wealth of data provides a valuable resource for training AI models and has been instrumental in the success of Machine Learning

C. Braun (✉) · M. F. Huber
Institute of Industrial Manufacturing and Management IFF, University of Stuttgart, Stuttgart, Germany
e-mail: christopher.braun@iff.uni-stuttgart.de

M. F. Huber
e-mail: marco.huber@ieee.org

Fraunhofer Institut for Manufacturing Engineering and Automation IPA, Stuttgart, Germany

© The Author(s) 2025
U. Schmuntzsch et al. (eds.), *New Digital Work II*,
https://doi.org/10.1007/978-3-031-69994-8_7

(ML) and Deep Learning (DL) algorithms, which rely on large data sets for training and generalisation. In addition, the open source community, industry investment and the development of the hardware and software infrastructure required for the efficient and scalable training and deployment of AI models have contributed significantly to the vibrant development and widespread adoption of AI.

The prerequisites for implementing ML and DL-based applications are being met in the context of Industry 4.0 due to increasing digitalisation and connectivity, resulting in the generation of vast amounts of data, coupled with advancements in hardware technology facilitating high computing power (Gauger et al. 2022). Consequently, Industry 4.0 stands to gain significant benefits from the ongoing advancements in AI. By integrating AI into production systems, Industry 4.0 aims to create highly flexible, efficient, and sustainable manufacturing processes (Wagner et al. 2020). It revolutionises traditional production methods and paves the way for smart, interconnected factories of the future. AI algorithms can be employed to analyse vast amounts of data collected from various sources, enabling Predictive Maintenance (PdM), quality control, and efficient resource allocation. Based on a survey conducted by the German digital association Bitkom, the companies surveyed consider PdM to be the second most important advantage of AI next to the increase in productivity in the context of Industry 4.0 (Berg 2019).

According to Ran et al. (2019), the intention of PdM is to shift from traditional reactive and preventive maintenance approaches to a more data-driven and condition-based methodology. By continuously monitoring and analysing the real-time data collected from various sensors, devices, and machines, PdM aims to identify early signs of equipment degradation or failure, allowing for timely intervention before any significant problems occur. By implementing PdM, organisations can schedule maintenance activities based on actual equipment condition rather than fixed schedules, thus reducing unnecessary maintenance actions and associated costs. It enables the identification of potential faults or anomalies before they lead to critical failures, ensuring maximum uptime and operational efficiency. Additionally, PdM helps optimise spare parts inventory management by accurately predicting the need for replacements or repairs.

The field of Prognostics and Health Management (PHM) can be considered the driving force for the wide accessibility of PdM in terms of technical realisation. PHM enables assessing the health of a system as well as predicting its future state based on up-to-date information on actual operating conditions, empowering engineers to leverage information from the system's health state to enhance their understanding and implement strategies for maintaining the system in its originally intended function (Kim et al. 2017). Nevertheless, modelling systems that experience trend-based degradation is considered one of the major challenges inherent to health modelling in manufacturing environments (Toothman et al. 2023). According to Hagmeyer et al. (2022), there exist three primary approaches for implementing diagnostic and prognostic applications including data-driven approaches, physical model-based approaches, and hybrid approaches. However, Kim et al. (2017) advise to focus on the development of novel hybrid approaches to compensate for the limitations of pure data-driven and model-based approaches.

Data-driven approaches are limited by the availability of run-to-failure data and the fact that causal relationships are not learned, possibly leading to implausible predictions, whereas model-based approaches are mainly limited by the significant complexity of the degradation process to be modelled (Hagmeyer et al. 2022). A promising approach to overcoming these limitations, which is in line with the recommended development of hybrid approaches, comes from the field of Theory-Guided Data Science (TGDS). Proposed by Karpatne et al. (2017), TGDS focuses on harnessing the extensive scientific knowledge available to enhance the efficacy of data science models. The similar research area of Informed Machine Learning (IML) (von Rueden et al. 2021) describes learning from a hybrid information source consisting of data and prior knowledge, where the "prior knowledge comes from an independent source, is given by formal representations, and is explicitly integrated into the machine learning pipeline". The authors state that integrating prior knowledge into learning systems enables performance improvements by reducing the amount of training data needed, increasing the interpretability as well as enhancing the generalisation capabilities of models. With IML considering all sources of knowledge, i.e. scientific, general and expert knowledge, it intuitively points to the possibility of productively engaging workers by providing such knowledge, which will be discussed again in the concluding considerations. TGDS, however, solely focuses on leveraging scientific knowledge such as physical laws, simulations and established models or frameworks within the relevant field. Accordingly, TGDS serves as a means to fully exploit the potential of PHM through the development of hybrid models, which ultimately enables the successful implementation of PdM strategies.

2 Predictive Maintenance

Maintenance is essential to ensure operational continuity, prevent costly downtime, and safeguard a company's reputation and competitiveness. The evolution of maintenance strategies has been driven by the necessity to mitigate the consequences arising from the inevitable degradation of engineering systems, transitioning from Reactive Maintenance (RM) to Preventive Maintenance (PM) and ultimately to PdM approaches (Ran et al. 2019; Baur et al. 2020; Huber 2021). RM is characterised by addressing equipment failures and performing repairs in response to their occurrence, whereas PM focuses on scheduled maintenance tasks performed to prevent failures and maintain equipment reliability. As a result, the strategy employed may lead to downtime or potentially unnecessary maintenance if failures do not occur within the planned intervals. This has led to the development of Condition-based Maintenance (CbM), a special case of PM in which the necessary maintenance activities are scheduled based on the current state of the system. However, CbM lacks the ability to anticipate incipient faults. By leveraging data-driven insights to predict equipment failures and schedule maintenance activities proactively, PdM builds on CbM and enables minimising both unplanned downtime and unnecessary preventive maintenance tasks, counteracting the shortcomings of RM and PM, respectively.

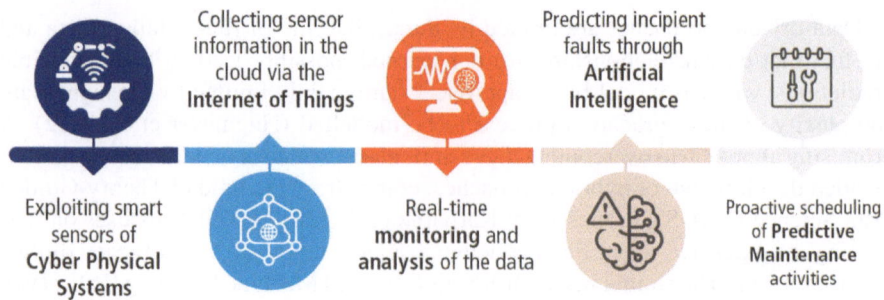

Fig. 1 Smart sensor technology (CPS) and its inter-connectivity (IoT) enable collecting and analysing vast amounts of data with the aid of learning algorithms (AI) to derive information about incipient faults, which allows to proactively schedule PdM before a catastrophic failure occurs (own illustration)

Cyber Physical Systems (CPS), the Internet of Things (IoT) and AI are key technological drivers contributing to the rise of Industry 4.0 (Wang 2016) and have enabled the long-standing concept of PdM (Fig. 1) to become widely accessible by enhancing this maintenance strategy in the following aspects (Ran et al. 2019): CPS involve the integration of physical systems with digital systems, allowing for real-time data exchange and control of sensors and actuators, while IoT facilitates extensive data acquisition by connecting a multitude of devices to the internet, enabling seamless interaction and communication among them. This, in turn, provides the necessary infrastructure for gathering and processing data, while AI leverages this data to predict and identify potential equipment failures. By employing sophisticated learning algorithms, AI analyses the data and discovers patterns that can indicate impending equipment issues, enabling proactive maintenance and optimisation of maintenance operations. However, despite the availability of these technological drivers, companies encounter a range of hurdles when implementing advanced PdM applications, including cost-related issues and the need for skilled human resources (Baur et al. 2020).

Recognising the potential benefits associated with the adoption of PdM, its implementation faces new challenges due to the increasing complexity and automation of modern industrial systems. According to Ran et al. (2019), key considerations include designing compatible system architectures, defining the specific purposes of PdM for cost and reliability trade-offs, and developing tailored fault diagnosis and prognosis approaches based on industry-specific needs. Accordingly, employing PdM requires to gain a comprehensive understanding of the overall PdM system, followed by clarifying the specific purpose or objective, and finally choosing a suitable approach that aligns with the specified goals. This demonstrates the significance of taking maintenance considerations into account during the early stages of component or machine design in order to leverage the advantages of predictive analysis (Chukwuekwe et al. 2016). The successful development of PdM applications necessitates a collaborative effort that brings together expertise from various fields and disciplines. Pooling the knowledge and perspectives of engineers, data scientists, domain experts, IT professionals, cybersecurity experts, and business management professionals allows to effectively address the multifaceted challenges of PdM implementation.

3 Prognostics and Health Management

By integrating fault detection and diagnoses, as well as prognosis of future system degradation, PHM enables the real-time health assessment based on current operational data and additional (historical) knowledge about the system (Braig and Peter 2022). This allows health management processes to be improved, as diagnostics refers to the process of identifying faults and determining their root causes, while prognostics involves assessing health and detecting incipient failure, facilitating the prediction of failure time and the determination of the Remaining Useful Life (RUL) (Lee et al. 2014). Upon closer examination, PHM can be further subdivided into four distinct tasks (Jia et al. 2018; Hagmeyer et al. 2022):

Fault detection   involves the identification and recognition of abnormal conditions, deviations, or anomalies in the behaviour or performance of a system.
Diagnosis   aims to determine the specific nature and root causes of the fault by accurately identifying the underlying issue.
Health assessment   focuses on evaluating and quantifying the risk of failure or health status of the system based on its current condition.
Prognosis   refers to the prediction of the future state of health and remaining useful life of the system.

PHM is a comprehensive concept applicable in various sectors that serves as a basis for advanced maintenance strategies such as PdM, but is to be separated from maintenance in general. In this context, however, an effective PHM system is expected to detect and isolate incipient faults, facilitate monitoring and prediction of fault progression, and support decision-making based on current assessment of system health, thereby enhancing maintenance strategies and overall system reliability (Lee et al. 2014).

3.1 Benefits and Challenges

The holistic discussion of the various benefits of PHM is distinguished by considering the positive impact on the system itself, the organisation as a whole, and the economy in general. Afterwards, the challenges related to implementing PHM are explained, specifically focusing on the approaches for prognostics applications in the following subsection. The overview given is based on Kim et al. (2017).

Benefits PHM offers benefits in terms of improved operational reliability and reliability prediction (Braig and Peter 2022), particularly for safety-critical systems that rely on accurate damage assessment and RUL estimation. The monitoring capabilities enable the collection of data that accurately reflects the actual life cycle conditions of the system. This mitigates the risk of failure in scenarios where the system is subjected to environmental and operational loads that surpass the allowed limits for its intended operating conditions. In addition, the use of PHM in production enables improved process quality control by providing comprehensive insights into the status of manufacturing equipment by means of monitoring and prognostics, including factors such as vibration, temperature and load, thus offering more information compared to traditional quality control methods. By effectively detecting incipient faults and preventing catastrophic failures, the accurate management of system health is achieved, leading to an overall enhancement in system safety.

The adoption of PHM allows organisations to optimise their maintenance efforts, allocate resources more efficiently, and reduce the overall maintenance costs due to the predictive nature of PdM. Continuously monitoring the condition of a system enables early detection of potential faults prior to their progressing to catastrophic failure, minimising unscheduled maintenance. With the ability to predict when a component is likely to fail, unnecessary inspections can be avoided, saving time and resources. Leveraging RUL information to increase the operational uptime of a component allows for maximising component utilisation and ultimately leads to substantial cost reduction for the component. As a result, PHM offers more cost-effective planning of maintenance activities by being based on the actual condition and health of the system. Hence, proactive scheduling facilitates optimising the availability of the system. By reducing the frequency of scheduled maintenance, the risk of maintenance-induced faults is lowered, where unintentional human-caused damage to the system may lead to unexpected downtime and operational disruptions (Chukwuekwe et al. 2016).

PHM contributes to an optimised logistics supply chain by providing an accurate RUL estimate, allowing replacement of components to be planned in advance and thus improving inventory management. Consequently, spare parts can be ordered at short notice as needed, which mitigates the risk of stockouts or excess inventory, leading to a significant simplification of the supply chain for spares (Sun et al. 2010). By minimising the frequency of repairs, a reduced demand for a large workforce emerges, resulting in notable cost savings in both system maintenance and training (Sun et al. 2010). A decrease in the need for repairs implies a lower workload for maintenance personnel, leading to reduced labour expenses and more efficient allocation of resources.

Original Equipment Manufacturers (OEMs) providing PHM-equipped products can increase their revenue by offering customers a more reliable product, which in turn benefits the economy. This requires a collaborative effort between OEMs and system developers, continually improving system design by leveraging the prognostics insights derived from real-time operational data of machines currently in use. In addition, Maintenance-as-a-Service (MaaS) is gaining popularity in the industrial maintenance industry as MaaS allows organisations to outsource their maintenance activities to a service provider (Zoll et al. 2018). Instead of handling maintenance tasks in-house, companies can rely on external experts to manage the maintenance of their equipment, infrastructure, or facilities. By adopting MaaS, organisations can focus on their core business activities while benefiting from specialised maintenance experts and reduced operational costs.

Challenges To ensure the successful implementation of prognostics applications, it is essential to conduct a preliminary analysis that primarily focuses on identifying the critical components of the system. According to Baur et al. (2020), eligible components for prognostics applications are distinguished by their substantial impact on the system, both in terms of performance and cost of downtime, coupled with the fact that degradation is observable over time through sensor technology. Consequently, rarely failing components that are responsible for major downtime require the implementation of sophisticated maintenance strategies to minimise the potential impact.

The identification of critical components is followed by the optimal selection and positioning of sensors aimed at acquiring representative data that accurately reflect the environmental, operational, and performance-related characteristics of a system (Cheng et al. 2010). This process is subject to high requirements as the data gathered form the foundation for the development of prognostics applications. Neglecting proper sensor selection and positioning or disregarding sensor reliability may result in inaccurate data negatively impacting the prognostic performance. Furthermore, it is essential to consider data compatibility as the integration of data from various sources may be challenging. In addition, managing and storing large amounts of data can be expensive, especially in real-time monitoring scenarios, requiring dedicated infrastructure and storage solutions to be defined in advance.

Collecting representative data related to degradation processes is challenging as directly observing damage is difficult or impossible in certain scenarios. Consequently, it becomes necessary to indirectly determine the extent of damage by measuring system responses, which are then used to estimate the level of damage. During this process, it is essential to extract features that accurately represent the damage that is hampered by the presence of noise in the system responses and may therefore require post-processing or filtering techniques (Toothman et al. 2023). Moreover, sufficient data resolution is crucial. Insufficient resolution may result in missed events, the inability to capture rapid changes in system health, or difficulties in accurately identifying precursor patterns. The data labelling process is particularly difficult due to the degrading nature of the data. Accurately annotating the relevant failure events requires domain expertise, making the process both costly and time-consuming.

Data-driven approaches for prognostics are based on the premise that failure prediction can be directly inferred from data (Baur et al. 2020), requiring rather small effort in terms of implementation as opposed to model-based approaches that involve complex modelling of machinery degradation processes. However, the latter allow for a more comprehensive understanding of the system behaviour, accommodating different operating conditions, and requiring less data for effective implementation, whereas purely data-driven methods generally lack proper interpretability. But explainability and interpretability are crucial for building trust and confidence in the predictions made by the models on part of the end users (Burkart and Marco 2021). The ability to understand and interpret the reasoning behind these predictions, and how they align with domain knowledge, is of paramount importance to stakehold-ers including maintenance personnel, operators, and decision-makers, representing a fundamental contribution to the concept of digital sovereignty. If models fail to offer transparent explanations, it becomes challenging to gain acceptance and trust in the predictions, hindering their practical adoption. When deciding between a data-driven or model-based approach for a PHM application, key factors to consider include the availability of data, the complexity of the system, and specific application require-ments. Hybrid methods, on the other hand, can partially mitigate the shortcomings of both approaches and will be the focus of Sect. 4.

Accurately predicting the RUL or health of a system relies on the capability of the model to effectively manage real-world uncertainties, which arise due to var-ious factors, including but not limited to insufficient data, varying environmental and operational conditions and model uncertainty. Lack of representative data can compromise the model's ability to derive a complete understanding of the system behaviour, leading to inaccurate predictions. Furthermore, different operating con-ditions, such as temperature, loads, and vibration levels, impact the degradation process and potentially affect the accuracy as well. For example, high temperatures may accelerate the aging and deterioration of components, while excessive vibra-tions can lead to mechanical stresses and increased wear. Lastly, the process of model simplification to ensure computational feasibility comes at the cost of intro-ducing uncertainty (Dewey et al. 2019). By making assumptions, the model may not fully capture the complexities of the system, resulting in inherent uncertainties in its predictive capabilities.

3.2 Data-Driven and Model-Based Approaches

Data-driven approaches enable handling vast amounts of data by means of processing and analysis, facilitating the identification of hidden patterns and the detection of previously unknown correlations that may have an impact on system performance. This allows leveraging the power of data without the need for pre-existing models or extensive system knowledge. Using statistical analysis and ML techniques that derive predictions directly from data also offers high flexibility and generalisability.

However, the effectiveness of data-driven approaches relies heavily on the quantity and quality of available data, which is limited due to the lack of efficient procedures for obtaining proper training data sets (Baur et al. 2020). Especially the need for run-to-failure data hinders data acquisition. In the machinery field, equipment is typically intended to remain operational, which poses a challenge in obtaining the necessary data that captures the complete progression of fault to failure, leading to inherently imbalanced data sets. In addition, dealing with heterogeneous data stemming from a large number of different types of sensors is challenging, as sensor information is partly redundant and not all signals contain information on specific fault types due to the varying impact of different fault types on different signals (Fink et al. 2020).

Data-driven prognostic approaches lack the ability to comprehend the underlying system dynamics, which potentially leads to predictions being physically inconsistent or implausible (Karniadakis et al. 2021). Moreover, with the approaches solely relying on data, the ability to extrapolate is constrained by the extent to which the data covers the entire state space. The inherent lack of explainability and interpretability hinders the task of tracing the factors or variables influencing the predictions, as well as understanding and providing a rationale for the underlying reasoning. Overcoming these limitations is particularly important in the context of PHM.

Following Baur et al. (2020), model-based approaches involve leveraging a detailed model of the system's degradation process, including first principles, system identification, and state estimation techniques. The authors argue that by incorporating domain knowledge and physical understanding, these approaches can capture the underlying system dynamics and degradation mechanisms more precisely, enhancing the accuracy and reliability of prognostic predictions. As a result, this capability enables the models to effectively account for various operating conditions, thereby facilitating the overall robustness.

Due to the complexity and automation of modern industrial systems, the adoption of model-based approaches is limited by the significant complexity of the degradation process to be modelled (Hagmeyer et al. 2022). As the complexity of the system increases, the number of variables and parameters in the model also grows, leading to increased computational requirements and potential challenges in parameter estimation and model identification. This can impact the practicality and scalability of the model, especially in real-time applications where computational efficiency is crucial. Balancing the level of detail and accuracy in the model while ensuring computational feasibility becomes essential in addressing this limitation and achieving reliable prognostic predictions.

Model-based approaches in prognostics are generally highly specific with respect to the application and therefore limited in their transferability and generalisability (Hagmeyer et al. 2022). Given the significant amount of time, effort, and expertise required to develop accurate and correct models, such a capability is desirable. However, a model that is tailored to one system may not accurately capture the complexities and variations of a similar system, making it less effective or even invalid when applied to a different context. As a result, the developed model cannot be readily applied without further adaptation or modification. In some cases, this process may even necessitate the development of entirely new models.

4   Theory-Guided Data Science

TGDS aims to overcome the limitations of data-driven and model-based approaches by effectively integrating scientific knowledge into data science models, which enables learning dependencies that have sufficient grounding in physical principles (Karpatne et al. 2017). Embedding physics into ML is defined by Karniadakis et al. (2021) as "the process by which prior knowledge stemming from our observational, empirical, physical or mathematical understanding of the world can be leveraged to improve the performance of a learning algorithm". The authors argue that enforcing physical consistency enables generalisable and scientifically interpretable models to be learned that remain robust in the presence of imperfect data. However, successfully implementing this approach requires substantial efforts in handling the diversity of forms in which scientific knowledge is represented and incorporated across various disciplines (Karpatne et al. 2017).

4.1 Integration of Knowledge

The hybrid approach of integrating scientific knowledge into data-driven methods enables developing models that provide accurate and physically consistent predictions (Karniadakis et al. 2021), with the ML pipeline offering diverse ways for incorporation. According to Kim et al. (2021), approaches for knowledge integration mainly refer to three components of the pipeline (Fig. 2), i.e., feature engineering, designing and regularisation, which are defined hereafter. Depending on the nature of scientific knowledge and the ML method employed, various combinations of approaches can be utilised for a particular problem (Karpatne et al. 2017).

Feature engineering refers to obtaining additional attributes through the use of informative priors such as simulation results or governing equations with the aim of improving the adequacy of the data.

Designing involves embedding the physical knowledge directly into the network architecture or model structure to enforce theoretical constraints.

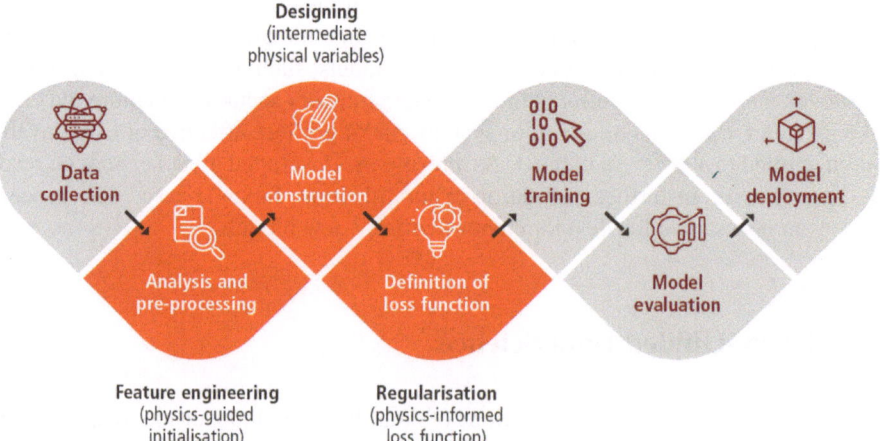

Fig. 2 Depiction of a typical ML pipeline, where integrating prior knowledge into the pipeline is mainly related to three of its components through feature engineering, designing and regularisation (own illustration)

Regularisation focuses on modifying the objective function of the learning algorithm to ensure physical conformance, typically by supplementing the loss term with physical laws.

4.2 Approaches for Prognostics and Diagnostics

The approaches presented below consider prior knowledge that is insufficient for holistic modelling but possesses the potential of increasing the predictive accuracy of data-driven methods. The non-exhaustive overview given is structured according to the components of the ML pipeline that enable knowledge integration. Moreover, references are provided demonstrating the successful adoption of TGDS in the field of PHM.

Integrating prior knowledge by simulating the system behaviour enables generating synthetic data, referred to as physics-guided feature engineering. Due to insufficient knowledge for holistic modelling, however, it is necessary to make simplified assumptions regarding the physical process to be modelled. Consequently, the generated data may not effectively represent the true underlying system dynamics, preventing its use as a supplement to the existing training data. Yet, fundamentally reflecting the governing physics by means of simulated data still enables improving the generalisation capabilities and robustness of the model (von Rueden et al. 2022). For this purpose, the data is utilised for pre-training the ML model on a rather simple problem in order to obtain initial parameters (Hagmeyer et al. 2022). The actual training of the model then involves fine-tuning based on the physics-guided initialisation, which saves computation time and addresses the shortcomings of insufficient real data. Dourado and Felipe (2019) propose utilising Physics-Informed Neural Networks (PINNs) (Raissi et al. 2019) for corrosion-fatigue prognosis of aircraft wings. The hybrid approach accurately compensates for the lack of information in corrosion-fatigue damage accumulation. Integrating basic understanding of the underlying physics in the form of a purely mechanical fatigue model enables leveraging the network to learn the corrosion effects, after being pre-trained on synthetic data in order to derive appropriate initial parameters.

The construction of neural networks is influenced by various design decisions such as the number of neurons per layer, the number of hidden layers, the initialisation of weights and the selection of activation and loss functions. These considerations play a crucial role in facilitating effective learning. Prior physical knowledge offers several ways to inform these design considerations in order to obtain both generalisable and scientifically interpretable results (Karpatne et al. 2017). A potential approach to utilising physical knowledge in guiding the architecture involves assigning physical meaning to certain outputs of neurons in the hidden layers (Huang and Jianhui 2022). These intermediate physical variables enable the explicit encoding of constraints and essentially enforce physical conformity by confining the hypothesis space. Cheng et al. (2022) accurately predict RUL in the context of bearing prognosis with multisensory signals by incorporating the knowledge of monotonic degradation, i.e., that components cannot heal without repair. In light of this, the network must factor in the irreversibility of the degradation process to ensure consistency with the evolution of the bearing's health condition. Therefore, a positive increment recurrence relationship is introduced to maintain the monotonicity, effectively modelling the degradation process as an intermediate physical variable.

Integrating differential equations into neural networks facilitates regularising the network's complexity and ensures physical plausibility by imposing constraints derived from the governing equations. This regularisation encourages the network to learn physically meaningful representations and helps in obtaining accurate and consistent solutions that respect the underlying physics, such as conservation laws, boundary conditions, or other physical principles (Raissi et al. 2017). Solving differential equations with PINNs (Fig. 3) requires derivatives of the network's output with respect to the inputs (Cuomo et al. 2022). By incorporating the governing equations as additional terms in the loss function, PINNs enforce the network to satisfy the underlying physics. Wen et al. (2023) propose a model fusion scheme based on PINNs, enabling PHM of Lithium-ion batteries. For this purpose, a semi-empirical semi-physical partial differential equation is developed to model the degradation dynamics of Lithium-ion batteries, which allows capturing the fading trends of the battery state of health. However, the authors emphasise that since the proposed loss function represents multi-task learning, an appropriate weighting method is compulsory for balancing the different loss terms in order to effectively improve the performance of PHM for Lithium-ion batteries.

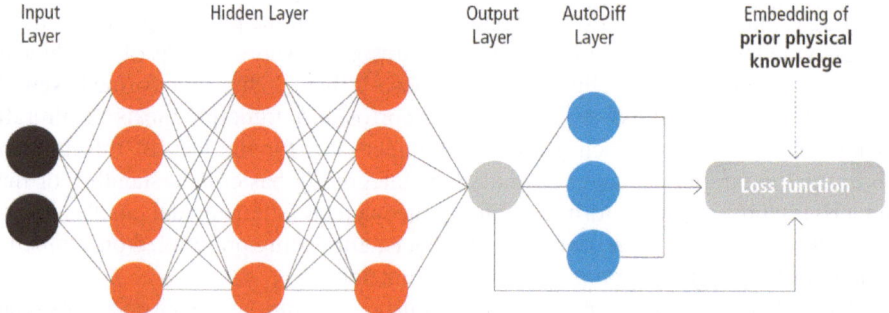

Fig. 3 Fundamental illustration of Physics-Informed Neural Networks (PINNs), which enable embedding prior physical knowledge into the loss function. Solving differential equations with PINNs requires derivatives (AutoDiff layer) of the network's output with respect to the inputs (own illustration)

5 Conclusion

Industry 4.0 continues to evolve and transform the manufacturing landscape, driven by the ongoing efforts of various fields and disciplines. An essential part of this transformation is the integration of PdM, which offers companies the opportunity to improve operational efficiency, increase safety measures and minimise overall maintenance costs. However, the implementation of PdM faces new challenges due to the increasing complexity and automation of modern industrial systems. Fortunately, the rapidly advancing field of PHM plays a vital role in making PdM more accessible in terms of technical realisation. The development of novel hybrid approaches enables effectively tackling prognostics and diagnostics while compensating for the limitations of data-driven and model-based approaches. In this process, scientific consistency is ensured by integrating prior knowledge about the underlying physics of the system to be modelled directly into data-driven methods. Nevertheless, the successful implementation of this approach necessitates substantial efforts in handling the diversity of forms in which scientific knowledge is represented. In this context, the question arises of how non-formalised domain knowledge can be used to improve model performance by means of IML. The idea of incorporating workers providing domain knowledge is directly related to the goal of digital sovereignty with regard to AI adoption, as it would allow workers to be actively involved in AI development by providing expertise and general feedback, improve the traceability and verifiability of AI algorithms by integrating domain knowledge, and increase acceptance and trust on the part of end users. Continuously promoting research activities aimed at improving PHM through the development of hybrid methods is compulsory to meet the emerging challenges related to the ongoing evolution of Industry 4.0.

References

Baur, M., Albertelli, P., Monno, M.: A review of prognostics and health management of machine tools. Int. J. Adv. Manuf. Technol. **107**, 2843–2863 (2020)

Berg, A.: Industrie 4.0 - jetzt mit KI. bitcom.org. https://www.bitkom.org/sites/default/files/2019-04/bitkom-pressekonferenz_industrie_4.0_01_04_2019_prasentation_0.pdf (2019). Accessed 03 June 2023

Braig, M., Zeiler, P.: Using data from similar systems for data-driven condition diagnosis and prognosis of engineering systems: a review and an outline of future research challenges. IEEE Access (2022)

Burkart, N., Huber, M.: A survey on the explainability of supervised machine learning. J. Artif. Intell. Res. (JAIR) **70**, 245–317 (2021). DOI: 10.1613/jair.1.12228 10.1613/jair.1.12228

Chen, X. et al.: Physics-informed deep neural network for bearing prognosis with multisensory signals. J. Dyn. Monit. Diagn. 200–207 (2022)

Cheng, S., Azarian, M.H., Pecht, M.G.: Sensor systems for prognostics and health management. Sensors **10**(6), 5774–5797 (2010)

Chukwuekwe, D.O. et al.: Reliable, robust and resilient systems: towards development of a predictive maintenance concept within the industry 4.0 environment. In: EFNMS Euro Maintenance Conference (2016)

Cuomo, S. et al.: Scientific machine learning through physics–informed neural networks: where we are and what's next. J. Sci. Comput. 92(3), 88 (2022)

Dewey, H.H., DeVries, D.R., Hyde, S.R.: Uncertainty quantification in prognostic health management systems. In: 2019 IEEE Aerospace Conference. IEEE, pp. 1–13 (2019)

Dourado, A., Viana, F.A.C.: Physics-informed neural networks for corrosion-fatigue prognosis. In: Proceedings of the Annual Conference of the PHM Society, vol. 11, p. 1 (2019)

Fink, O. et al.: Potential, challenges and future directions for deep learning in prognostics and health management applications. In: Engineering Applications of Artificial Intelligence, vol. 92, p. 103678 (2020)

Gauger, I., Nagel, T., Huber, M.: Hybrides Maschinelles Lernen im Kontext der Produktion. In: Digitalisierung souver ä n gestalten II: Handlungsspielr ä ume in digitalen Wertsch ö pfungsnetzwerken. Springer, pp. 64–79 (2022)

Hagmeyer, S., Zeiler, P., Huber, M.: On the integration of fundamental knowledge about degradation processes into data-driven diagnostics and prognostics using theory-guided data science. In: PHM Society European Conference, vol. 7. 1, pp. 156–165 (2022)

Huang, B., Wang, J.: Applications of physics-informed neural networks in power systems-a review. IEEE Trans. Power Syst. (2022)

Huber, M.: Data Science: Grundlagen, Architekture und Anwendungen. In: Haneke, U. et al. (Ed.) Data Science: Grundlagen, Architekturen und Anwendungen, 2nd edn. dpunkt.verlag. Chap. Predictive Maintenance, pp. 255–274 (2021)

Jia, X. et al.: A review of PHM data competitions from 2008 to 2017: methodologies and analytics. In: Proceedings of the Annual Conference of the Prognostics and Health Management Society, pp. 1–10 (2018)

Karniadakis, G.E. et al.: Physics-informed machine learning. Nat Rev Phys 3(6), 422–440 (2021)

Karpatne, A. et al.: Theory-guided data science: a new paradigm for scientific discovery from data. IEEE Trans. Knowl. Data Eng. 29(10), 2318–2331 (2017)

Kim, N.-H., An, D., Choi, J.-H.: Prognostics and Health Management of Engineering Systems. Springer International Publishing, Switzerland (2017)

Kim, S.W., et al.: Knowledge integration into deep learning in dynamical systems: an overview and taxonomy. J. Mech. Sci. Technol. 35, 1331–1342 (2021)

Lee, J. et al.: Prognostics and health management design for rotary machinery systems—reviews, methodology and applications. Mech. Syst. Signal Process. 42(1–2), 314–334 (2014)

Raissi, M., Perdikaris, P., Karniadakis, G.E.: Physics informed deep learning (part i): data-driven solutions of nonlinear partial differential equations (2017). arXiv:1711.10561

Raissi, M., Perdikaris, P., Karniadakis, G.E.: Physics-informed neural networks: a deep learning framework for solving forward and inverse problems involving nonlinear partial differential equations. J. Comput. Phys. 378, 686–707 (2019)

Ran, Y. et al.: A survey of predictive maintenance: systems, purposes and approaches (2019). arXiv:1912.07383

Sun, B. et al.: Benefits analysis of prognostics in systems. In: 2010 Prognostics and System Health Management Conference. IEEE, pp. 1–8 (2010)

Toothman, M. et al.: Overcoming challenges associated with developing industrial prognostics and health management solutions. Sensors 23(8), 4009 (2023)

von Rueden, L., Houben, S., et al.: Informed pre-training on prior knowledge (2022). arXiv:2205.11433

von Rueden, L., Mayer, S., et al.: Informed machine learning–a taxonomy and survey of integrating prior knowledge into learning systems. IEEE Trans. Knowl. Data Eng. 35(1), 614–633 (2021)

Wagner, P. et al.: KI-Anwendungsfälle in der Produktion". In: ten Hompel, M., Vogel-Heuser, B., Bauernhansl, T. (eds.) Handbuch Industrie 4.0: Produktion, Automatisierung und Logistik. Springer, Berlin, pp. 1–24 (2020)

Wang, K.: Intelligent predictive maintenance (IPdM) system-Industry 4.0 scenario. WIT Trans. Eng. Sci. **113**, 259–268 (2016)

Wen, P. et al.: Fusing models for prognostics and health management of lithium-ion batteries based on physics-informed neural networks (2023). arXiv:2301.00776

Zoll, M., Jäck, D., Vogt, M.W.: Evaluation of predictive-maintenance-as-a-service business models in the internet of things. In: 2018 IEEE International Conference on Engineering, Technology and Innovation (ICE/ITMC). IEEE, pp. 1–9 (2018)

Christopher Braun is a research associate at the Institute of Industrial Manufacturing and Management (IFF) at the University of Stuttgart as well as the Fraunhofer Institute for Manufacturing Engineering and Automation (IPA). Within the departments of Cognitive Production and Cyber Cognitive Intelligence, respectively, his focus lies on leveraging the potential of artificial intelligence in an interdisciplinary fashion. Specifically, his research interests revolve around the integration of domain-specific knowledge and prior information into machine learning methodologies, aiming to enhance algorithmic capabilities and robustness within the realm of informed machine learning.

Prof. Dr. Marco F. Huber is Professor for Cognitive Production Systems at the University of Stuttgart. In addition, he is director of the Machine Vision and Signal Processing Department and the Cyber Cognitive Intelligence (CCI) Department of the Fraunhofer Institute for Manufacturing Engineering and Automation (IPA) in Stuttgart. His research focuses on machine learning, explainable artificial intelligence, machine vision, and robotics. Prof. Huber received his Ph.D. degree and habilitation degree in computer science from the Karlsruhe Institute of Technology (KIT), Germany, in 2009 and 2015, respectively. From 2009 to 2011 he headed the research group 'Variable Image Acquisition and Processing' at the Fraunhofer IOSB in Karlsruhe. He then worked as a Senior Researcher at AGT International in Darmstadt, Germany, until 2015. Before joining the University of Stuttgart and Fraunhofer IPA, Prof. Huber was responsible for product development and data science services at USU Software AG in Karlsruhe. He has authored or co-authored more than 150 publications in various high-ranking journals, books, and conferences.

Analyzing and Designing Sociodigital Sovereignty on Individual and Organizational Levels—A Case Study

Ulrike Schmuntzsch⊙ and Ernst Andreas Hartmann⊙

Abstract Rapidly evolving digital transformation processes, emerging technologies such as artificial intelligence systems, and constantly changing market conditions have a significant impact both on employees at the individual level in the workplace and on companies at the organizational level. The preservation, protection, and further development of knowledge and intellectual property for both individuals and organizations are becoming increasingly important to enable self-determined interaction—meaning sovereignty—in the market. However, a key question is often how such digital transformation processes can be successfully managed for both entities in their daily business. This article introduces the concept of Sociodigital Sovereignty as two pillars—the individual and the organizational level—and presents classification matrices as tools for industrial application. These tools can be used to analyze and design transformation process and improve Sociodigital Sovereignty at the workplace and enterprise levels. The industrial use case presented focuses on a software launch. In a three-part workshop with company employees, the situation was analyzed at the individual workplace level—first without the new software and then with it—using the respective classification matrix. After prioritizing key challenges and proposed solutions, the third part of the workshop used the classification matrix on the organizational level to structure the developed recommedations for action. To summarize, the results show the strong connections between individual and organizational levels and confirm the applicability of the concept of Sociodigital Sovereignty as a tool for sociotechnical analysis and design.

Keywords Digital sovereignty · Sociodigital sovereignty · Sociotechnical analysis · Sociotechnical design · Digital transformation · Industrial use case

U. Schmuntzsch (✉) · E. A. Hartmann
Institute for Innovation and Technology (IIT), Steinplatz 1, 10623 Berlin, Germany
e-mail: schmuntzsch@iit-berlin.de

© The Author(s) 2025
U. Schmuntzsch et al. (eds.), *New Digital Work II*,
https://doi.org/10.1007/978-3-031-69994-8_8

1 Introduction

Digitalization and transformation are creating new challenges for companies and employees around the world. Rapidly changing market conditions as well as the latest technologies such as artificial intelligence (AI) systems or more generally algorithmic systems affect both the individual level of the employee at the workplace and the organizational level of the company. In these times of change and transformation, knowledge and intellectual property are becoming increasingly important assets for interacting with others, adapting to constantly changing conditions, and asserting oneself as independent and self-determined entities in the market. Accordingly, in this initial situation, employees and companies face similar challenges and share common aims: both seek to preserve, protect, and further develop their knowledge and competencies in order to maintain their sovereignty, which critically depends on these assets. This means that they want to maintain control over their environment and their situation (Couture and Toupin 2019; Hartmann 2021a), but moreover to successfully pursue their own aims and objectives (Hartmann 2021a, b, 2022; Shajek and Hartmann 2023). In order for this to be successful for both sides, both entities—individuals and organizations—need to be taken into account when launching transformation and digitalization processes. The question often arises: how can this be implemented in practice? Which concepts and theories from the past can be used and which are (modified) relevant for the future and applicable in practice?

Bearing in mind that even the most highly automated system can still be seen as a sociotechnical system consisting of human, technology, and organization, the concept of Sociodigital Sovereignty with its two pillars—individual level and organizational level—was applied as an analytical tool to an industrial use case. During the process, the concept and the classification matrices developed were used to analyze the situation within a company during a software launch and to elaborate design suggestions in order to improve Sociodigital Sovereignty within the company.

This article first presents the theoretical background of the concept of Sociodigital Sovereignty with its two pillars—the individual level and the organizational level—and the respective classification matrices as analytical tools. Subsequently, the proof of applicability to a specific industrial use case is described and reflected upon. This is followed by a summary and a conclusion in which the interplay between theory and practice is analyzed. Finally, an outlook on the further development of the concept and its classification matrices is provided and directions for future activities are given.

2   Theoretical Background

2.1   *Sociodigital Sovereignty on an Individual Level*

The project described here is part of a more comprehensive endeavor to understand, discuss, and develop Sociodigital Sovereignty, which is the ability of individuals and organizations to successfully pursue their aims and objectives by employing digital technology (Hartmann 2021a, b, 2022; Shajek and Hartmann 2023). In this context, Sociodigital Sovereignty was conceptualized to be applicable on an individual level—humans—as well as an organizational level—companies and other organizations (Hartmann 2021a). The specific project described in this chapter has links to both levels—individual and organizational. Within this chapter and the scope of the publication in general, the organizational level will be emphasized.

In the following, however, both aspects, Sociodigital Sovereignty on individual and organizational levels, respectively, will be described one after the other.

To capture Sociodigital Sovereignty on an individual level, concepts from work psychology (Hacker 2005; Oesterreich 1981) and sociotechnical systems theory (Cherns 1976; Hartmann 2005; Mühlbradt et al. 2022; Trist and Bamforth 1951) were used to construct a conceptual matrix, consisting of three columns and three rows (Hartmann 2021a, 2022; Hartmann and Shajek 2023).

The three columns describe three aspects of Sociodigital Sovereignty at the individual workplace level[1]:

- Transparency and Explainability: Transparency of the work system as a whole—consisting of technological, human, and social subsystems—is a prerequisite for humans being able to exercise control. Complex algorithmic and AI-based systems, however, are inherently complex and intransparent. In these cases, transparency must be provided with extra effort. These are aspects within the domain of Explainable AI, or XAI (High-Level Expert Group on AI [AI HLEG] 2020; Mueller et al. 2019; Pentenrieder et al. 2023). Similarly, transparency must also refer to human (e.g., qualifications) and organizational (e.g., tasks and roles) phenomena.
- Confidence of action—or efficiency (*Effizienz*) in the sense of Oesterreich (1981)—refers to the fact that humans, when acting in sociotechnical systems, can be confident that the effects of their actions will be consistent with what they intended when choosing these actions (Hartmann 2021a).
- Freedom of action—divergence (*Divergenz*) as it is called by Oesterreich (1981)—describes environments offering humans a range of different courses of action from which they may choose with discretion. This is close to the concepts of

---

[1] The following, already published research—as cited—is presented to provide readers with an understanding of the background, context, and theoretical foundations of the original work to be presented in Section 3 of this contribution.

'degrees of freedom' (*Freiheitsgrade*) and 'scope of action' (*Handlungsspiel-raum*) as used in action regulation theory (Hacker 2005; Hartmann 2005, 2021a).

The guiding questions in the nine cells of Fig. 1 illustrate which facets of Sociodigital Sovereignty are represented by the respective cells of the matrix.

Fig. 1 Conceptual matrix for sociodigital sovereignty on the individual or workplace level (own illustration, Hartmann and Shajek 2023)

2.2 Sociodigital Sovereignty on an Organizational Level

To measure Sociodigital Sovereignty on an organizational level, a slightly different approach was pursued (Hartmann 2021a; Hofmann et al. 2023). As rows of the matrix, confidence of action and freedom of action were chosen, similar to the matrix on the individual level (Fig. 1), but here, these concepts refer to actions on an organizational level rather than to individuals (Luhmann 1982; Škrinjarić 2022).

For the columns of the matrix, an approach from the domain of organizational knowledge, intellectual capital, and knowledge balance sheets was adopted (Edvinsson and Malone 1997; Hartmann et al. 2014; Mertins 2005). Here, three pillars of capital—from a perspective of organizational knowledge—are distinguished:

- Human Capital: The knowledge, skills, competencies, and qualifications of the company's employees.
- Structural Capital: The company's ability to utilize and develop knowledge internally. This is closely related to theories of organizational learning (Argyris 1976; Argyris and Schön 1996) and the conduciveness of organizations—their structures, processes, culture and values—to learning and innovation (Cedefop 2012).
- Relational Capital: The company's ability to leverage and enhance knowledge through collaboration with external entities such as educational or research institutions, associations, public bodies, and other companies. Relational capital is strengthened through close collaboration between companies and other organizations within regional or industry-specific clusters (Furman et al. 2002).

Figure 2 below illustrates this matrix for Sociodigital Sovereignty on the organizational level.

Fig. 2 Facets of sociodigital sovereignty of organizations (own illustration, Hartmann 2021a)

The individual aspects of Sociodigital Sovereignty on the organizational level are (Hofmann et al. 2023):

- Human Capital: Here, the depth of relevant (digital) knowledge is the aspect associated with confidence of action, whereas the variety of relevant (digital) knowledge refers to freedom of action on an organizational level.
- Structural Capital: Single-Loop-Learning means that companies improve the existing ways of doing things; this increases confidence of action. In Double-Loop-Learning, companies invent new ways of doing things, which increases their freedom of action (Argyris 1976; Argyris & Schön 1996).
- Relational Capital: In terms of confidence of action, the relations between the company and its external partners should be reliable. For freedom of action, the

network of external relations should provide degrees of freedom for alternative actions.

2.3   Interim Summary of the Theoretical Background

As presented above, the concept of Sociodigital Sovereignty includes both levels—an **individual** one on the level of the workplace and an **organizational** one on the level of companies and organizations (Hofmann et al. 2023). On both levels, confidence of action (of individuals or organizations) and freedom of action (of individuals and organizations) are regarded as aspects of Sociodigital Sovereignty. On the individual level, an additional aspect is the transparency and explainability of the respective working environment.

The constituting elements of Sociodigital Sovereignty are conceived of somewhat differently between individual and organizational levels. On the individual level, according to sociotechnical systems theory (Cherns 1976; Mühlbradt et al. 2022; Trist and Bamforth 1951), people, organization, and technology are considered to be the core pillars on which Sociodigital Sovereignty must rest. On the organizational level, the three components of intellectual capital (Edvinsson & Malone 1997; Hartmann et al. 2014; Mertins 2005)—human, structural, and relational capital—were selected as core elements of Sociodigital Sovereignty, because developing Sociodigital Sovereignty implies innovative capabilities, for which these three types of intellectual capital are crucially important (Cedefop 2012; Hartmann et al. 2014; Hofmann et al. 2023). Taken together, at both the individual and organizational level, Sociodigital Sovereignty is about achieving self-determination by employing and developing resources that are important for both—individual as well as organizational—competence development (Škrinjarić 2022).

The concept of Sociodigital Sovereignty with both levels was conceptualized to apply to different company use cases. The different facets of each matrix allow for an individual and detailed assessment on the level of the individual workplace as well as on an organizational level. Such holistic sociotechnical analysis offers the possibility to evaluate the effects of digitalization and transformation on the employees as well as on the company itself. Furthermore, these instruments allow to actively involve employees and to include their experiences and knowledge in such complex change processes, which then hopefully leads to better solutions and ensures more willingness, commitment, and engagement throughout the company. Moreover, evaluation results as well as the concept of Sociodigital Sovereignty and its facets itself can then serve as a basis and guideline to support such digital transformation processes in a way that is beneficial for both sides—the employees and the company.

3 Proof of Applicability of the Concept of Sociodigital Sovereignty on an Industrial Use Case

The following section presents an industrial use case in which the concept of Sociodigital Sovereignty and its classification matrix were applied for the first time to analyze the challenging situation of a software implementation within a company as well as to help find solutions to the problems identified. As already described in the introduction (Chap. 1), constantly changing working and market conditions require constant adaptation from companies and organizations as well as from employees themselves in order to keep up to date and to be able to assert themselves on the market. In this specific use case, a company—providing highly specialized, knowledge-based technical services to the manufacturing industry—wants to strengthen its competitiveness by improving, among other factors, its internal resource management and external delivery times. A new planning software should be launched for this purpose. Accordingly, this process not only requires employees to familiarize themselves with the new tool but also includes an organizational restructuring and necessarily an adaptation to it. Generally, and also in this specific case, such transformation processes entail multiple challenges for the company and its employees which sometimes provoke discord and resistance. In order to systematically evaluate and address such challenges as well as to find practicable solutions for both sides—employees and company—the concept of Sociodigital Sovereignty and its classification matrix was used as an assessment tool from the Institute for Innovation and Technology (iit) within a workshop with three participants of the industrial company, being responsible for the software development and implementation project in question.

3.1 *Workshop Design and Implementation*

As mentioned above, the company's initial situation relates to the launch of a new software that also requires a restructuring of different work routines. These work routines constitute the core business processes of the company. By the time of the workshop, the whole system had not yet been implemented, but a test run had already taken place. The new software and the whole process were discussed emotionally among employees. With the help of the developed classification matrix and a first questionnaire, it was possible to structure the debate and its multiple arguments from supporters and opponents. As already described in a previous publication (Schmuntzsch and Hartmann 2023), during the workshop, the classification matrix developed for the individual workplace level was used for a stepwise consideration of both sides and to first analyze the nine facets of the concept of Sociodigital Sovereignty on the level of the individual workplace.[2]

---

[2] The already published article—as cited—deals with the theoretical background of the concept of Sociodigital Sovereignty on an individual level at the workplace and its classification matrix. Based

Table 1 Overview of the three parts of the workshop

	First Part	Second Part	Third Part
Focus of Analysis	Strengths and weaknesses *without* new planning software	Opportunities and threats *with* new planning software	Key challenges and solutions
Level of Sociodigital Sovereignty	Individual level at the workplace	Individual level at the workplace	Organizational level at the company

Guiding questions were developed for each facet, resulting in a total of 40 questions. These questions were case-specific adaptations and specifications of the more general guiding questions as shown in the nine cells of Fig. 1.

Following the introduction of the matrix and questions, the work routine before launching the new software was analyzed in the first part. The second part focused on the situation where the new software has already been launched. In both parts, participants were asked to note down all positive and negative experiences and associations with respect to each work situation on paper cards. In both cases, all cards were sorted into the nine facets of the classification matrix on the level of the individual workplace.

In the subsequent third part of the workshop, participants were asked to prioritize the main challenges of launching the new software along with restructuring the work routines. This prioritization focused on identifying the main challenges that the company needed to address and resolve. In the following step, a brainstorming session was conducted where participants were asked to propose different solutions for each of the prioritized challenges. After the workshop, in the subsequent analysis of the findings, the classification matrix on an organizational level was used in order to categorize and reflect the results. The three parts of the workshop and the respective focus are presented in Table 1.

3.2 Results of the First and Second Workshop Part—From Strengths and Weaknesses to Opportunities and Threats

Based on the results of the first and second part of the workshop, a SWOT analysis was carried out to evaluate aspects of both sides with the help of the classification matrix on the **level of the individual workplace** (for theoretical background see Sect 2.1). The answers from the first part of the workshop referring to the work routines without new software represent the strengths (S) (positive) and weaknesses

on this, the first and the second part of the workshop as well as its results are presented in detail. However, the present publication addresses mainly the concept of Sociodigital Sovereignty on an organizational level and describes in detail the third part of the workshop and its subsequent analysis with the respective classification matrix and its results. Moreover, both pillars of Sociodigital Sovereignty—the individual one and the organizational one—were discussed and reflected with regard to their interconnections.

(W) (negative). In contrast, predicted associations and experiences during the test run of the new software represent the opportunities (O) (positive) and threats (T) (negative). As already mentioned, the workshop results of this analysis refer to the individual level of the workplace and have already been described in detail in a previous presentation (Schmuntzsch and Hartmann 2023). Summarizing the results of this part of the workshop, participants made it very clear that transparent and comprehensible tasks, technologies, and working environments are of huge importance for employees (**transparency/explainability**). Likewise, work routines were favored in which employees feel confident, so that their actions will efficiently lead to the intended outcome (**confidence of action/efficiency**). In contrast, they feared situations where employees are responsible for the outcome, but it is unclear to them what the software is doing and why. Moreover, participants appreciate a work environment that can be described as holistic and diverse in its tasks and routines. This also means having all the information and being able to personally apply work routines to different tasks (**freedom of action/divergence**) (Schmuntzsch and Hartmann 2023).

Generally, it has to be stated that all three aspects of the concept of Sociodigital Sovereignty on the individual level of the workplace play an important role to employees. However, it is emphasized that sufficient knowledge in depth and width about technologies, tasks, work routines, and the organizational structure and processes is seen as a prerequisite for the sociotechnical system in which the employees work to be perceived as transparent and explainable. In turn, the perceived transparency and explainability are then considered as preconditions to develop confidence of action during the work routines as well as to make use of the freedom of action offered. However, the emphasis is not only on the depth and breadth of knowledge among staff but also on its accessibility and the ability to utilize it within the company when needed. Various concerns and wishes expressed and prioritized can be grouped around two main aspects which were classified as the key challenges of the company: The first one relates to the preservation and further development of **employees' expertise**. The second one encompasses all aspects related to the creation of technical, but more importantly, **organisational prerequisites** for the optimal sharing of knowledge within the company. Both categories were selected as prioritized challenges to be further discussed in the third part of the workshop.

3.3 Results of the Third Workshop Part—From Challenges to Solutions

Looking at the results of the third part of the workshop, the prioritized challenges and the proposed solutions are described in more detail below. Largely, most of the concerns relate, on the one hand, to the preservation and further development of **employees' expertise** and, on the other hand, to a well-balanced **adjustment of the organizational structure** to a knowledge-sharing culture.

Focusing on the **employees' expertise**, participants of the workshop express their wishes for a strategic human resource development including individual offers for training courses based on an individual and confidential feedback on work results as well as individual development paths for each employee. Furthermore, the work routine should be designed in an interesting and varied way, so that the staff's expertise in depth and width is preserved and further developed. Especially, with respect to digitalization and automation, participants expressed the wish to maintain a critical mindset in order to question software-generated results and to intervene in case of failure and exceptional events.

In order to preserve and further develop **employees' expertise** in times of digitalization, participants proposed to establish a strategic human resource management with mentoring program, competence analysis, and customized simulator training. The mentoring program for newcomers should be designed and conducted with experienced experts in order to reduce training periods. Individual competence analysis, to qualify employees individually according to their deficits and goals, also aims to provide an organizational overview of the different competence profiles and bottlenecks within the company, and how to close such gaps. With the help of such a competence matrix, it is then possible to show individual development pathways. Therefore, a comprehensive and recurrent simulator training program should be established with the help of experienced experts. For the implementation of that and the documentation of practical expert knowledge, one option suggested was to release employees from at least 10 percent of their working time for knowledge documentation (in the first step foreseeable retirees, then all). The training program as well as the software itself should provide individual and confident feedback and suggestions for further training options. Apart from that, learning texts should be included in the system. Generally, the whole system should contain low-threshold training offers that are individually tailored and completely confidential. Nevertheless, it was also suggested that manual testing procedures should be carried out either randomly or at regular intervals in order to maintain practical skills and abilities.

Looking at the challenges expressed referring to the **adjustment of the organizational structure,** one focus is on changing the mindset within the company from solitary expertism to a culture of shared knowledge. Therefore, the goal is to drive a cultural change that encourages knowledge sharing and fosters active participation with appreciation and incentives. Moreover, participants considered it necessary to create semi-autonomous working groups in order to establish job rotation and division of tasks. Another main challenge refers to the digitalization and standardization of work routines across different groups without destroying well-functioning individual processes within each group. In order to digitalize structures and processes to build up a database documentation, participants find it important to keep documentation effort low.

The proposed solutions to adjust the **organizational structure** aimed at changing mindset by appreciating knowledge sharing in the virtual and real world. Therefore, teambuilding measures to foster motivation and spirit as well as semi-autonomous working groups, and temporary project responsibility as well as regular role changes were among the proposed solutions. The main objective of these proposed solutions is to establish a new culture of teamwork in which the division of labor, mutual support, and knowledge sharing are normal. Nevertheless, it was stated that work routines should continue to be organized in an interesting and varied way including simpler routines as well as more challenging tasks. Another focus of the proposed solutions is the user-friendly design of the user interface in order to minimize documentation effort and, above all, to motivate employees to actively participate. Therefore, technical solutions such as gamification elements and other motivational elements should be integrated into the new software. Moreover, organizational changes such as specifically established documentation teams for each topic and different types of compensation for members of such teams, e.g., reduced turnover targets, were proposed to encourage motivation.

In summary, various solutions are proposed to address the two key challenges. Apart from a more user-friendly design of the software and corresponding documentation, most of the concerns and wishes address the human and organizational level and not primarily the technical level. In order to build up more expertise among the staff, a strategic human resource management program with different elements should be established. Moreover, the aim is to evoke a change of mindset from solitary expertism to a knowledge-sharing culture. Therefore, an adjustment of the organizational structure including different elements should be initiated in order to spread and share employees' knowledge in depth and width within the company. Since most of the proposed solutions have to be tackled by the company as a whole on an organizational level and not solely on the individual workplace level, the following categorization and reflection of these findings will be carried out by using the classification matrix on an organizational level (cf. Sect. 2.2).

3.4 Reflection of Key Challenges and Solutions by Using the Classification Matrix on an Organizational Level

Looking at the theoretical background presented in Sect 2.2, the concept of Sociodigital Sovereignty on the organizational level refers to everything that is associated with the company itself as an entity and not (directly) with the individual level of the workplace. Nevertheless, the reciprocal interaction between both levels has to be considered since decisions and adjustments on the organizational level have a huge impact on the individual level of the workplace. Otherwise, as seen in the specific use case, complex problems often emerge on the individual level of the workplace, but also have an organization-wide impact. Often, they can only be tackled holistically with solutions addressing the organizational level.

Since the key challenges and solutions described in this specific industrial use case affect the complete organization, a change in perspective from the individual level of the workplace—where the various problems emerged and were gathered—to the organizational level—where the solutions have to be found organization-wide—seems to be useful. Hence, for classifying and reflecting the described challenges and posed solutions, the classification matrix on an organizational level will be used further on.

The classification matrix on the organizational level consists of three aspects: human capital, structural capital, and relational capital (cf. Sect. 2.2). Both aspects, human capital and structural capital, refer to internal matters of the company or organization. Human capital is related to the depth and variety of the relevant knowledge within a company, which addresses the two aspects, confidence of action and freedom of action. Structural capital encompasses the two aspects of single-loop and double-loop learning within a company itself. At the single-loop level, an organization improves the existing ways of doing things which leads to an increase in the confidence of action. At the double-loop level, new ways of doing things are invented which then hopefully lead to an increase in the freedom of action. In contrast to human and structural capital, relational capital refers to all external connections and relationships of a company or organization. Basically, it can be stated that all three types of capital can be seen as essential resources of the company in order to achieve and maintain sovereignty during digital transformation processes. With the help of these different facets of the concept of digital sovereignty, the nature of an organization can be described, evaluated, and designed internally and externally: internally according to such aspects as the quality and quantity of its knowledge within the company as well as the quality of its learning behavior. In addition, the quality of the relationships with external partners can also be part of the assessment. It is therefore crucial to keep all facets in mind and to evaluate on a case-by-case basis which of these are relevant. Based on such an assessment, a picture of the organization is composed which highlights the need for action in order to strive for more sovereignty in times of digital transformation processes.

Looking at the concrete challenges and solutions worked out in our specific industrial use case and described in the section above, it has to be stated that the first main focus is on the preservation and further development of employees' expertise, and the second one on the organizational adjustment to a change of mindset. Hence, on the organizational level, most of the prioritized challenges and proposed solutions refer to the aspects of human capital and structural capital. Thus despite the fact that this industrial use case deals with the launch of a new software tool, the prioritized challenges relate, on the one hand, to the depth and width of staff knowledge (human capital). On the other hand, the expressed challenges refer to the structural capital associated with the topic of single- and double-loop-learning on an organizational level (structural capital). As in this specific use case, the industrial company is characterised by the absence of external partnerships, e.g. to suppliers or other organizations, aspects referring to the relational capital were not mentioned during the workshop.

In contrast, a number of aspects were mentioned relating to the human capital including both the depth as well as the width of the knowledge within the company. Accordingly, employees' knowledge shall be preserved and further developed mainly with the help of a newly established strategic human resource management which includes a mentoring program, individual competence analyses, and individually tailored simulator training. Therewith, new employees should be familiarized faster with the organization's working routines and older employees should have the opportunity to specialize in certain areas as well as to pass on their knowledge in mentoring programs. To foster the expansion of knowledge in depth and variety, the launched software should be complemented by a documentation part with learning texts explaining directly on the spot where a certain use case is being handled. Furthermore, a training software should also offer opportunities to preserve and further develop employees' knowledge. Generally, employees should be given a variety of individual low-threshold opportunities to close individual knowledge gaps as well as to expand in other areas according to their interests.

Apart from these offers to foster individual learning, another important aspect mentioned is the improvement of the organizational adjustment to (digital) transformation processes (structural capital). The proposed solutions address the aspects of single- and double-loop learning. Since in this specific use case not only is a new software tool launched, but also a comprehensive organizational restructuring takes place, nearly all of the proposed solutions can be categorized as double-loop learning. Looking at the set of aspects mentioned by the participants, it is striking that the company invents or plans to invent a variety of new ways of doing things and not only wants to do the existing things in a new way. For example, semi-autonomous working groups, job rotation, and division of tasks shall be newly established within the company. Furthermore, active participation shall be fostered by establishing incentives and appreciation as well as by a user-friendly design of the documentation software. Therewith, the company wants to initiate a change of the organizational and individual mindset from solitary expertism to a culture of common knowledge sharing. With the help of these various measures which address the aspect of double-loop learning, the company aims to redefine work practices to promote more participation, cooperation, and critical thinking.

Finally, it has to be stated that both aspects—preservation and further development of employees' knowledge in depth and width (human capital) as well as the adjustment of the organizational culture to a knowledge-sharing culture (structural capital)—are intertwined. Hence, employee-oriented mentoring programs, individual competence analyses, and individually tailored simulator training designed to address human capital can be seen as a prerequisite for achieving a positive change in the organizational mindset. Conversely, changing the organizational structure can be seen as a necessary prerequisite for employees to actively live the desired cultural change and to take part in human-resource management programs offered.

4 Summary and Conclusion

Looking at the social and technological challenges posed by the emergence of artificial intelligence systems, or more generally more algorithmic systems, and the ever-increasing global competition described in the introduction in Chap. 1, market conditions for companies and working conditions for employees are changing rapidly. Both, companies and employees, are affected by these digital transformation processes. As a result, both are striving to assert themselves under these conditions and to act in a self-determined and/or sovereign manner. But what does this mean, and how can Sociodigital Sovereignty be achieved both on the individual level at the workplace as well as on the organizational level of the company? Taking into account different theories from work psychology as well as from sociotechnical system theories, the concept of Sociodigital Sovereignty and its classification matrices were conceptualized with its two pillars to make it applicable to industrial use cases on an individual as well as on an organizational level. With its theoretical roots in different theories, Sociodigital Sovereignty can generally be defined as the ability of individuals and organizations to successfully pursue their aims and objectives by employing digital technology (Hartmann 2021a, b, 2022; Shajek and Hartmann 2023).

Looking at the classification matrix on the **individual level** at the workplace, Sociodigital Sovereignty refers to the three aspects **transparency/explainability**, **confidence of action**, and **freedom of action** within a sociotechnical system that includes the different subsystems of human, technology, and organization. According to this nine-cell matrix, it is important for the individual employees that their technical and organizational working environment is perceived as comprehensible and predictable, providing a certain scope for action. With regard to the classification matrix on an **organizational level**, the three defining aspects are **human capital**, **structural capital**, and **relational capital**. Human and structural capital relate to the internal quality and quantity of knowledge as well as to the internal learning behavior of the company or organization, whereas relational capital refers to the external partnerships of the organization, e.g. to suppliers, customers, and educational and research institutions. Both matrices are conceptualized to be applicable to industrial use cases in order to support companies and other organizations on an individual level at the workplace as well as on an organizational level to achieve more Sociodigital Sovereignty.

Based on the theoretical background of the concept of sociotechnical sovereignty and its classification matrices on an individual and on an organizational level presented in Chap. 2, an industrial use case in the form of a workshop as proof of applicability was presented in Chap. 3. In conclusion, the workshop demonstrated that the concept of Sociodigital Sovereignty and its classification matrices are useful tools for structuring debates and supporting companies in implementing new technologies and work routines. In the first and second parts of the workshop, the classification matrix on the individual level at the workplace was used to systematically analyze the positive and negative aspects that were raised regarding the situation within the company without and with the implementation of the new planning software. Generally, the nine facets of the matrix reflected almost all the concerns, fears, wishes and hopes expressed by the workshop participants. As a result, two comparable pictures of the working situation with and without the new planning software served as a basis for further analysis. Overall, most of the concerns and wishes address the human and organizational level and not primarily the technical level, although a software launch served as the specific use case here. One of the key challenges identified was building up more expertise among the staff in depth and in width, so that employees have the opportunity to perceive their working environment as transparent and explainable as well as predictable in its actions, enabling them to take advantage of the freedom of action offered. Another key challenge identified is to evoke a change of mindset from solitary expertism to a knowledge-sharing culture.

The third part of the workshop focuses on finding solutions to the two prioritized key challenges. Looking at the proposed solutions, apart from a more user-friendly design of the software and corresponding documentation, most of the proposed solutions address the human and the organizational level. With regard to the human level, a strategic human resource management program with different elements should be put in place to promote the development of employees' expertise. In order to bring about a change of mindset from solitary expertism to a knowledge-sharing culture, an adjustment of the organizational structure including different elements should be initiated. Since most of the proposed solutions need to be addressed not only on an individual level at the workplace, but by the company on an organizational level, the classification matrix on an organizational level was used. In conclusion, it proved to be a useful tool for categorizing the findings of the third part of the workshop and reflecting on them afterwards, as the need for action became apparent. Looking at the classified results, human capital and structural capital are the most dominant forms mentioned, while relational capital played a subordinate role. Fostering the increase of employees' expertise in depth (human) and width as well as evoking a change of mindset to a knowledge-sharing culture (structural), accordingly, the company pursues different goals. For example, shared knowledge should allow for a more flexible deployment of staff in order to better distribute human resources. In addition, the stimulated knowledge exchange should lead to a better familiarization of new employees as well as to secure expert knowledge so that it is not lost with the employees once they retire. Looking at the goals and the challenges, it turns out that both types of capital are intertwined and need to be addressed holistically. Developing the knowledge of employees can only be successful and sustainable if a change of mindset can be successfully evoked on the structural level. Otherwise, a change of mindset to a knowledge-sharing culture requires that various and individual training opportunities as well as comprehensive and explainable information material has to be accessible to everyone and not limited to just a few. In this way, the desired depth and width of expertise can be built up among employees.

Having a closer look at this particular use case, it is striking that hardly any challenges and related suggestions were mentioned belonging to the relational capital. On the one hand, it seems obvious since this specific company—providing highly specialized, knowledge-based services—has (almost) no external relationships to suppliers or other network partners. On the other hand, in addition to the various aspects of human and structural capital mentioned above, this could be another important lever for tackling complex problems. Apart from an external software company that develops the planning software and a consulting firm that assisted in advance to plan the roll-out of the software, there was no external support during the introduction or in daily business. Although it was not specifically mentioned as a finding by the participants of the workshop, this lack of relational capital, which was revealed by structuring all the workshop findings using the classification matrix, could be another important aspect to tackle in order to solve the company's problems.

Acting as individuals and organizations in a complex world with growing competition and ever-changing conditions, as well as additionally facing a digital transformation process, can be exhausting and overstraining without external partners or a reliable external network. Thus, even though not mentioned explicitly, building up an external network and reaching out for external support could be another important outcome of the evaluation process for the company. With the help of such external networks, it could be possible to benefit from other expert knowledge, to relieve the daily workload and to better manage the digital transformation process. Such external partner(s) could, for example, help to address the two key challenges mentioned above as well as to support the software launch within the company. This again shows the connection between the three types of capital and the need to evaluate a situation holistically.

In conclusion, the application of the concept of Sociodigital Sovereignty with both classification matrices as evaluation tools to an industrial use case provided on the one hand well-structured workshop results and also revealed unconscious fields of action. On the other hand, for further activities, it can also serve as a guideline to evaluate and implement new solutions in order to address the mentioned and revealed challenges within the digital transformation process in a holistic and human-centered way. Considering that even the most highly automated systems are still sociotechnical systems consisting of humans, machines, and organizations, newly introduced technologies not only require adaptations at the workplace level but also lead to restructuring within the company, impacting the sovereignty of both individuals and the organization. This in turn requires holistic evaluation tools and design guidelines that consider the three aspects human, technology, and organization as different, but equally important and intertwined parts of a sociotechnical system. Therefore, the concept of Sociodigital Sovereignty with its two pillars and its classification matrices has proven to be helpful in a first industrial use case.

The next final chapter gives an outlook on further developments of the method and possible next steps.

5 Further Development of the Method and the Next Steps

After successfully applying the concept of Sociodigital Sovereignty and its classification matrices on a first industrial use case, it became clear that workshops such as those described in this article can only be one part of a whole toolbox to continue our support for the specific company and for such digital transformation processes in companies in general. Since employees and their knowledge are the most important factors within a company, especially when it comes to successfully managing digital transformation processes, workshops with only a few and selected participants can be a good point of reference to reveal the most pressing challenges of a company. However, putting these findings on a broader footing as well as ensuring the commitment of employees in the initiation of the changes, it is essential to enable employees to participate in a variety of ways. Looking at the transformation process as a whole, it seems to be useful to enable broad employee participation both in the analysis of the challenges as well as in the development of possible solutions. Thus, focusing on the analysis phase, the guiding questions for each facet of the matrix used in the first and second part of the workshop have already been developed into a questionnaire that can now be used and validated in an **employee survey**. This should make it possible to broaden the initial findings of a workshop and to identify potential additional or different challenges and initial approaches to solving them. Ensuring employee participation in the next phase, when possible solutions have to be worked out in more detail, we have had good experience in our research project with so-called **co-creation workshops**. Co-creation workshops are a form of inclusive participation in which drafters accompany a workshop and visualize system design drafts from scratch according the participants' discussions (Hofmann et al. 2023; Pentenrieder et al. 2023). Such direct collaboration with designers allows for immediate visualization of employees' ideas and opens up discussion of different possible solutions. By enabling broad employee participation in a variety of ways, the internal expertise of the staff can flow into the solution-finding process, and with that, secure commitment during complex and lengthy (digital) transformation processes.

However, in addition to the important factors of employee participation and inclusion of internal knowledge, the analysis using the classification matrix on an organizational level also reveals the importance of relational capital for obtaining external support and other specific external knowledge from suppliers and network partners. Examples of such relational capital can be specific software development companies as well as consulting firms that accompany (digital) transformation processes with their specific advice and external expertise. Looking at our particular industrial use case and the developed concept of Sociodigital Sovereignty and its classification matrix in general, it can be stated that also our involvement within the industrial company can be seen as a form of external consulting service and respectively relational capital. By developing a whole package with different tools for analyzing and designing digital transformation processes on an individual level at the workplace as well as on an organizational level at the company level, we aim to provide companies with a holistic and human-centered external consultancy service. Accordingly, depending on the individual requirements of the specific use case, our service consists of different tools such as employee surveys, evaluation, and co-creation workshops as well as topic-specific presentations in order to deepen the knowledge in certain thematic areas. In the future, it is therefore planned to further develop and validate the framework as well as the different parts of the toolbox in other use cases.

In conclusion, the challenges of digitalization and transformation today affect both the organizational level of companies as well as the individual level of employees at the workplace. Since both parties have the similar goal of interacting in a self-determined and independent way, as well as to assert themselves in the market, it seems to be useful to apply the concept of Sociodigital Sovereignty and its classification matrices on both pillars—on an individual and on an organizational level—in order to analyze as well as to design such (digital) transformation processes within companies. From a sociotechnical systems perspective, holistic and user-centered evaluation tools and design guidelines that take into account the three aspects human, technology and organization as well as their interaction, have proven their usefulness in this first industrial use case.

References

Argyris, C.: Single-loop and double-loop models in research on decision making. Adm. Sci. q. **21**(3), 363 (1976). https://doi.org/10.2307/2391848

Argyris, C., Schön, D.A.: Organizational Learning II: Theory, Method, and Practice. Addison-Wesley, Organization development series (1996)

Cedefop (Ed.): Learning and innovation in enterprises. Publications Office of the European Union (2012)

Cherns, A.: The principles of sociotechnical design. Hum. Relats. **29**(8), 783–792 (1976). https://doi.org/10.1177/001872677602900806

Couture, S., Toupin, S.: What does the notion of "sovereignty" mean when referring to the digital? New Media Soc. **21**(10), 2305–2322 (2019). https://doi.org/10.1177/1461444819865984

Edvinsson, L., & Malone, M.S.: Intellectual Capital: Realizing Your Company's True Value by Finding its Hidden Roots, 1st edn. HarperBusiness (1997). http://www.loc.gov/catdir/descri ption/hc041/96051533.html

Furman, J.L., Porter, M.E., Stern, S.: The determinants of national innovative capacity. Res Policy: Policy Manag Econ Stud Sci Technol Innov **31**(6), 899–933 (2002)

Hacker, W.: *Allgemeine Arbeitspsychologie* (2., vollständig überarbeitete und ergänzte Auflage). Huber (2005)

Hartmann, E.A.: Arbeitssysteme und Arbeitsprozesse. Mensch - Technik - Organisation: Bd. 39. vdf Hochschulverl. an der ETH (2005)

Hartmann, E.A.: Digitale Souveränität in der Wirtschaft – Gegenstandsbereiche, Konzepte und Merkmale. In: Hartmann, E.A. (Ed.), Digitalisierung Souverän Gestalten: Innovative Impulse im Maschinenbau. Springer (2021a). https://doi.org/10.1007/978-3-662-62377-0_1

Hartmann, E.A. (Ed.): Digitalisierung souverän gestalten: Innovative Impulse im Maschinenbau. Springer (2021b). https://doi.org/10.1007/978-3-662-62377-0 https://doi.org/10.1007/978-3-662-62377-0

Hartmann, E.A.: Digitale Souveränität: Soziotechnische Bewertung und Gestaltung von Anwendungen algorithmischer Systeme. In: Hartmann, E.A. (Ed.), Digitalisierung souverän gestalten II: Handlungsspielräume in digitalen Wertschöpfungsnetzwerken. Springer (2022).

Hartmann, E.A., Shajek, A.: New Digital Work and Digital Sovereignty at the Workplace – An Introduction. In A. Shajek & E. A. Hartmann (Eds.), New Digital Work: Digital Sovereignty at the Workplace, 1st ed., pp. 1–15. Springer International Publishing; Imprint Springer (2023). https://doi.org/10.1007/978-3-031-26490-0_1

Hartmann, E.A., von Engelhardt, S., Hering, M., Wangler, L., Birner, N.: Der iit-Innovationsfähigkeitsindikator: Ein neuer Blick auf die Voraussetzungen von Innovationen. Institut für Innovation und Technik (iit) (2014). http://www.iit-berlin.de/de/indikator/downlo ads/iit_perspektive_innovationsfaehigkeitsindikator.pdf

High-Level Expert Group on AI: The Assessment List For Trustworthy Artificial Intelligence (ALTAI) for self assessment (2020). https://digital-strategy.ec.europa.eu/en/library/assessment-list-trustworthy-artificial-intelligence-altai-self-assessment

Hofmann, A., Hartmann, E.A., Shajek, A.: Digitale Souveränität in soziotechnischen Systemen— KI-Nutzung und Krisenbewältigung. Gruppe. Interaktion. Organisation. Zeitschrift Für Angewandte Organisationspsychologie (GIO) **54**(1), 95–105 (2023). https://doi.org/10.1007/s11612-023-00674-9

Luhmann, N.: Autopoiesis, Handlung und kommunikative Verständigung. Z. Soziol. **11**(4), 366–379 (1982). https://doi.org/10.1515/zfsoz-1982-0403

Mertins, K. (Ed.): Wissensbilanzen: Intellektuelles Kapital erfolgreich nutzen und entwickeln; mit 16 Tabellen. Springer (2005). https://doi.org/10.1007/3-540-27519-3

Mueller, S.T., Hoffman, R.R., Clancey, W., Emrey, A., Klein, G.: Explanation in Human-AI Systems: A Literature Meta-Review, Synopsis of Key Ideas and Publications, and Bibliography for Explainable AI (2019). http://arxiv.org/pdf/1902.01876v1

Mühlbradt, T., Shajek, A., Hartmann, E.A.: Methoden der Analyse und Gestaltung komplexer soziotechnischer Systeme: trends in der Forschung. In Gesellschaft für Arbeitswissenschaft (Chair), *Frühjahrskongress 2022, Magdeburg: Technologie und Bildung in hybriden Arbeitswelten* (2022)

Oesterreich, R.: Handlungsregulation und Kontrolle. Urban & Schwarzenberg (1981)

Pentenrieder, A., Hahn, P., Schaffrath, S., Krieger, B., Brzoska, S., Peters, R., Künzel, M., Hartmann, E.A.: Designing explainable and controllable artificial intelligence systems together: inclusive participation formats for software-based working routines in the industry. In: Shajek, A., Hartmann, E.A. (Eds.), New Digital Work: Digital Sovereignty at the Workplace. Springer (2023)

Schmuntzsch, U., Hartmann, E.A.: Highly automated and master of the situation?! Approach for a human-centered evaluation of AI systems for more sociodigital sovereignty. In: Stephanidis, C., Antona, M., Ntoa, S., Salvendy, G. (Eds.), Communications in Computer and Information Science. HCI International 2023 Posters, vol. 1832, pp. 494–501. Springer Nature Switzerland (2023). https://doi.org/10.1007/978-3-031-35989-7_63

Shajek, A., Hartmann, E.A. (Eds.): New Digital Work: Digital Sovereignty at the Workplace. Springer (2023)

Škrinjarić, B.: Competence-based approaches in organizational and individual context. Humanities Soc Sci Commun **9**(1) (2022). https://doi.org/10.1057/s41599-022-01047-1

Trist, E.L., Bamforth, K.W.: Some social and psychological consequences of the longwall method of coal-getting. Hum. Relat. **4**(1), 3–38 (1951). https://doi.org/10.1177/001872675100400101

Dr. Ulrike Schmuntzsch studied 'Business Psychology' at the former University of Applied Science in Lüneburg and 'Human Factors' at the Technical University (TU) Berlin. Focusing on Work and Engineering Psychology, she worked as a research associate at the Chair of Human-Machine Systems at the TU Berlin from 2011 until 2022. As part of several application-oriented research projects with various industry partners, she was responsible for human-centered evaluation and design. Her doctoral thesis, which she completed in 2014, focuses on the development and evaluation of a warning glove as a means of user support during maintenance work in industrial applications. Since 2022, Ulrike Schmuntzsch has been a research associate at the Institute for Innovation and Technology (iit) and a consultant at VDI/VDE Innovation + Technik GmbH. In this role, she is part of the project team 'Digital Sovereignty in Business'.

Dr. Ernst Andreas Hartmann —after studying psychology, specialising in work and organisational psychology—obtained his doctorate as Dr. rer. nat. at the Faculty of Mechanical Engineering at RWTH Aachen University in 1995. In the 1990's, he worked at the Hochschuldidaktisches Zentrum/Lehrstuhl Informatik im Maschinenbau (University Teaching Centre/Chair of Information Technology in Mechanical Engineering) at RWTH Aachen. In this context, he engaged in projects on academic reform and took part in the development of new forms of academic teaching/learning. Furthermore, he carried out research on the design of man-machine systems, and issues of industrial work organisation. In the mid-1990's, Ernst A. Hartmann was an internal consultant for organisation and process development at John Deere Werke Mannheim. In 2002, he qualified as lecturer (habilitation) in psychology and received the 'venia legendi' for Work and Organisational Psychology; since then, he has been a private lecturer for work systems and process design at RWTH Aachen. From 2001 to 2004, he was responsible for the scientific coordination of the program 'Lernkultur Kompetenzentwicklung' ('competence development and learning cultures') of the German Federal Ministry of Education and Research at the 'Arbeitsgemeinschaft Betriebliche Weiterbildungsforschung ABWF e.V.' ('Association for Research in Continuing Education').

From 2004 to 2016, Ernst A. Hartmann was head of the Socioeconomic Department at VDI/ VDE Innovation + Technik GmbH in Berlin; since 2016, he has been head of the Education, Science, and Humanities Department. Since 2007 he has functioned as one of the directors of the Institute for Innovation and Technology (iit).

Wikipedia's Atypical Oganizational Model: Digital Sovereignty 20 Years in the Making

Arne Klempert and Delphine Ménard

Abstract This article explores the history of Wikipedia and Wikimedia as a unique example of the evolving concept of digital sovereignty. With roots in the open, free, and non-profit model, Wikipedia has become a global phenomenon. Based on the authors' extensive involvement since the early years, this article examines Wikipedia's journey of over two decades to unravel relevant aspects of sovereignty within its unconventional organizational framework. The concept of digital sovereignty was nascent when Wikipedia emerged in 2001. In its 24-year evolution, Wikimedia's atypical organizational model, shaped by a mix of intent and happenstance, fostered digital independence while unintentionally creating pockets of dependence. Looking at the origins and the foundational principles, this article sheds light on various aspects of dependence, brought about in the areas of content, collaboration, governmental influence, legal framework and funding models. The article concludes by highlighting the cultural dimension of digital sovereignty, emphasizing how Wikipedia's digital sovereignty is the result of foundational philosophies, open ideation spaces, and iterative adaptation.

Keywords Wikipedia · Wikimedia · Digital sovereignty

1 Introduction

The Free Encyclopedia Wikipedia has grown to be one of the ten most visited websites in the world and the only very large online platform based on an open, free, and non-profit model. It is the starting point of the Wikimedia movement, a complex constellation of organizations, online projects, and communities.

A. Klempert (✉) · D. Ménard
Grüner Weg 6, 61462 Königstein Im Taunus, Germany
e-mail: arne@klempert.de

D. Ménard
e-mail: dmenard@notafish.com

© The Author(s) 2025
U. Schmuntzsch et al. (eds.), *New Digital Work II*,
https://doi.org/10.1007/978-3-031-69994-8_9

We both became volunteers in the early years of Wikipedia: Arne joined the German community in 2003, and Delphine started contributing to the French Wikipedia in 2004. We then went on parallel journeys in the Wikimedia movement and served on Wikimedia boards and committees on a national and global level. We both held paid staff roles, Arne at Wikimedia Deutschland and Delphine at the Wikimedia Foundation.

This article, drawing on our personal experience in the Wikimedia movement, studies the Wikipedia phenomenon 20 years in the making in the hope of providing insight into what digital sovereignty looks like in an atypical organization. The concept of digital sovereignty was not what the founders had in mind when Wikipedia was born in 2001—it barely existed at the time.

Wikimedia is an atypical organizational model. Based on a blend of intentionality and chance, its foundational culture as well as structural and technical developments led to independence in the digital space while creating some unintended level of dependence.

2   Origins

When thinking about the genesis of a successful online platform, one imagines a brilliant mastermind with a revolutionary idea, a lot of work to build a startup, and then cunning strategies to draw users to the platform to scale it. This is not how Wikipedia came to be. The collaborative free encyclopedia owes its existence, and large parts of its current essence, to a few decisions made in the early days rather than any advanced planned roadmap, and to developments that took place by chance or oversight and emerged out of the intrinsic dynamics of the project.

2.1   Genesis: Establishing the Digital Space

Jimmy Wales had intended to create an online encyclopedia with free content, and his initial project was set up in a traditional way, rooted in academics. Volunteer experts were to write encyclopedic articles and make them publicly accessible after having undergone successful peer review. Wikipedia's predecessor was called Nupedia and it advanced rather sluggishly.

In looking for ways to improve Nupedia, Larry Sanger, who was hired as editor-in-chief to oversee its development, became aware of a technology called "wiki". Wikis—websites powered by software allowing instant collaboration with multiple users—had been around for several years, created in 1995 by Ward Cunningham.

In January 2001, when Sanger introduced the wiki concept to the Nupedia community, he framed it as "an idea to add a little feature to Nupedia" (Sanger 2003). Following the hesitation among the members of the Nupedia Advisory Board to integrate it into their encyclopedia project, Wales and Sanger moved the wiki to its own domain and called it "wikipedia" (Sanger 2005), a portmanteau made from the words wiki and the suffix pedia, from encyclopedia. Wikipedia was intended to be separate from Nupedia, and was to be used to feed it with articles.

The platform turned out to be an instantaneous success. On 12 February 2001, it counted 1,000 articles and increased that number tenfold within the following six months. Nupedia was closed two years later, in September 2003, with about 100 articles, counting those published and those in various stages of development (Jemielniak 2012).

2.2   People Occupying the Space: Birth of a Community

Wikipedia grew exponentially but it was far from being a mainstream phenomenon, and it was far from being useful as an encyclopedia. Participants became attracted through mentions in tech-focused online media and word of mouth. Wikipedia's presence in search engine results also sent new visitors to the site.

Kurt Jansson, later a founding member of Wikimedia Deutschland, joined Wikipedia in the summer of 2001 and described these early years as a "wonderfully anarchic time of trying things out and finding oneself". From the very beginning, he was "fascinated by Wikipedia as a social experiment" (Der Spiegel 2016). In the media reports of these years, the Wikipedia phenomenon was viewed with awe, but with no chance of eventually rivaling Britannica or Encarta. (Benjakob and Harrison 2020). MIT Technology Review wrote in September 2001 "it [Wikipedia] will probably never dethrone Britannica, whose 232-year reputation is based upon hiring world-renowned experts and exhaustively reviewing their articles with a staff of more than a hundred editors." (Helm 2001). Wikipedia was being observed but hardly considered.

Each media report, however wrong their assessment was, brought new interested readers to the project, each of them a potential author that would help improve the content. More and more people were turning to Wikipedia for information, even though there was still concern about its reliability. Public skepticism eased considerably in 2005 when an expert-led investigation carried out by the scientific journal Nature concluded that "Wikipedia [came] close to Britannica in terms of the accuracy of its science entries" (Giles 2005).

The massive influx of new contributors in those early years and the increased expectations of the public had a significant impact on shaping the community. With an increased focus on maintaining quality and efficiency, the participants developed stricter rules and started using algorithmic tools to deal with contributions.

In 2007, after years of massive growth, participation came to a halt. The number of editors—56,000 at the peak—entered a phase of decline. The decline in participation stopped around 2014 and editor numbers seem stable since then. Wikimedia today benefits from a large community of contributors who write and add content to the projects, as well as curate the encyclopedia and its sister projects, and organizers, who work to support the contributing communities (Fig. 1).

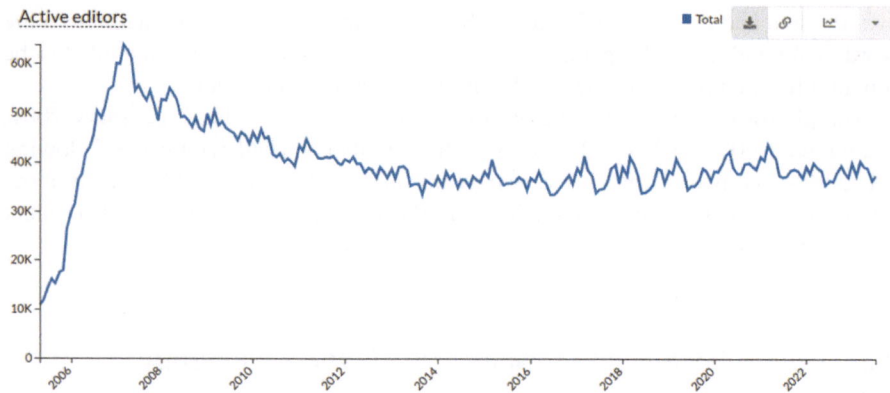

Fig. 1 Number of active contributors on the English Wikipedia April 2005—September 2023 (https://stats.wikimedia.org/#/en.wikipedia.org/contributing/active-editors/fulll|line|2005-05-01~2023-09-01|(page_type)~content*non-content|monthly)

2.3 Anchoring the Digital Space: Development of an Organizational Framework

Wikipedia was originally funded by Bomis, a private company owned by Jimmy Wales. The content was intended to be available free of charge from the beginning. After conversations within the community about how Wikipedia should be financially supported and heated debates around adding ads on the website, Jimmy Wales announced the formation of the Wikimedia Foundation, a US non-profit, in June 2003.

In the early years, the Wikimedia Foundation reflected the Bomis influence: the first Board of Trustees consisted of Jimmy Wales and his two Bomis Co-Founders Tim Shell and Michael Davis. In 2004, Florence Devouard and Angela Beesley, who were both active contributors to Wikipedia, were appointed to the Board of Trustees. By 2006, a majority of the board members had a community background.

The fact that Wikipedia needed financial and organizational support sparked initiatives to found other Wikimedia organizations. In 2004, a group of Wikipedians in Germany founded Wikimedia Deutschland, a non-profit membership organization. Other communities followed. Wikimédia France, Wikimedia Serbia, Wikimedia Italia, and Wikimedia Polska were founded in 2005. Non-European chapters only came to be several years later, Wikimedia Argentina, and Wikimedia Taiwan in 2007, Wikimedia Australia in 2008, Wikimedia South Africa in 2013, establishing Wikimedia's presence on five continents. Wikimedia counts about 40 chapters today.

After the creation of a new affiliation model in 2012, the number of affiliates grew, with a growing number of User Groups, some focusing on countries, others on languages or shared interests. Wikimedia counts 146 User Groups in 2023.

The Wikimedia Foundation—largely relying on volunteers in its early years—went through organizational growth after 2007, when the Board of Trustees hired the first full-time Executive Director, Sue Gardner. When she joined, the Wikimedia Foundation counted less than ten employees. In 2014, there were around 200, and by the end of 2022, the Wikimedia Foundation had grown to about 700 employees.

Some Wikimedia chapters and User Groups moved to a professionalized model. Wikimedia Deutschland paved the way and hired its first employee in 2008. Today, it has 160 employees and 100.000 supporting members and is the largest Wikimedia organization after the Wikimedia Foundation. Other movement affiliates stayed significantly smaller, most of them remaining purely volunteer-driven.

Wikipedia is considered a very large online platform, along with YouTube, Facebook, or TikTok. The organizations that support it count at most 1000 employees worldwide, a fraction of the companies operating other websites of this magnitude. It is important to note that none of the employees are involved in editorial work on Wikipedia in their professional capacity. This remains the sole responsibility of the volunteer editor communities, with about 300,000 contributors around the world. Most employees are working on software development and infrastructure. Other staff resources are working on global development, community support, fundraising, legal and community advocacy.

Wikipedia is a grassroots project. It was fostered by a community that started to organize on wiki and eventually embraced the real world. Remembering the fact that the Wikimedia organizations came after Wikipedia as a project is indispensable to understanding the boundaries of the Wikimedia ecosystem. In 2023, it is impossible to dissociate the organizational framework from the online wiki projects and the communities that keep them alive. Independence stems from all of them.

3   Foundational Principles as the Seat of Independence

3.1   Copyleft: Free Licenses and the Right to Fork

Wikipedia was created under a free license, the GFDL (GNU Free Documentation License), meant to ensure that the content was free to use and share. In 2009, Wikipedia migrated under a Creative Commons license, the CC-BY-SA (Attribution Share-Alike), which was more flexible and easier to use.

A free license allows for reuse and expansion, fosters collaborative authorship, and ultimately serves public interest (Kim 2007). The encyclopedia is thus there to be used and reused by all, commercially or not. One can take all the content (not just an article) and start their own encyclopedia, or sell it as a book, or use it in another context.

This right was exerted as early as 2002 with the departure of contributors to Wikipedia in Spanish, who, frustrated by talks of adding advertisements to the website, decided to go and start a fork of Wikipedia. They migrated the Spanish language content to another site and founded the Enciclopedia Libre Universal. While the license allowed it, the technical migration was a bit difficult (Tkacz 2011), but

Enciclopedia Libre Universal still exists today and is a good early example of how a free license fosters independence.

Ease of use granted by the license is strengthened by the fact that MediaWiki, the piece of wiki software that powers Wikipedia, is also available under the GNU General Public License (GPL 2.0), another free license. Anyone with the capacity to host a website can host a MediaWiki instance and import Wikipedia's content. In theory at least, Wikipedia is reproducible to infinity and designed that way.

Choosing free licenses made the content independent from its authors and even from Wikipedia itself. They strengthened the overall independence of Wikipedia from influence and allowed for maximum reach and reuse possibilities, making Wikipedia a common good. With those, Wikipedia becomes the model of a world where "citizens must have access to a public space built on public digital infrastructures governed by them and not by private actors" (Keller 2022).

3.2   Wiki Philosophy: Anyone Can Edit

In hindsight, the experience of Nupedia shows that a closed, peer-reviewed model probably would not have worked. Opening Wikipedia as a wiki was an efficient way to funnel in as many contributors as possible. "Anyone can edit" is one of the features of Wikipedia which makes it most resilient. Producing and curating content does not depend on any one person, or even on any one group of people, but rather on the goodwill of hundreds of thousands of people across the years who have given their time, be it minutes correcting a spelling mistake, or days and years writing featured articles.

Wikipedia does not have one editorial board that decides what goes to print or what does not, it is a living ecosystem that regenerates as people come and go, and contribute to expanding articles. The strength of the collective over time is best illustrated when one looks at the history of Wikipedia articles. The first version of the French Wikipedia article "Pomme" (apple) started with the words: "La pomme est un fruit"[1] (The apple is a fruit). Today, it is a multi-page article that has been edited by more than 1,400 editors (people with any kind of edit), and counts about 395 authors (people who have added substantial information).

Collective authorship and the idea that anyone can edit allows a form of independence that has been necessary for Wikipedia's growth, lowering the barrier of entry for people to contribute to the repository of the world's knowledge.

---

[1] https://fr.wikipedia.org/w/index.php?title=Pomme&oldid=18085.

3.3   Wikipedia Pillars and Core Content Policies

Wikipedia's second pillar: "Wikipedia is written from a neutral point of view" states "We strive for articles with an impartial tone that document and explain major points of view, giving due weight for their prominence" ("Wikipedia: Five Pillars" 2005–2023). The basic idea behind Neutral Point of View (NPOV) is that contributors should aim at writing articles with objectivity in mind, presenting information to the reader that can be verified and emanates from reliable sources. NPOV is one of the three Core Content Policies of Wikipedia ("Wikipedia: Core Content Policies" 2022), along with Verifiability and No Original Research.

NPOV differs significantly from the approach of traditional encyclopedias, which focused on general doctrine, probably because they relied on individual experts as authors and were limited in the length of their articles. Wikipedia's approach opens the door to a plurality of opinions and hopefully keeps the path open to a critical approach to any subject. One might question the validity of NPOV as a successful (or not) way to write objective articles, but NPOV constitutes a clear element of independence for Wikipedia.

By wanting to gather and present all research and all major points of view on a subject, Wikipedia fosters independence from one "client": a person, a company, and even a school of thought that would want to tell their version of any one story and have it be the only public version.

3.4   Non-profit Status and Small Donor Funding Model

From the very beginning, Wikipedia needed financial support to exist. Even though the Wikipedia authors are volunteers, hosting the website (and facilitating volunteer community initiatives) has a cost. Bomis took care of these costs until 2003. With Wikipedia's growth, it became clear that Bomis would not be able to foot the bill forever. Bomis was also not going to be the entity to bring Wikipedia forward.

The non-profit model would guarantee independence from the for-profit and was deemed more in line with the work volunteers were doing and the fact that the content was both free to reuse and free of charge. The Wikimedia Foundation was founded as a US non-profit (501(c)3) in June 2003 and, following a server outage in late 2003, started its first fundraiser, to raise 20,000 US dollars (Feddern 2003).

The target amount came in within days, but it was already clear back then that the exponential growth of Wikipedia would require future fundraisers to bring in more money. When a dedicated fundraising team was established at the Wikimedia Foundation, the question of how to get revenue became more strategic. After some experiments with alternative revenue streams in 2008 and 2009, the "many-small-donors model" was intentionally prioritized: "because it aligns fundraising with the rest of the Wikimedia movement: it makes it global and empowers ordinary people. It also enables us to stay focused on our own mission and strategy, rather than being

pulled off-course by large funders' needs and desires. It makes us independent."
(Gardner 2010). This small donor model has evolved again over the years, integrating
email campaigns beside the banner campaigns and it is largely what has been keeping
Wikipedia alive.

The values behind a non-profit model and the commitment to the support of many
rather than dependence from a few felt like a natural path of growth for Wikimedia, in
line with the Open Source and open-to-all philosophies that its growing communities
had championed in the early days.

4 (In)dependence

The foundational principles that gave birth to and accompanied the growth of
Wikipedia and the Wikimedia movement have been solid grounds for independence,
and have played an important role in establishing sovereignty. The content is free, it
is powered by free software, curated and produced by volunteers, and anyone can edit
the website, making free and neutral knowledge available. Many of these grounds
for independence, however, come with their share of unintended dependence.

4.1 Content Dependence

Wikipedia is open for anyone to edit, inviting amateur authors to write about any
subject. The culture of amateur writers, however, discredits the Wikipedia authors
by design: they are considered non-reliable, and their work can only be accepted
if they can provide sources to back their writing. One of Wikipedia's fundamental
rules is "No original research": contributors can only add content if they can cite
other people's work: academics, researchers or journalists, and any other respected
sources.

The rules enacted through the core content policies posit a problematic dependence
on external sources. External sources are sometimes obsolete (outdated research)
and/or structurally biased, which skews the creation of user-generated information Wikipedia relies on (Graham et al. 2014). They are also inexistent in a form
Wikipedia would accept—for example, oral citations are not allowed—which hinders
the capacity of contributors to report on much of the world's reality (Kaffee 2019).

The "Neutral Point of View" then becomes the point of view of educated males
from the northern hemisphere. This notably influences the way a certain corpus
of knowledge is represented in Wikipedia, among this knowledge from the Global
South and Indigenous Knowledge (Gallert 2013). In its 2017–2018 Annual Plan,
Wikimedia Argentina supported an initiative developed to counter this bias, working
with journalists to raise awareness about the lack of sources that Wikipedians could
work with, urging them to research important subjects that needed more local sources
(Wikimedia Argentina 2018).

To help address this problem, the Wikimedia Foundation has been working on mapping the Knowledge Gaps along three dimensions: Readers, Contributors, and Content and for each has been studying representation (gender, age, geography…) and interaction gaps (Redi et al. 2021) in order to provide a base of where to start to remedy some of this dependence.

4.2 Barriers to Contribution: Not Anyone Can Edit

Wikipedia relies on the general public's contribution to grow its content. One way for new editors to emerge is them visiting the website and clicking on the edit button to participate. This makes the "anyone can edit" wiki principle's success dependent on traffic to the website. Knowledge hosted by the Wikimedia websites is increasingly integrated in other tools such as knowledge graphs, or personal virtual assistants, and is consumed outside of Wikipedia itself, which makes the edit button harder to discover.

The decline of the number of editors, which worried Wikimedia for a long time, stopped around 2014. The reason for the decline and stop in growth has been studied by various people, and a number of barriers to contribution have been identified. In 2013, Halfaker notes: "Wikipedia has changed from the encyclopedia that anyone can edit to the encyclopedia that anyone who understands the norms, socializes himself or herself, dodges the impersonal wall of semi-automated rejection, and still wants to voluntarily contribute his or her time and energy, can edit." (Halfaker et al. 2013).

Ford and Wacjman, in a 2017 study, analyze how Wikipedia's infrastructure and policies favor male contributors, making Wikipedia a less than inclusive community for women (Ford and Wajcman 2017). Structural barriers to contribution, including lack of access, narrow the wiki principle of anyone can edit and make Wikipedia dependent on only a fringe of the potential contributors it could benefit from by favoring the dominance of "white male, Northern Hemisphere" editors.

Wikimedia local or thematic affiliates and communities, with support from the Wikimedia Foundation, have developed initiatives for targeted recruitment of editors and programs to empower contributions from underrepresented groups and geographies, which have somewhat moved the needle on who gets to participate, but imbalance is still great when it comes to how many women or editors from the Global South are active on the Wikimedia projects (Wikimedia Foundation 2021a).

4.3 Freedom of Speech and State Censorship

The term digital sovereignty has been used to talk about how national states control both the flow of data and the technology that enables that flow, or how social movements and individuals counter government surveillance through the use of Open Source software or privacy tools (Couture and Toupin 2019).

In order to make the sum of human knowledge accessible to all, Wikipedia depends on governments and national infrastructures which are, on occasion, denied. Wikimedia sites have repeatedly been severely affected by government measures in various countries. In some cases, measures were directed against specific content, the retrieval of which was prevented locally, others were directed at Wikipedia as a whole, and some also at contributors. Those blocks are probably a result of Wikimedia communities' tireless work in tackling fake news and propaganda.

One of the first countries to restrict access to Wikipedia was China. After an initial block of Chinese-language Wikipedia in the run-up to the 15th anniversary of the massacre in Tiananmen Square, Wikipedia has been blocked intermittently for years. Since 2019 the block has been permanent, affecting all versions of Wikipedia. (Wikimedia Foundation 2019). China has moreover actively fought Wikimedia's presence in the World Intellectual Property Organization, laying its veto to the Wikimedia Foundation obtaining an accreditation as a permanent observer for the third time in 2023.

In 2017, the Turkish government invoked a law to protect national security and implemented a block of all versions of Wikipedia. The Wikimedia Foundation went to court against this in Turkey and later took it to the European Court of Human Rights (ECHR). It took more than two and a half years until the Constitutional Court of Turkey found that the block was unconstitutional, which then led to the lift of the ban by the Turkish government (Tecimer 2020).

The Wikimedia Foundation regularly comments on such blocking with a reference to its mission, "a world in which every person can freely share in the sum of all knowledge". It stands to reason that this also includes the defense against censorship and the protection of user privacy. In 2021, the Wikimedia Foundation implemented its own Human Rights Policy, aiming to establish Wikimedia's commitment "to protect and respect the full range of human rights across all of our projects, while serving as a foundation for our broader work in advocating for policies and technologies that enable our global movement to thrive." (Wikimedia Foundation 2021b).

4.4 Legal and Organizational Frameworks

The Wikimedia Foundation is not only strictly against censorship by authoritarian states. Wikimedia has also made it its mission to defend the work of volunteer authors against other forms of interference. The way the organizational model grew is a testament to a focus on legal independence.

As chapters came to be, many aspects of the roles and responsibilities of those local organizations vis-à-vis the Wikimedia Foundation remained unclear for a long time. The earliest document of their relationship is probably the initial bylaws of Wikimedia Deutschland, which stated clearly its independence from the Wikimedia Foundation. Not every chapter was founded in the early days with this independence clause, but distance from the Wikimedia Foundation was enforced by the chapters

Committee as it became clear that tying Wikipedia to a non-US entity might make Wikipedia liable to non-US law.

Despite the global ambition and reach of its mission, the Wikimedia Foundation limits its legal presence to the US. Apart from the generally free speech-friendly legal framework granted by the US Constitution, Section 230 of the Communications Decency Act (CDA) provides a safe harbor provision, a "bargain struck between the legislator and the services that enable the free flow of information on the internet" and grants immunity to websites from claims made against what their users post (Walsh and Oh Lam 2010). This legal framework is clearly of advantage for Wikipedia's self-regulating principle.

The declared "independence" of Wikimedia Deutschland became important when the German chapter faced several legal issues. Since 2006, there have been repeated legal disputes in Germany in which the German association was involved. Publicly known cases include disputes over the naming of the late hacker "Tron" and the German comedian Atze Schröder. In both cases, the plaintiffs failed (Kleinz 2007).

Interestingly enough, the community sometimes follows local laws that are more restrictive than US law. An example of this is the case of two men convicted for the murder of German actor Walter Sedlmayr in 1990. Around the time of their release from prison in 2007/2008, both used legal means to remove their names from the internet in general, based on German law which allows criminals a right to privacy once they have served their sentence. The German-speaking community initially followed this request because they understood the German legal framework, and did not name the two men. The English-language community and the Wikimedia Foundation had a different take, and, based on the First Amendment of the US Constitution, the names stayed in the English version of the article (Schwartz 2009).

In its Transparency Report, the Wikimedia Foundation states how many requests it has received to disclose user data, and change or remove content. In relation to the enormous size of Wikipedia, the number of requests received is extremely low and the number granted is almost zero—an indication of how much the Wikipedia community strives to respect the law but also how Wikimedia is mindful of its users' privacy and rights.

Thanks to advantageous US laws, and the concentration of liability in the US organization, Wikimedia reduces its exposure to more constraining laws in other jurisdictions. At the same time, it increases its dependence on the US legal framework, which might be problematic should this framework come to change drastically.

4.5   Funding Models for Independence

In 2021–2022, Wikimedia has received 165 million USD from over 13 million donors. The small donor model represents more than 80% of Wikimedia's revenue. However, while the model ensures independence from a unique source of funds, it increases dependence on visits to the website. With increasing direct integration of

Wikipedia content in other platforms, it is unclear whether the website will continue to provide the necessary traffic to the donation pages.

The global fundraising campaigns are undoubtedly successful from a financial point of view, but they have been regularly criticized—both for being overly dramatic in tone, and for being aimed at the growing size of reserves. Keeping large reserves and emphasizing the urgency are common non-profit tactics, but the community has been vocal in keeping the Wikimedia Foundation accountable. As Dewey points out in the Washington Post, "[The Wikimedia Foundation] grew out of the near-anarchic online community surrounding the wiki movement, and is still beholden to its ethics." (Dewey 2015).

To reduce its reliance on annual donation drives, the Wikimedia Foundation started an Endowment, as a "permanent safekeeping fund" (Dredge 2016). Announced in 2016 with a goal to reach a $100 million through fundraising within ten years, the Endowment reached its target ahead of schedule in 2021. In the 2021/2022 fiscal year, the Wikimedia Foundation raised 165 million USD, coming from 13 million donations.

In 2021, the Wikimedia Foundation also started a for-profit venture, Wikimedia Enterprise, to offer API services to large business users of Wikimedia content. Wikimedia Enterprise aims to give its client a more reliable and sustainable API and push them to invest in the future of Wikimedia by supporting efforts to share the content that they use for their own for-profit ventures.

Wikimedia Deutschland's funding structure is an example of a different take on the small donor model, as it leverages its status as a membership organization and converts donors into supporting members. In 2022, Wikimedia Deutschland exceeded the 100,000-member mark with an annual contribution volume of 5 million euros. In the same year, 400,000 individual donors donated 15 million euros.

Funding is a critical aspect of sovereignty, as it is an enabler for the development of strategies that ensure Wikipedia's and Wikimedia's future and independence.

5   Conclusion

Twenty-two years after starting as what could have remained a marginal experiment, Wikipedia has established a completely new approach to knowledge, how it is gathered, distributed and curated. Wikipedia's content is still free, maintained by a large self-organizing community of volunteers, and is hosted by a non-profit entity. It provides free and verifiable information to the benefit of individuals and organizations.

All these characteristics are aspects that illustrate digital sovereignty. When looking at Wikipedia's and Wikimedia's development, it is clear that their independence was not engineered. On the contrary, it results from a complex mixture of decisions, non-decisions, and organic developments. Some decisions were made intentionally by individuals in the beginning, others stemmed from the wiki culture, and others again were pragmatic ways to adapt to Wikipedia's development and its evolving environment. As the Wikimedia Ecosystem grew and matured, it revealed unexpected pockets of dependence.

What we are learning from the case of Wikipedia and Wikimedia is to see digital sovereignty as a cultural approach rather than a deliberate organizational or technological design. Wikipedia's history and impact illustrate digital sovereignty in the organizational space as a polarity between intent and happenstance.

Establishing foundational philosophies, and providing individuals and their collectives with an open space to ideate and iterate around what happens to them have worked for Wikipedia. This liminal space was needed for actors of the system to see where the foundational philosophies fell short, and to work to avoid the dependence they might create.

The concept of digital sovereignty has a deterministic aspect to it: self-determination, clear delineation, clear boundaries. Advocating for an open space seems to be counterintuitive because the nature of an open space is that you don't see the end of it, but then, twenty years ago, nobody imagined that Wikipedia would get to where it is today... At least, we didn't.

References

Benjakob, O., Stephen, H.: "2 From Anarchy to Wikiality, Glaring Bias to Good Cop: Press Coverage of Wikipedia's First Two Decades." *Wikipedia @ 20*, October (2020). https://wikipedia20.mit press.mit.edu/pub/u1f6cq5i/release/2.

Couture, S., Toupin, S.: What does the notion of 'sovereignty' mean when referring to the digital? New Media Soc. **21**(10), 2305–2322 (2019). https://doi.org/10.1177/1461444819865984

Der Spiegel: "Wikipedia wird 15: Unverzichtbar, allgegenwärtig, in Lebensgefahr," March 16, 2016, sec. Netzwelt (2016). https://www.spiegel.de/netzwelt/web/wikipedia-wird-15-unverzichtbar-allgegenwaertig-in-lebensgefahr-a-1082465.html

Dewey, C.: "Wikipedia Has a Ton of Money. So Why Is It Begging You to Donate Yours?" *Washington Post* (2015). https://www.washingtonpost.com/news/the-intersect/wp/2015/12/02/wik ipedia-has-a-ton-of-money-so-why-is-it-begging-you-to-donate-yours/. Accessed 2 Dec 2015

Dredge, S.: "Wikipedia Launching $100m Fund to Secure Long-Term Future as Site Turns 15." *The Guardian*, January 15, 2016, sec. Technology (2016). https://www.theguardian.com/techno logy/2016/jan/15/wikipedia-fund-future

Feddern, B.: "Wikimedia startet Spendenaufruf." heise online. https://www.heise.de/news/Wikime dia-startet-Spendenaufruf-90891.html (2003). Accessed Accessed 29 Dec 2003

Ford, H., Wajcman, J.: 'Anyone can edit', not everyone does: wikipedia's infrastructure and the gender gap. Soc. Stud. Sci. **47**(4), 511–527 (2017). https://doi.org/10.1177/0306312717692172

Gallert, P.: "Reliable Sources for Indigenous Knowledge: Dissecting Wikipedia's Catch–22," (2013)

Gardner, S.: "Why Wikimedia's New Revenue Strategy Makes Me Happy." *Sue Gardner's Blog* (blog). https://suegardner.org/2010/07/17/hello-world/ (2010). Accessed 17 July 2010

Giles, J.: Internet encyclopaedias go head to head. Nature **438**(7070), 900–901 (2005). https://doi.org/10.1038/438900a

Graham, M., Bernie, H., Ralph, S., Ahmed, M.: "Uneven Geographies of User-Generated Information: Patterns of Increasing Informational Poverty." SSRN Scholarly Paper. Rochester, NY. https://papers.ssrn.com/abstract=2382617 (2014).

Halfaker, A., Stuart Geiger, R., Morgan, J.T., Riedl, J.: The rise and decline of an open collaboration system: how Wikipedia's reaction to popularity is causing its decline. Am. Behav. Sci. **57**(5), 664–688 (2013). https://doi.org/10.1177/0002764212469365

Helm, J.:. "Free the Encyclopedias!" MIT Technology Review. https://www.technologyreview.com/2001/09/04/235538/free-the-encyclopedias/ (2001). Accessed 4 Sept 2001

Jemielniak, D.: "Wikipedia: An Effective Anarchy," November. https://depot.ceon.pl/handle/123 456789/315 (2012)

Kaffee, L.-A.: "The Sum of All Knowledge? Oral Citations on Wikipedia." *Medium* (blog). https://medium.com/@lucie.kaffee/the-sum-of-all-knowledge-oral-citations-on-wikipe dia-abaad65c5b0c (2019). Accessed 17 Sept 2019

Keller, P.: "Digital Commons Are a Pillar of European Digital Sovereignty." Open Future. https://openfuture.eu/blog/digital-commons-are-a-pillar-of-european-digital-sovereignty. (2022). Accessed 21 June 2022

Kim, M.: The creative commons and copyright protection in the digital era: uses of creative commons licenses. J. Comput.-Mediat. Commun. **13**(1), 187–209 (2007). https://doi.org/10.1111/j.1083-6101.2007.00392.x

Kleinz, T.: "Lobbygruppe scheitert mit Verfügungsantrag gegen Wikimedia." heise online. https://www.heise.de/news/Lobbygruppe-scheitert-mit-Verfuegungsantrag-gegen-Wikimedia-132 430.html (2007). Accessed 24 May 2007

Redi, M., Gerlach, M., Johnson, I., Morgan, J., Zia, L.: "A Taxonomy of Knowledge Gaps for Wikimedia Projects (Second Draft)." (2021). https://doi.org/10.48550/arXiv.2008.12314

Sanger, L.: "[Nupedia-l] Let's Make a Wiki." https://web.archive.org/web/20030414014355/http:/ /www.nupedia.com/pipermail/nupedia-l/2001-January/000676.html (2003). Accessed 14 Apr 2003

Sanger, L.: "The Early History of Nupedia and Wikipedia, Part II - Slashdot." https://slashdot.org/story/05/04/19/1746205/the-early-history-of-nupedia-and-wikipedia-part-ii (2005). Accessed 19 Apr 2005

Schwartz, J.: "Two German Killers Demanding Anonymity Sue Wikipedia's Parent." *The New York Times*, November 13, 2009, sec. U.S. https://www.nytimes.com/2009/11/13/us/13wiki. html (2009)

Tecimer, C.: "Why the Turkish Constitutional Court's Wikipedia Decision is No Reason to Celebrate." *Verfassungsblog*, January. https://doi.org/10.17176/20200120-172218-0(2020)

Tkacz, N.: "The Spanish Fork: Wikipedia's Ad-Fuelled Mutiny." *Wired UK*. https://www.wired.co. uk/article/wikipedia-spanish-fork (2011). Accessed 20 Jan 2011

Walsh, K.M., Oh Lam, S.: "Self-Regulation: How Wikipedia Leverages User-Generated Quality Control Under Section 230." SSRN Scholarly Paper. Rochester, NY. https://papers.ssrn.com/abstract=1579054 (2010)

Wikimedia Argentina: "Grants:APG/Proposals/2017–2018 Round 1/Wikimedia Argentina/Progress Report.". https://meta.wikimedia.org/wiki/Grants:APG/Proposals/2017-2018_round_ 1/Wikimedia_Argentina/Progress_report_form (2018)

Wikimedia Foundation: "Wikimedia Foundation Urges Chinese Authorities to Lift Block of Wikipedia in China." Wikimedia Foundation. https://wikimediafoundation.org/news/2019/ 05/17/wikimedia-foundation-urges-chinese-authorities-to-lift-block-of-wikipedia-in-china/ (2019). Accessed 17 May 2019

Wikimedia Foundation: "Community Insights Report." 2021. https://meta.wikimedia.org/wiki/Com munity_Insights/Community_Insights_2021_Report/Thriving_Movement (2021a)

Wikimedia Foundation: "Wikimedia Human Rights Policy." Wikimedia Foundation Governance Wiki. 2021. https://foundation.wikimedia.org/wiki/Policy:Human_Rights_Policy. (2021b)

"Wikipedia:Core Content Policies." 2022. In *Wikipedia*. https://en.wikipedia.org/w/index.php? title=Wikipedia:Core_content_policies&oldid=1124190755

"Wikipedia:Five Pillars." 2023. In *Wikipedia*. https://en.wikipedia.org/w/index.php?title=Wikipe dia:Five_pillars&oldid=1179858344.

Wikimedia Material

"Copyright - MediaWiki." 2001. 2023 2001. https://www.mediawiki.org/wiki/Copyright.

"History of the Wikimedia Foundation Board of Trustees." 2023. Meta Wiki. 2023. https://meta.wikimedia.org/wiki/History_of_the_Wikimedia_Foundation_Board_of_Trustees/en

"History of Wikipedia." 2023. In Wikipedia. https://en.wikipedia.org/w/index.php?title=History_of_Wikipedia&oldid=1181728644#Founding_of_Wikipedia

Iskander, M.: "Love Wikipedia? Get to Know the Nonprofit behind It." Wikimedia Foundation. https://wikimediafoundation.org/news/2023/10/30/love-wikipedia-get-to-know-the-nonprofit-behind-it/. (2023). Accessed 30 Oct 2023

"Pomme." 2002. In Wikipédia. https://fr.wikipedia.org/w/index.php?title=Pomme&oldid=18085.

"Pomme - Page History - XTools." 2023. X-Tools Wikimedia Cloud. 2023. https://xtools.wmcloud.org/articleinfo/fr.wikipedia.org/Pomme

Wales, J.: "[Wikipedia-l] Announcing Wikimedia Foundation". https://lists.wikimedia.org/pipermail/wikipedia-l/2003-June/010743.html. (2003) 20 June 2003

Wikimedia Deutschland: "Jahresbericht 2022 Wikimedia Deutschland e.V." (2023a)

Wikimedia Deutschland. "Wikimedia Deutschland: Über uns." Wikimedia. 2023. https://www.wikimedia.de/ueber-uns/. (2023b)

Wikimedia Foundation:. "Resolution:Licensing Update Approval - Wikimedia Foundation Governance Wiki." https://foundation.wikimedia.org/wiki/Resolution:Licensing_update_approval (2009)

Wikimedia Foundation. 2014. "Wikimedia Foundation 2013–2014 Annual Report." https://annual.wikimedia.org/2014/ (2014)

Wikimedia Foundation: "Wikimedia Foundation Reaches $100 Million Endowment Goal as Wikipedia Celebrates 20 Years of Free Knowledge." Wikimedia Foundation. https://wikimediafoundation.org/news/2021/09/22/wikimedia-foundation-reaches-100-million-endowment-goal/ (2021). Accessed 22 Sept 2021

Wikimedia Foundation: "Fundraising Report 2021–2022." Meta. 2022. https://meta.wikimedia.org/wiki/Fundraising/2021-22_Report (2022)

Wikimedia Foundation: "Transparency Report - January to June 2023." Wikimedia Foundation. https://wikimediafoundation.org/about/transparency/2023-1/ (2023a)

Wikimedia Foundation: "Wikimedia Foundation Annual Plan - 2023–2024." https://meta.wikimedia.org/wiki/Wikimedia_Foundation_Annual_Plan/2023-2024/Foundation_Details (2023b)

"Wikimedia Movement Affiliates." 2023. Meta Wiki. 2023. https://meta.wikimedia.org/wiki/Wikimedia_movement_affiliates

Arne Klempert is an independent management consultant and primarily advises medium-sized and large companies on Strategic Communication and Transformation. He joined the German Wikipedia in 2003 and was involved in the Wikimedia movement in many capacities: he was a spokesperson for the German Wikipedia, co-founded Wikimedia Deutschland in 2004 and, as Managing Director of Wikimedia Deutschland (2006 until 2008), became the first paid employee of a Wikimedia section. He was also a member of the Board of Trustees of the Wikimedia Foundation from 2009 to 2012.

Delphine Ménard currently advises organisations as a Human Resources and Change Expert, and serves as the People and Culture Manager at the non-profit Association for Progressive Communications (APC). She graduated in 2024 from INSEAD's Executive Master in Change (EMC) programme.

Delphine Ménard joined the French Wikipedia in 2004. She served as a volunteer on the boards of Wikimédia France, Wikimedia Deutschland and Wikifranca. From 2016 to 2022, she worked for the Wikimedia Foundation, first as a Programme Officer for the Annual Plan Grants Program, before moving to Human Resources in 2020. She is still an active Wikimedia Volunteer today, advising on HR, strategy and governance in her spare time.

Why Trustworthiness is the Cornerstone of Digitalization

Ulla Coester and Norbert Pohlmann

Abstract The importance of trust is generally recognized. Yet the principle of "everything in moderation" also applies, meaning, in short: trust is good, but control is better. This insight has wide-ranging implications, especially in the current discussion surrounding digitalization. But what happens when it is no longer possible to control an IT or AI solution due to its complexity? At first glance, there are only two options: either refrain from using the technology altogether or simply trust the provider. But neither alternative is suited to the task of advancing digitalization in a meaningful way. On the contrary, providers and users must cooperate on equal terms in order to shape the digital transformation together in a responsible manner. This requires users to trust the technology and the provider. Due to the strong interdependence between trust and trustworthiness, trustworthiness is the cornerstone of the digital transformation.

Keywords Trust · Trustworthiness · Digitalization · Complexity

1 Building Trust in Digitalization

1.1 Building Trust: Reducing Complexity

In the context of digitalization, a new form of interdependency has arisen between manufacturers or provider and users and also user companies due to the increasing complexity intrinsic to the use of new technologies. This results in consequences for both parties: On the one hand, it affects the users' competence in decision-making, and on the other, it limits the options available to the providers. Neither of the two

---

U. Coester · N. Pohlmann (✉)
Institute for Internet Security—if(is), Westphalian University of Applied Sciences, Neidenburger Strasse 43, 45897 Gelsenkirchen, Germany
e-mail: pohlmann@internet-sicherheit.de

U. Coester
e-mail: coester@internet-sicherheit.de

© The Author(s) 2025
U. Schmuntzsch et al. (eds.), *New Digital Work II*,
https://doi.org/10.1007/978-3-031-69994-8_10

parties has full sovereignty in the sense of freedom of action, because every action of the respective party has consequences. In practical terms, this means that users may be required to disclose more data when using services, even if they do not want to, and their data may be used in ways they cannot control. Conversely, users now have more opportunities to monitor activities of providers.

Efforts to improve the quality of this interrelationship are therefore essential, especially those focused on the acceptance of innovative technologies in general and AI solutions in particular. The necessity of such efforts becomes particularly clear in light of the fact that innovative technologies—and thus the entire internet/IT infrastructure—have not only become more and more complex but also increasingly opaque. This results in a serious dilemma: Increasing use is enforced—whether intended or involuntary—while knowledge about the background of and interrelationships within these structures is decreasing. This dilemma has the potential to produce behavioral dichotomy in users: either disproportionate rejection of the technology and corresponding services or blind trust. Both behaviors are counterproductive in terms of value-creating digitalization. Although the latter does not preclude use in general, it prevents meaningful use of new applications or innovative services, since use is neither based on a high level of decision-making competence nor on the sovereignty, or autonomy, of the user.

However, since trust has fundamentally positive connotations—according to sociologist Niklas Luhmann, trust is a mechanism for reducing complexity (Luhmann 1968), i.e., it makes life easier—providers should direct their activities toward building a relationship of trust with their users. At first glance, this idea might seem trivial. Yet while trust may reduce complexity, the construct of "trust" is in itself complex, since the collective conditions under which it manifests are highly exacting. Thus, it is necessary to examine this concept in more detail.

1.2   Building Trust: The User's Perspective

According to Luhmann (Luhmann 2000), trust enables an optimistic view of the future, although individuals possess neither sufficient information nor the necessary control to justify this optimism. Trust can therefore be described as an adaptive strategy that helps individuals retain their capacity for action in a world characterized by uncertainty. Put simply, according to Luhmann (Luhmann 2000), trust is a mechanism for reducing complexity. Generally, human beings possess varying capacities for trust, but this capacity differs among individuals. In principle, there are various concepts that can be used to explain the emergence of trust, such as "trust based on routine" (Pohlmann 2022). In the context of digitalization, the model of "trust based on reason" is particularly pertinent. Establishing this type of trust depends on the one hand, on a person's benefit, interests, and preferences and, on the other hand, on their ability to process information and to recognize trustworthy interaction partners based on certain criteria. According to a widely used model (Möllering 2006), these criteria include, for example, competence, benevolence, and integrity of the trusted party.

Rational trust thus constitutes a person's attribution of reasons for trustworthiness to, for example, a particular provider. In the context of digitalization, this concept could be applied as follows: The user must receive reliable signals of trustworthiness regarding the provider's competence, goodwill, and integrity.

1.3 Building Trust: Change of Perspective to AI Provider

Trust—also in the context of digitalization, as mentioned above—can only be understood as a positive behavior if it is not blind and naive. This assumption is complicated by the fact that users must first trust in outside expertise because they are unaware of the capabilities, knowledge, and intention of the provider in question. Therefore, first and foremost, it is crucial to dispel the user's belief that the provider's interests have been placed above their own, or that a provider could behave opportunistically. Providers should recognize this as their duty—not least because this allows them to develop a relationship of trust with their users. In the context of digitalization, we could thus posit: The provider must prove that the user can trust them.

1.4 Building Trust: Law Versus Moral Philosophy/Ethics

In principle, it could be argued that between existing rules and regulations (laws) and those that will come into force in the foreseeable future, enough is already being done to nurture trust. But upon closer inspection, laws alone are insufficient because it may be possible for providers to develop and deploy AI that, while technically legal, is nevertheless illegitimate, because it ignores the interests of the user, serving only one-sided corporate interests.

In order to build trust, it is crucial to preclude mistrust. For providers, this necessitates increased transparency, especially relating to AI. In other words, providers must provide users with all the information they need to be able to build trust and thus achieve digital sovereignty as described above. Therefore, it is particularly important for AI provider to openly address how their actions relate to the interests of their customers—above and beyond any legal requirements (Fig. 1). Self-limitation, defined as "the ability to limit one's own freedom, whether in the form of external expectations or in the form of one's own actions, in such a way that the use of this freedom does not result in harm" (Suchanek 2021), is one method by which providers express respect for their customers. In terms of digitalization, this requires a consistent sense of responsibility on the part of the provider—starting with management strategy and ending with the developer, who is thus obliged to deliver an AI solution that is fair, transparent, and explainable. This rule could be expressed as follows: To build trust, providers should exercise transparency by openly presenting their organizational modalities and corresponding AI solutions within the framework of a "provider commitment" and by agreeing to adhere to these principles. As a caveat

Fig. 1 Law and moral
philosophy (own illustration)

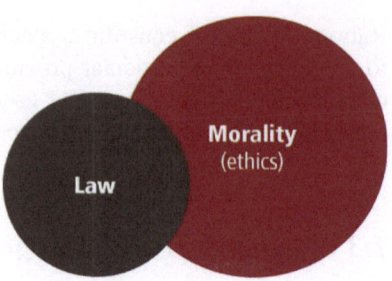

to this rule, however, it is essential to maintain a degree of moderation with regard to transparency, because "transparency describes only the provision of information and does not necessarily result in understanding. In fact, too much information can limit comprehension." (DENKIMPULS DIGITALE ETHIK).

2 Approach: Trustworthiness Platform—Building Trust to Leverage the Potential of Digitalization

2.1 Structure of the Model of Trustworthiness

The relationships for building trust are illustrated in the model of trustworthiness (Fig. 2) and elucidated below based on each individual aspect (Coester and Pohlmann).

2.2 Definitions

Trust, trust giver, and trust receiver: Among other factors, trust relies on a subjective conviction of the correctness of actions. In principle, trust is necessary to reduce complexity and is thus required whenever the user is confronted with an uncertain or insecure situation or when the outcome of an action may involve risk. The trust giver's (user's) "ability to allow" themselves to trust a trust receiver (AI company) therefore equates to their willingness to refrain from questioning the respective trust receiver and, correspondingly, to expose themselves to a certain degree of risk.

Institutional trust: A basic requirement in encouraging people to use AI solutions is the promise of added value. Conversely, this means that if users do not obtain added value from using the AI solution, they are more critical in their assessment and thus less willing to adopt the respective solution. AI providers must also take additional steps to encourage users to extend their capacity to trust AI solutions. This can be achieved by transferring interpersonal trust—i.e., the relationship of trust that develops between people based on individual criteria—to AI solutions. The basic

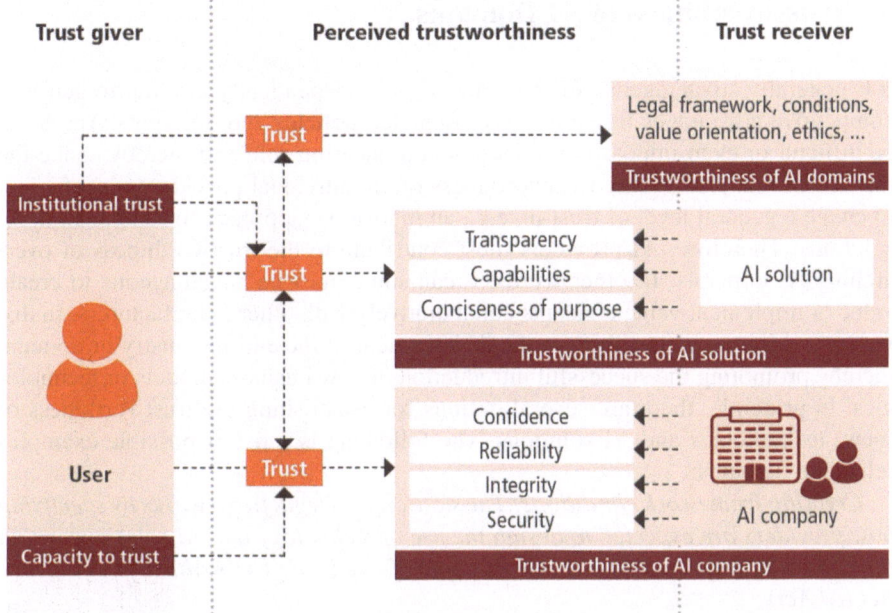

Fig. 2 Model of trustworthiness (own illustration)

idea behind this is that the providers present themselves as trustworthy so that the users can transfer their trust to them based on an assumed similarity.

Perceived trustworthiness: The concept of trustworthiness is based on the assumption that it is possible to rely on some specific aspect. As a rule, perceived trustworthiness is based on the apparent functionalities of the AI solution and measures taken by the AI provider (trustworthiness aspects). This can be demonstrated by not asking for all of the information about a person, but only the details that are necessary.

Trustworthiness: Trustworthiness is based on the assumption that it is possible to rely on given information, solutions, or actors. Presenting the trustworthy aspects of AI solutions as well as those of the provider company as a whole fosters user trust. Evidence of this can be provided in the form of a company commitment.

AI company: An AI company is a manufacturer or provider of AI technologies, products, or services. These individual categories are referred to collectively as AI solutions.

Users: Users include all parties who use AI technologies, products, or services, including user companies.

AI solution: In the context of artificial intelligence, an AI solution is any application that utilizes artificial intelligence mechanisms or an artificial intelligence system.

3   Trustworthiness of AI Domains

It is generally advantageous for providers to act independently in order to generate competitive advantages based on user-orientated policies and decisions. Yet it can be difficult or even impossible to assume a pioneering role and thereby shape the market. In many cases, the trustworthiness of an individual provider is insufficient to create a general level of trust in, e.g., an innovative approach to fundamental AI solutions. Therefore, AI providers must contribute to the trustworthiness of overarching AI domains. In other words, it can sometimes be advantageous to create value or implement value concepts collaboratively with other manufacturers. In this way, providers may contribute to the development of the entire industry or domain, thereby promoting the successful introduction of new business models or technologies. In principle, there are several options for establishing the trustworthiness of (new) technologies and AI solutions. The following is a list of possible examples related to domains:

Creating framework conditions: The state creates legal frameworks by specifying how providers are expected to design the use of technology and AI solutions within the context of a given domain. One example of this is the EU Artificial Intelligence Act (AI Act).

Motivating ecosystems: One example in this area is the Gaia-X industry consortium. Demonstrating trustworthiness is not only a relevant issue for new IT technologies; even for established technologies, standards must be redefined periodically to align with shifting values, and new standards must be implemented accordingly. According to Gaia-X principles, providers must guarantee that their AI solutions comply with European law and ensure data portability while adhering to the strictest IT security standards and offering transparency about data usage.

4   Trustworthiness of AI Solutions

For AI providers to transfer trust to their AI solutions in a dedicated manner, they must account for other aspects—apart from artificial intelligence—that play a role in determining perceived trustworthiness, including transparency, performance, and conciseness of purpose. Only by considering all these aspects can providers nurture user trust in their AI solutions.

4.1   Trustworthiness Aspect: Transparency of an AI Solution

The principal benefit of embracing transparency is that this practice demonstrates the AI providers' willingness to seriously consider their users' needs and to communicate openly. By no means does this involve disclosing every detail of the AI solution or the associated business activities. Rather, it means that users are provided with all the relevant information they need to make a sound decision about the trustworthiness of the AI solution in question. Consequently, quality of information plays a decisive role: It should be participatory and balanced, meaning that it should account for the interests of all parties in equal measure. In the past, this form of communication was not necessary. However, due to the growing complexity of current solutions, such transparency is now essential to increase the users' willingness to adopt AI solutions. This situation illustrates the interplay between trust and trustworthiness: An AI provider depends on the acceptance of its users. At the same time, due to the potential of smart applications to influence users, it becomes increasingly important to demonstrate respect for the users' need for sovereignty and privacy.

In depth instruction for use for an AI solution: One way of ensuring transparency is, e.g., to provide—preferable as a package insert—an instruction leaflet detailing the potential effects of the AI-based application and how it can be controlled. This insert should also include information on the risks of use and how users can avoid or mitigate potential problems.

4.2 Trustworthiness Aspect: Capabilities of an AI Solution

The capabilities of an AI solution are those parameters that the user can immediately comprehend and control. Therefore, the measurable criteria by which the user evaluates the AI solution are a product of the extent to which users feel supported in achieving their intended purpose and the suitability of the AI solution for their needs. Reliability and predictability are among the relevant evaluation criteria here. It is also important for the AI provider's expertise to be reflected in the capabilities of its AI solution. If a solution lacks capabilities, this ultimately reflects strategic errors or inadequate expertise on the part of the provider that produced it. This example demonstrates both the connection and interplay between the trustworthiness of the AI solution and that of the AI provider itself. The usability or performance of the AI solution, among other factors, can serve as an evaluation benchmark for users.

Performance of an AI solution: How precise are the results, how high is the quality of the respective analyses, and how quickly can models be adapted to ensure that they remain up to date.

4.3 Trustworthiness Aspect: Conciseness of Purpose of an AI Solution

Whether or not an AI solution can be said to have conciseness of purpose becomes apparent when examining the intended use of the AI solution. To ensure its solution has a conciseness of purpose, AI providers must precisely define the function and intention of the AI solution during development. Accordingly, the intended use of the AI solution should be clear to the user. For this reason, it is important to ensure that the purpose of the AI solution and the use of its characteristic features can be easily and immediately understood. Yet this does not imply that achieving conciseness for purpose requires a low level of functionality. Furthermore, providers must openly communicate any relevant changes or extensions to the AI solution, especially if these alterations affect the users' ability to identify the originally intended use. If, in addition to the actual application, an AI solution includes supplemental functions designed solely for the interests of the AI provider or third parties, these functions must also be clearly presented and described.

Disclosure of the business model: If sensitive user data is collected under the "pay with personal data" business model and used for individualized advertising and/ or sold to third parties for profit, providers must clearly communicate this intended use.

5   The Trustworthiness of AI Providers

When users weigh up whether or not to use new IT technologies, the features of the respective AI solution are not the only factors involved in their decision. The AI providers' reputation also plays an important role. At present, it is apparent that user trust in IT technologies, applications, and services is not (yet) fully justified. Therefore, providers are required to meet additional conditions to increase their level of trustworthiness. To this end, AI providers must disclose their strategies to the outside world. In practical terms, this means aligning their actions with four trustworthiness aspects: confidence, reliability, integrity, and security. These factors allow users to rationally assess a provider's trustworthiness and quickly and easily evaluate the associated parameters.

5.1   Trustworthiness Aspect: Confidence in an AI Provider

Confidence is one of the key factors to trustworthiness. Generally speaking, with regard to functionality, confidence can be generated by providers with the capability and means to provide reliable, secure AI technology, services, and applications. It is important for providers to develop a strategy to address this aspect and then document it in a general confidence guideline that apply to all employees. For this purpose, among others, providers must create a concept defining the parameters to be fulfilled. The following parameters are of key importance here:

Employees: Parameters pertaining to employee education, certification, and further training: Have the employees studied mathematics, computer science with a focus on artificial intelligence, or data science, or have they received further training in those fields; what experience do they have in the area of AI; what additional skill do they have?

Quality standards: *Parameters for development and production: Description of the development process, definition of the accompanying quality assurance process, including implementation and specification of life cycle management criteria.*

5.2 Trustworthiness Aspect: Reliability of the AI Provider

Reliability means that AI solutions execute only those processes that users desire or expect and that they do so with as close to 100% reliability as possible. Reliability thus implies that AI providers are fundamentally benevolent, meaning that they act in the best interest of their users, i.e., they focus primarily on their users' needs rather than on their own provider interests. One example of benevolent behavior would be for a provider to refrain from instrumentalizing obvious weaknesses of their users—and thus causing harm—to gain a (financial) advantage. For instance, a provider might choose to exploit a customer's preference for high-value branded products: Based on a user's buying behavior, he is placed in a particular category and regularly offered expensive products without a discount. However, given that no provider is perfect, cultivating reliability requires a willingness on the part of the provider to continue developing. In order to compensate for their existing deficits, providers must implement mechanisms that allow them to continuously and proactively improve their reliability and demonstrate this development to users. Ideally, user confidence should precisely mirror the actual reliability of the AI provider or the AI solution. Conversely, the AI providers risk damaging or losing their trustworthiness entirely if their actions are inconsistent with their public image. How will AI providers have to address this issue in the future?

What criteria should they include in their reliability management policy? The following parameters are key to reliability management:

Act cooperatively *in order to more effectively identify the real needs of the user and provide individual support when problems arise. Assuming overall responsibility in the event of damage or issuing recalls if problems are identified are examples of cooperative behavior. The AI provider must inform its users immediately—by direct means, if possible—should serious vulnerabilities be identified. When such information is preemptively published by third parties, for example* via *social media or the trade press, this reduces the trustworthiness of the AI provider.*

Act responsibly *to create added value for users through the correct use of functions that benefit the user. In general, in the context of AI solutions, acting responsibly means fulfilling all specifications required to ensure that input data is of high quality. Data must be complete, representative, and correct. The AI provider should have a single person in the organization whose responsibility it is to ensure that all necessary measures are taken to guarantee that these criteria are met* (Coester and Pohlmann 2020).

Yet, data selection is not the only key factor; responsible handling is also essential. Some AI solutions may require it to start a debate as to whether the data analysis involved is truly in society's best interest. For example, what if Google Street View were used to predict the likelihood of people becoming involved in accidents based on where they live in order to derive more cost-effective insurance policies? Indeed, as the research team under Łukasz Kidziński of Stanford University and Kinga Kita-Wojciechowska from the University of Warsaw discovered, the location variable turned out to be a surprisingly good indicator of accident likelihood (Coester 2020).

5.3   Trustworthiness Aspect: Integrity of the AI Provider

AI providers demonstrate integrity by considering all influencing factors that are relevant to trustworthiness, paying particular attention to the ethical dimensions. This means that, as a trust receiver, an AI provider must be capable of fulfilling all promises made, willing to consistently keep them, and prepared to observe social norms and values.

The ethical orientation of AI providers will be subject to even greater scrutiny in the future. Various studies corroborate this point. In one such example, it was found that 93% of users in Germany demand ethically responsible use of IT technology. Therefore, integrity should be viewed as an essential tenet of all business activities. One-dimensional, purely technically-oriented mindsets that disregard ethical considerations and values are poised to become less and less profitable—or may only prove profitable at all in the short term. This may be explained by the volatility of user behavior, which is subject to rapid influence by negative events or social media posts. Another key aspect to consider in this context is the variability in trust between individual users, a factor that renders general attitudes relatively difficult to assess. Therefore, one of the most important steps for AI provider is to draft an integrity maxim that includes a clear affirmation of their business model and other factors specific for provider. This maxim must address all relevant ethical considerations. Some applicable considerations to keep here in mind are:

Privacy protection: *This includes, on the one hand, responsible handling of customer data, such as immediate deletion when it is no longer needed, and protecting this data through encryption. In addition, it encompasses the commitment to refrain from exploiting user data for additional commercial purposes.*

Accountability: *One of the key aspects of accountability is ensuring the verifiability of the quality of AI solutions. In addition, it should be mandatory for providers to review the technologies they use and disclose any issues with the potential to negatively affect society.*

Responsible use of AI: *This includes, among other things, that providers do everything in their power to protect both individual users and society from harm, i.e., providers must not prioritize profit over human wellbeing. In other words, providers must avoid the behavior attributed to Facebook by a former employee during a hearing before the U.S. Congress: "The providers' leadership knows how to make*

*Facebook and Instagram safer but won't make the necessary changes because they have put their astronomical profits before people." She further emphasized that Facebook's actions carry major ramifications for democracy and society, claiming that the provider relaxed its misinformation filters following the 2020 US election to attract more users (*Tagesschau*).*

5.4   Trustworthiness Aspect: Security of the AI Provider

Recognizing the importance of cybersecurity and implementing the appropriate measures theoretically ensures that AI solutions can be used on the internet at low risk. Yet, this scenario remains aspirational, given that ransomware and DDoS, among other security issues, are commonplace. Some AI solutions or services used today do not offer the level of trustworthiness required to handle critical business processes securely. AI providers therefore require an adequate and well-defined IT security policy to guarantee the best possible protection—during the development process and later when the customers are using the AI solution. Because users (as well as user companies) are often not equipped to protect themselves adequately, AI providers must continuously ensure that their AI solutions are up to date concerning security. The following measures, among others, are critical in the context of security:

Presentation of cybersecurity measures used: The AI provider must demonstrate the actions they are taking to protect both the AI solution and their own company from cybersecurity risks.

Regular review of products and the AI provider itself: The aim of this measure is to actively and continuously identify vulnerabilities with the help of penetration tests, red teams, and bug bounty programs designed to eliminate security vulnerabilities as quickly as possible. Systems are then updated before these vulnerabilities can be leveraged for attacks. This applies to both proprietary AI solutions as well as to the AI providers themselves and their suppliers. This makes it possible to maintain a high level of security—which is also verifiable for the user—throughout the ongoing development process.

6   Conclusion: Trustworthiness as a Competitive Advantage

There is a crisis of trust—this fact is undeniable: "Eighty percent of Germans currently have little or no trust in the major digital corporations. This was the finding of a survey of 5,000 internet users in Germany conducted by the opinion research institute Civey on behalf of Next Conference" (Absatzwirtschaft).

The negative response to the Cambridge Analytica scandal, for example, or the current barrage of fierce media criticism leveled against Instagram clearly demonstrates the effects of negligent conduct. In other words, a responsible approach is key, especially in an international context. By demonstrating responsible behavior alone, European companies can stand out from Asian and American AI providers, who often neglect this responsibility because this approach does not (yet) fit their business model.

German companies can establish a market position based on their trustworthiness due to these conditions. Responsible ethical—and thus trustworthy—action is not at all incompatible with economic interests. On the contrary, value-oriented behavior is in the long-term economic interest of providers. In fact, both in Germany and abroad, users are increasingly demanding that AI solutions should be employed in an ethically justifiable manner.

As to the concept of digital sovereignty, it is important to consider this approach holistically for the community of states in Europe and promote it on a fundamental level, thus ensuring that the development and use of AI solutions align with prevailing morals. Moral aspects assume a high priority within the framework of the trustworthiness platform and are expressed through appropriate design, as AI providers must comprehensively state the values on which both their business activities and the design of their AI solutions are based. This enables users to make informed, sovereign decisions about which AI provider and solutions are trustworthy.

References

Absatzwirtschaft: Digitalkonzerne haben Faktor Mensch aus den Augen verloren. https://www.abs atzwirtschaft.de/digitalkonzerne-haben-faktor-mensch-aus-den-augen-verloren-vertrauensve rlust-von-facebook-google-co-immer-staerker-221217/

Coester, U.: Digitale Ethik—ein Problem in der Marktforschung? In Marktforschung für die Smart Data World. Springer Gabler (2020)

Coester, U., Pohlmann, N: Wie können wir der KI vertrauen?—Mechanismus für gute Ergebnisse, IT & Production—Zeitschrift für erfolgreiche Produktion, Technik-Dokumentations-Verlag (2020) Ausgabe 2020/21

Coester, U., Pohlmann, N.: Vertrauenswürdigkeit schafft Vertrauen—Vertrauen ist der Schlüssel zum Erfolg von IT- und IT-Sicherheitsunternehmen. DuD Datenschutz und Datensicherheit— Recht und Sicherheit in Informationsverarbeitung und Kommunikation, Vieweg Verlag, 2/2022

DENKIMPULS DIGITALE ETHIK: Transparenz und Nachvollziehbarkeit algorithmischer Systeme https://initiatived21.de/uploads/03_Studien-Publikationen/Denkimpulse-Ethik/08-Transparenz-Nachvollziehbarkeit/d21-denkimpuls-ethik08-Transparenz-Nachvollziehbarkeit-Algorithmen.pdf

Luhmann, N.: Vertrauen: Ein Mechanismus der Reduktion sozialer Komplexität, 2000, Lucius & Lucius, Auflage: 4. Auflage, Reihe: UTB: Soziologie fachübergreifend (2000)

Luhmann, N.: „Vertrauen: Ein Mechanismus der Reduktion sozialer Komplexität", 1968 Luhmann, N.: Vertrauen: Ein Mechanismus der Reduktion sozialer Komplexität. Originalausgabe, F. Enke Verlag, Stuttgart (1968)

Möllering, G.: Forschungsbericht 2006—Max-Planck-Institut für Gesellschaftsforschung. Why do we trust? A theretical approach to an everyday problem

Pohlmann, N.: Cyber-Sicherheit—Das Lehrbuch für Konzepte, Mechanismen, Architekturen und Eigenschaften von Cyber-Sicherheitssystemen in der Digitalisierung. Springer-Vieweg Verlag, Wiesbaden (2022)

Suchanek, A.: Ethik und Digitalisierung. In: Hackspiel-Mikosch, E./Neuhaus, R. (Hg.): Ethische Herausforderungen der Digitalisierung und Lösungsansätze der angewandten Wissenschaften (Wissenschaftliche Publikationsreihe der Hochschule Fresenius, Bd. 1), S.21–36, Open-Access-Publikation (2021). ISSN: 2749-4403
Tagesschau: Politiker fordern strengere Regulierung. https://www.tagesschau.de/wirtschaft/untern ehmen/socialmedia-facebook-eu-101.html

At the beginning of 2023, Ulla Coester began working as an academic assistant at the Institute for Internet Security, where she is the project leader for the research project 'Trustworthiness Plat-form for AI Solutions and Data Spaces'. In total, she has many years of professional experience, some international, in self-employment as well as in management posi-tions. Until 2022, she was member of the Standardization Evaluation Group in IEC (SEG 10) 'Ethics in Autonomous and Artificial Applications', In 2022, she was con-tributor to the 'Normungsroadmap KI|AG Grundlagen'. Since 2016, Ulla Coester also has been lecturer for 'Digital Ethics' at the Fresenius University of Applied Sciences, Cologne.

Prof. Dr. Norbert Pohlmann is Professor in the Computer Science Department for cyber secu-rity and the director of the 'Institute for Internet Security—if(is)' at the Westphalian University of Applied Sciences Gelsenkirchen. He is also chairman of the board of the IT Security Association TeleTrusT and member of the board of the Internet industry association eco.

Innovation Capacity in Manufacturing: A Question of Autonomy?

Gina Glock◉

Abstract This study examines the latent structures and interplay between the constructs employees' autonomy and innovation capacity in organizations. Using the example of the German manufacturing sector, the question is addressed as to which autonomy-related and innovation-friendly resources are available in companies and which organizational and interactional challenges workers have to face. Methodically, the contribution builds on the explorative approaches of factor and cluster analysis using the BIBB/BAuA employment survey. As a key result, this analysis isolates *innovation capacity clusters* that divide employees in manufacturing into three groups that give an impression of specific work resources and demands: the potentials (I), the passive balanced (II), and the autonomous frontrunners (III). Cluster I describes a group of employees who have little autonomy and face high organizational demands. Their innovation potential is nevertheless high but is hardly ever utilized. Cluster II has generally good working conditions and a balanced relationship between resources and demands, but the innovation-friendly resources are very limited. Cluster III describes the most innovative workers who are very autonomous. However, they run the risk of being crushed under the given organizational pressure. The three clusters provide companies with an opportunity to optimize their innovation capacity.

Keywords Innovation capacity · Manufacturing · Autonomy · Decision-making · Problem solving · AI skills

1 Introduction and Background

Autonomy is an essential resource for workers to cope with work demands. The ability to determine the method, prioritization, or scheduling of one's own work supports overcoming the daily challenges at work, dealing with unforeseen situations, or facing new topics, organizational structures, or technologies. Job resources

G. Glock (✉)
Dresden University of Technology, 01062 Dresden, Germany
e-mail: gina.glock1@tu-dresden.de

U. Schmuntzsch et al. (eds.), *New Digital Work II*,
https://doi.org/10.1007/978-3-031-69994-8_11

175

are fundamentally opposed to job demands. Their (im-)balance determines the setting and positive direction of organizational variables (see JD-R model, (Bakker and Demerouti 2007; Demerouti et al. 2001)). It is assumed that granting resources has a certain mitigating effect on demands (so-called "buffer theory"). Other authors, such as Karasek with his *job demand-control model* (Karasek 1979), have also established a connection between job demands and autonomy, which determines the stress situation of employees. From the perspective of both company and workers, exploring the appropriate level of autonomy involves finding a balance between freedoms and constraints (Moldaschl 2001)—whereby the opinions regarding the *right* degree of autonomy certainly diverge between the two parties.

Early approaches to the role and concepts of autonomy in the workplace date back to the 1970s and 1980s, when job redesign and reorganization were under consideration in organizations. In 1975, psychologist J. Richard Hackman and economist Greg R. Oldham put forward a widely cited definition of work autonomy: "The degree to which the job provides substantial freedom, independence, and discretion to the employee in scheduling the work and in determining the procedures to be used in carrying it out" (Hackman and Oldham 1975). In 1985, management professor James A. Breaugh expanded this approach, which is more related to work methods and work scheduling, by adding another dimension, the so-called criteria autonomy, which describes the extent to which employees can influence the aspects that are used to evaluate their performance (Breaugh 1985). Autonomy thus essentially refers to the question of how work tasks are carried out, prioritized, and evaluated. It is important to emphasize that it is not only the objective conditions of the job that shape the attitude, motivation, and behavior of employees, but also their subjective perception of their own degrees of freedom (Hackman and Lawler 1971).

The ability to act in a self-determined fashion at work is closely related to the innovative capacity of workers (Breaugh 1985; Sia and Appu 2015; Spiegelaere et al. 2014). Existing studies show that it is beyond question that there is a positive relationship between the two variables. What is needed in practice, however, appears to be a precise specification of which workers, with which autonomy-related resources, and under which organizational conditions really have innovation-friendly resources at their disposal. How can particularly innovation-strong workers be supported, and how can innovation-passive workers be activated? Which profile and which characteristics distinguish these workers in the first place?

This contribution thus explores the central question of which latent structures and linkages exist between employee autonomy and their innovation capacity for companies. The analysis follows an exploratory approach to answering this question by identifying *innovation capacity clusters* among workers. These are intended to provide guidelines for enabling workers to deal with (technological) change and, at the same time, serve as a proxy for the ability to deal with both the production and the use of technical and digital innovations. The development of innovation capacity is therefore discussed as a prerequisite for digital work. As a field of analysis, the manufacturing sector is chosen because it faces particular challenges in its dual role as supplier and provider of new technologies (Dispan and Schwarz-Kocher 2018).

The contribution proceeds as follows: Sect. 2 establishes the specific theoretical relationship between autonomy and innovation capacity and introduces the initial assumptions of the explorative, quantitative analysis. Section 3 uses a factor analysis to discuss two work scenarios in manufacturing according to the presence of autonomy, innovation-friendly resources, organizational requirements, and interactional conditions. Section 4 subsequently clusters three groups of employees according to their innovation capacity and specifies these based on selected structural, occupational, and technology-related variables. Section 5 closes the analysis with a summary and conclusion.

2 Autonomy and Innovation at the Organizational Level: An Approach Across Four Dimensions

The autonomy of employees describes self-determination at work, such as making decisions independently, planning and scheduling of work tasks, or selecting work methods (Breaugh 1985; Hackman and Oldham 1975). This includes the extent to which the working conditions of employees can be co-determined.

The studies on job redesign by the previously cited authors Hackman and Oldham are regarded as prominent examples of the establishment of a connection between autonomy, critical psychological states, and on-the-job outcomes: In their Job Diagnostic Survey (also Job Characteristics Model), Hackman and Oldham identify the positive expression of autonomy as a key success factor for the realization of positive personal and work outcomes such as good performance, low absenteeism, and turnover (Hackman and Oldham 1975). The positive influence of autonomy in the workplace has been proven for several factors: organizational commitment (Park and Searcy 2012), work engagement (Shin and Jeung 2019; Spiegelaere et al. 2016), creativity (Sia and Appu 2015), innovative work behavior (Spiegelaere et al. 2014), job satisfaction (Wheatley 2017), and performance (Langfred and Moye 2004), among others. In addition, granting autonomy to workers has a strengthening impact on well-being, as it can alleviate strain (Bakker and Demerouti 2007; Demerouti et al. 2001), work pressure (Carayon 2006), or delimitation (Voydanoff 2004). Embedded in the fields of occupational psychology and sociology, autonomy forms an essential work resource that regulates motivating or stress-reducing developments in the workplace.

Creativity and innovative work behavior are the factors that enable employees to deal with innovations and digital tools. Previous studies prove that autonomy at work affects both of these aspects: the opportunity to show creativity and innovative behavior at work. Breaugh's autonomy dimensions, i.e., work method, work scheduling, and work criteria autonomy (Breaugh 1985) have been shown to influence workplace creativity positively (Sia and Appu 2015). However, creativity only refers to the first, idea-generating phase of an innovation process (Spiegelaere et al. 2014). Innovative work, meanwhile, is intended to cover all phases of the innovation process and refers to "all employee behavior directed at the generation, introduction and/or application (within a role, group or organization) of ideas, processes, products or procedures, new to the relevant unit of adoption that supposedly significantly benefit the relevant unit of adoption" and is likewise positively related to the autonomy of workers (Spiegelaere et al. 2014).

Based on this insight, there is a need for a suitable strategy of analysis to specify this linkage between autonomy and the innovative capacity of employees more precisely and to place it in relation to organizational constraints at work. In particular, the following remarks address the question of *to what extent employee groups can be isolated that provide specific innovation capacities and what work resources and demands these groups face.*

Therefore, this contribution essentially follows the ideas and analytical approaches of Lorenz and Valeyre (2005), who attempted to examine organizational variety in companies by applying an explanatory factor analysis (EFA) and a hierarchical cluster analysis. Their main focus was on the impact of organizational innovation and human resource management on the labor market structure in the EU-15 countries.

In the first step of the analysis, an exploratory factor analysis is conducted. EFA is applied in analytical scenarios that aim to identify latent constructs of meaning within a certain selection of variables, i.e., nothing other than to examine the correlation structures between variables. EFA is thus based on the fundamental assumption that observed variables are influenced by a limited number of independent, not directly measurable, latent variables or factors. However, these factors can be estimated using linear combinations of the observed variables. In relation to the research question, the aim of EFA is to reveal the hidden connections between four dimensions of analysis: autonomy, innovation-friendly resources, organizational demands, and interactional resources.

In the concrete implementation of the EFA, five methodical aspects following Fabrigar, Wegener et al. are adhered to Fabrigar et al. (1999): item selection, appropriateness for the research objective, procedure selection, factor reduction, and rotating method. Based on the explorative factor analysis, a hierarchical cluster analysis is conducted to better distinguish between distinct analytical dimensions (see below).

The present study methodically takes up these approaches, uses the quantitative analysis of an employee survey, and describes the organizational structures with regard to autonomy and innovative capacity in the German manufacturing sector. In this process, clusters of innovation capacity are identified, which, based on their specific characteristics, provide information about the ability or inability of workers to innovate under the given organizational conditions. Thus, this contribution considers innovation capacity at the employee level and includes organizational conditions of work in the analysis.

The analysis is based centrally on the BIBB/BAuA Employment Survey (Erwerbstätigenbefragungen/ETB). This representative cross-sectional survey is conducted every six years as a telephone interview. The survey is carried out by the Federal Institute for Vocational Education and Training (BIBB) and the Federal Institute for Occupational Safety and Health (BAuA). The current survey dates from 2018 and includes 20,012 respondents, aged 15 and above, who are in paid employment for at least 10 h per week. The BIBB/BAuA-ETB provides extensive analysis options on working conditions, environments, and strains at the workplace from employees' perspectives. The Scientific Use File of the 2018 survey was acquired for the present analysis. The analysis included employees in the manufacturing sector and thus a total of 3,839 observations.

The sector definition for manufacturing is based on Section C of the Classification of Economic Activities (WZ 2008). In the following analyses, manufacturing is subdivided into six branches:

- Chemicals and pharmaceuticals (WZ 2008 = 20, 21)
- Metal processing (WZ 2008 = 24, 25)
- Electrical engineering and IT (WZ 2008 = 26, 27)
- Mechanical engineering (WZ 2008 = 28)
- Automotive (WZ 2008 = 29)
- Other manufacturing (remaining items of Section C).

The selection of variables for the analysis is related to the approach of Lorenz and Valeyre, whose work focuses on autonomy-related factors of work performance, organizational frameworks, and learning opportunities for employees. Using factor and cluster analysis, the authors have attempted to define the key forms and distribution of work organization within so-called *work organization clusters*. This contribution builds on this analytical approach and will focus on the interplay between autonomy at work and the innovation-promoting resources available to employees. Organizational requirements as well as interactional resources are taken into account (Table 1).

In reference to the definition of autonomy by Hackman and Oldham (1975) and with regard to the given items in the BIBB/BAuA-ETB, autonomy is essentially defined by the degree of freedom to determine the manner and planning of the execution given to a work task. Monotony and working on precise instructions have a negative impact on the autonomy of employees. A high value is placed on independent work and the possibility of making difficult decisions for oneself. The innovation-friendly resources at work are expressed through the opportunity to learn new things and to apply one's own problem-solving skills. Innovation possibilities are thus also determined by dealing with inconsistencies and an explorative approach to work.

In the following model, it is assumed that selected organizational requirements considerably influence and determine the possibility of utilizing autonomy and innovation-friendly resources. These include, in particular, working under strong performance pressure and with precisely quantified output figures, as well as keeping an eye on the parallelism of work processes as a proxy for the complexity of work. Interactional resources are likewise included in this analysis since cooperation and practices in teams also determine the possibility of independent work and the achievement of goals. Support from colleagues and superiors form positive interactional resources, while the constant shouldering and negotiation of responsibility and compromises are assumed to restrict individual freedom of action. In this description, special consideration is given to the fact that experience and judgment about

Table 1 Selection of BIBB/BAuA-ETB items for further analysis

Dimension	Variable	Variable description	Effect
Autonomy	F327_02	Making difficult decisions independently	+
	F411_02	Work execution precisely prescribed	−
	F700_02	Planning and scheduling work independently	+
	F411_03	Work processes precisely repeated	−
	F503	Completing tasks mainly independently	+
Innovation-friendly resources	F327_01	Responding to and solving problems	+
	F327_03	Identifying and closing knowledge gaps	+
	F411_04	Understanding and familiarizing with new tasks	+
	F411_05	Improving processes or trying something new	+
Organizational demands	F411_01	Working under deadline/performance pressure	−
	F411_07	Distinct minimum output or time prescribed	−
	F411_09	Tracking different processes simultaneously	−
Interactional resources	F327_04	Taking responsibility for other people	−
	F327_05	Convincing others/negotiating compromises	−
	F700_12	Receiving support from colleagues	+
	F700_13	Receiving support from superiors	+

Notes +/ − describes the assumed effect direction on the respective dimension. Variables marked with—are inverted
Source BIBB/BAuA-ETB 2018. Own categorization

processes, products, and social interaction are also necessary in many occupations in manufacturing (Hirsch-Kreinsen 2018).

3 Innovation-Enhancing Work in Manufacturing: A First Exploratory Review

EFA initially isolates two factors. The following characteristics of work in manufacturing become visible in a two-dimensional space (Fig. 1): Scenario 1 (x-axis) describes a work situation in which both autonomy and innovation-friendly resources are clearly positive. Employees are frequently able to make decisions independently, solve problems, learn new things, and try things out. At the same time, the organizational demands are relatively challenging. The pressure to perform and the complexity of the work are particularly negative. Interactional resources are also low; in particular, compromises often have to be made and responsibility for others has to be shouldered.

Scenario 2 (y-axis) paints a positive picture of work that allows for autonomy. Independent and plannable work is flanked here by little monotony or specification regarding the method and manner of performing the work. Organizational requirements, in the sense of pressure to perform and predefined output, seem less demanding. Higher interactional resources, especially through the support of colleagues and superiors, have a beneficial effect. The clearest difference to scenario 1 lies in the innovation-friendly resources, which hardly come into play and thus still show clear potential for improvement.

Thus, there appears to be autonomy and flexibility at work across the board in manufacturing, but the promotion of innovative capacity varies greatly. This difference is possibly related to organizational demands on employees as well as the dependence on and degree of support from third parties.

An attempt to transfer this scenario analysis to selected branches and professions in the manufacturing sector reveals a clear heterogeneity regarding autonomy and innovation opportunities (Fig. 2): At the branch level (orange), mechanical engineering, electrical engineering, and IT, as well as chemicals and pharmaceuticals,

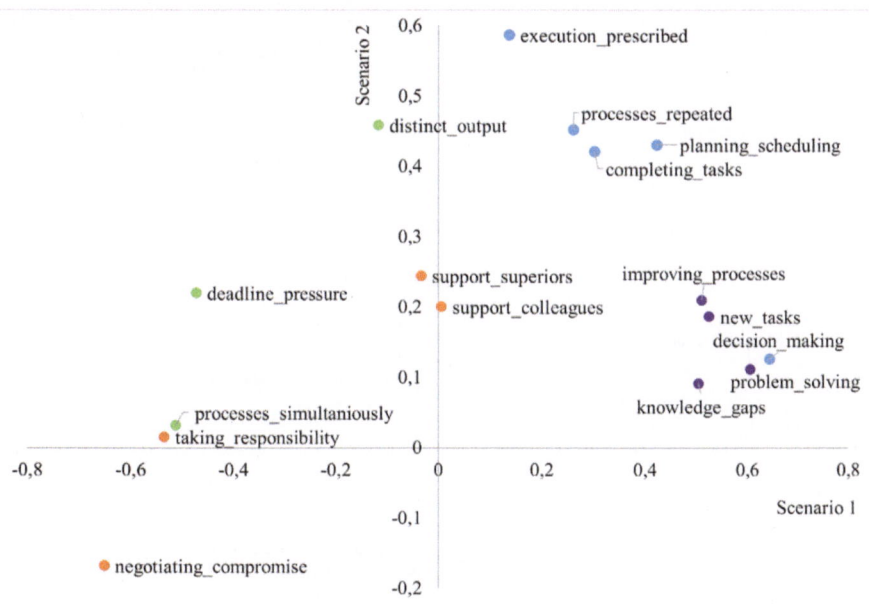

Fig. 1 Work scenarios in manufacturing (own illustration). *Notes* The dimensions of analysis are autonomy (blue), innovation-friendly resources (purple), organizational demands (green) and interactional resources (orange). *Source* Own figure based on BIBB/BAuA-ETB 2018

are in the positive ranges, which also indicates a basic innovation-friendliness and the granting of autonomy for employees. Chemicals and pharmaceuticals are more likely to belong to scenario 1, thus having more demanding organizational conditions and fewer interactional resources. Mechanical and electrical engineering and IT, which tend to move into scenario 2, benefit from organizational and social support but provide employees with fewer resources conducive to innovation. Other manufacturing and metal processing, on average, are far from these autonomy-promoting environments.

At the occupational level (blue), these results are partially confirmed: occupations in the R&D domain are clearly in the positive range, unsurprisingly more in scenario 1, in which employees benefit from autonomy and innovation-friendliness. Occupations in the fields of mechatronics, energy, and electronics tend to lie in scenario 2, i.e., they still have room for improvement, especially with regard to the granting of innovation-friendly resources. Surprisingly, occupations in mechanical engineering and automotive engineering are in the negative range. Thus, there still appears to be considerable potential for improvement here in terms of innovation-friendliness. Furthermore, in line with the overall metalworking branch, the corresponding occupations in metal production, construction, and processing are also located in a more autonomy- and innovation-unfriendly setting.

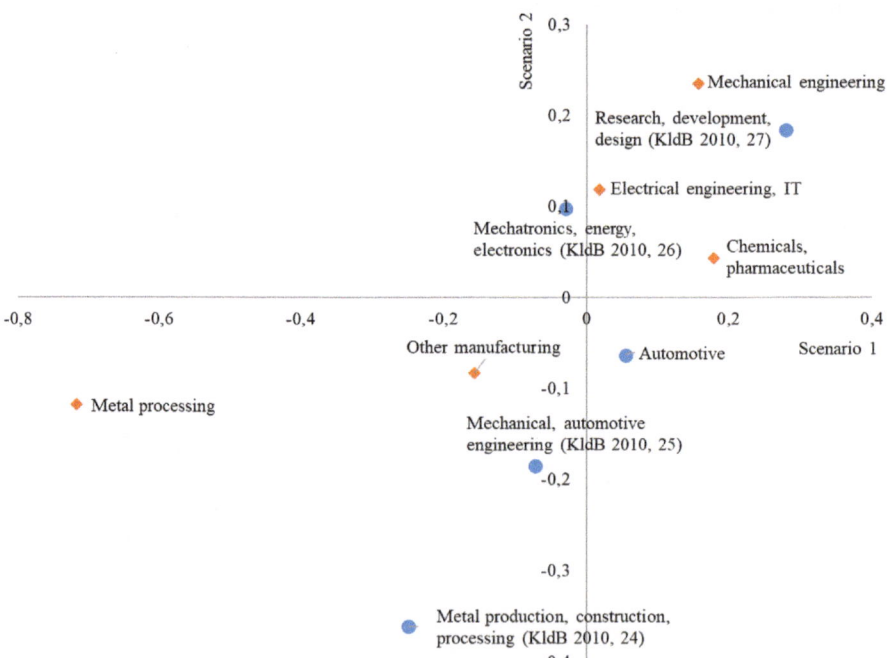

Fig. 2 Work scenario in manufacturing by branch and selected occupations (own illustration). *Notes* The dimensions of analysis are the branch level (orange) and the occupational level (blue). *Source* Own figure based on BIBB/BAuA-ETB 2018

4 The Relation Between Autonomy and Innovation Capacity: An Attempt at Clustering

4.1 Of Potentials, Balanced Employees, and Front Runners

Based on the EFA, a hierarchical cluster analysis is conducted to better distinguish the previously defined analytical dimensions of autonomy, innovation capacity, organizational demands, and interactional resources in manufacturing. This analysis identifies three clusters of employees with specific resources and requirements (Table 2):

- Cluster I: *The potentials*
- Cluster II: *The passive balanced*
- Cluster III: *The autonomous front runners.*

Cluster I—*the potentials*—describes a group of employees in the manufacturing sector (around 37%) who are equipped with relatively few resources and demanding organizational conditions. The cluster designation "potentials" is intended to indicate in particular that although there are clear innovation-friendly resources for this group, these are hardly used in everyday life due to little autonomous work and organizational requirements. Employees in this cluster show the lowest levels of autonomy at work. They rarely make difficult decisions, and according to their own statements, not even half of them are autonomous at work. Many of these employees experience monotony as well as precise instructions on how to do their work. However, the situation regarding innovation-friendly resources is hopeful: Almost two-thirds often react independently to problems. A good third frequently familiarize themselves with new tasks. However, the organizational demands appear to be high. Most employees regularly work under performance pressure and under the requirement of a targeted output. The interactional resources, however, suggest that they are often supported, at least by colleagues.

Cluster II—*the passive balanced* (around 20%)—refers to employees who show considerably higher levels of autonomy. The cluster designation "passive balanced" is intended to indicate that the working conditions for this group are fundamentally good. They are able to work independently and are relatively less challenged by organizational conditions. It remains to be discussed why possible innovation-related resources are not (or cannot) be used. Most employees in this cluster work predominantly independently and plan their work accordingly. They are seldom dictated how to do their work. Monotony and the rare opportunity to make difficult decisions independently are an issue for them. However, the group clearly falls behind the other clusters in terms of openness to innovation-friendly resources. New topics, tasks, or even trying out new things are very often out of the question for the balanced employees. Independent problem solving is ranked low in comparison to the other clusters. What is clearly visible, however, are the better organizational conditions. More than a third never work under time pressure. The majority do not have a quantified work goal. The complexity of the work also seems somewhat lower. This is

Table 2 Innovation capacity cluster by BIBB/BAuA-ETB items

Dimension		I	II	III	% of total
	Total	1,431	786	1,622	3,839
Autonomy	Making difficult decisions independently	29%	10%	63%	40%
	Work execution precisely prescribed (−)	13%	66%	76%	50%
	Planning and scheduling works independently	42%	73%	94%	70%
	Work processes precisely repeated (−)	9%	32%	61%	36%
	Completing tasks mainly independently	48%	80%	94%	74%
Innovation-friendly resources	Responding to and solving problems	63%	46%	94%	73%
	Identifying and closing knowledge gaps	28%	12%	52%	35%
	Understanding and familiarizing with new tasks	31%	25%	72%	47%
	Improving processes or trying something new	20%	13%	54%	33%
Organizational demands	Working under deadline/ performance pressure (−)	8%	35%	5%	13%
	Distinct minimum output or time prescribed (−)	23%	75%	58%	49%
	Tracking different processes simultaneously (−)	15%	38%	5%	15%
Interactional resources	Taking responsibility for other people (−)	30%	54%	15%	29%
	Convincing others/ negotiating compromises (−)	21%	30%	1%	14%
	Receiving support from colleagues	71%	86%	86%	81%
	Receiving support from superiors	44%	71%	65%	59%

Notes Variables with negative effect direction (−) reflect the response spectrum "rarely" or "never". All others reflect the response option "frequently"
Source BIBB/BAuA-ETB 2018. Own calculations

framed by interactional resources, which are not only expressed through the support of colleagues or superiors but are also reflected in a lower dependence on third parties, i.e., the need to negotiate compromises or even to take responsibility for others.

Cluster III—*the autonomous front runners*—ultimately describes a group (around 42%) with predominantly high resources but also strong work demands. The cluster designation "autonomous front runners" describes the high level of autonomy and innovation-related work resources available to this group. They have the basic prerequisites for working in a high-change environment. However, high work demands and dependencies in a complex environment clearly resist these resources. In terms of autonomy and innovation-friendly resources, employees in this cluster show the most promising conditions in the manufacturing sector: Two-thirds often make difficult decisions independently, and almost all of them generally work independently. It rarely happens that the details of the work are prescribed or that there is monotony in the work. Relatively often, they are confronted with new tasks, work their way into them, and fill gaps in their knowledge. The demands on the problem-solving ability of these employees are high. However, these resources are coupled with exceedingly high organizational demands, which lead to conceivably conflict-laden work. The complexity of their tasks is high, and they work under constant pressure to perform. These conditions are exacerbated by interactional dependencies. The autonomous front runners often take responsibility for others and have to constantly negotiate compromises in their work with others. At least they benefit relatively often from the support of colleagues and superiors.

Of course, these clusters are a simplification of reality. Nevertheless, they provide an opportunity to reflect on the interplay of work resources and demands. The employees in each cluster face different challenges. In all clusters, levels of conflict are visible that relate more to resources or requirements and open up opportunities for companies to act. It is now necessary to further specify these clusters.

4.2 Distinction by Structural and Occupational Characteristics

From a branch perspective, there is little difference in the distribution of the individual clusters (Table 3). Nevertheless, there are some striking features: employees in the metal processing industry tend not to be found among the autonomous front runners, while mechanical engineering is underrepresented among the potentials. With the exception of metal processing, cluster III, i.e., the particularly autonomous employees, is well represented among the selected industries.

The difference between the clusters is much clearer with regard to selected occupations. It is again confirmed that employees in occupations in metal production, construction, and processing are more likely to be found among the potentials, while occupations in research and development tend to be found among the autonomous front runners. As previously suspected, the more innovative professions in R&D tend

Table 3 Innovations capacity cluster by branch and selected occupations

	I (%)	II (%)	III (%)	% of total
Manufacturing sector (WZ 2008, C)				n = 3,839
Chemicals, pharmaceuticals (20, 21)	12	10	14	12%
Metal processing (24, 25)	17	15	11	14%
Electrical engineering, IT (26, 27)	15	18	17	16%
Mechanical engineering (28)	8	11	15	12%
Automotive (29)	18	15	20	18%
Other manufacturing (rest)	32	31	22	28%
Selected occupations (KldB 2010)				n = 1,594
Metal production, construction, processing (24)	32	23	11	22%
Mechanical, automotive engineering (25)	37	29	33	34%
Mechatronics, energy, electronics (26)	15	25	20	19%
Research, development, design (27)	15	22	36	25%

Source BIBB/BAuA-ETB 2018. Own calculations

to be among the latter. Mechanical and automotive engineers are distributed in relatively balanced proportions. Mechatronics engineers and occupations in the energy and electronics sectors tend to accumulate among the passive balanced.

Considering the level of qualification, it becomes clear that clusters I and II are predominantly made up of employees with vocational training, while almost two-thirds of the autonomous front runners are employees with a university or master's degree (Table 4). This tendency corresponds with the weekly working hours, which are predominantly up to 40 h for the potentials and the passive balanced. Cluster III, however, has a high density of employees who regularly work over 40 h. This is an alarming indication that the work of the autonomous front runners is particularly stressful and demanding. There are hardly any differences in terms of company size. Cluster III tends to move towards larger companies.

Thus, the picture of the employment clusters in manufacturing is completed as follows: Employees in the potentials cluster tend to have vocational training and, in some cases, work well over 30 h a week. The passive balanced cluster paints a rather homogeneous picture with a more stable ratio between vocational training and higher-level qualifications. Also, in terms of working hours and company size, the ratio is quite different compared to the other groups. The autonomous front runners collect many highly qualified employees, the majority of whom work overtime and are employed in large companies. They show signs of the concept of an entreployee (Pongratz and Voß 2003), i.e., a type of employee whose performance is achieved within the context of a critical personal process of delimitation. Parallels to leadership and management responsibility are also conceivable (Ribbat et al. 2021).

Table 4 Innovation capacity cluster by selected personal and organizational variables

	I (%)	II (%)	III (%)	% of total
Level of education				n = 3,832
Without vocational qualification	7	8	2	5%
Vocational training	72	63	33	54%
Upgrading training	10	9	16	12%
University	11	20	49	29%
Working hours per week				n = 3,839
10 to 20	4	10	2	5%
21 to 30	6	10	4	6%
31 to 40	58	58	39	50%
More than 40	32	22	55	39%
Company size				n = 3,787
Micro company (less than 10)	6	8	3	5%
Small company (10 to 49)	17	21	11	15%
Medium company (50 to 249)	27	26	24	26%
Large company (more than 250)	50	45	62	54%

Source BIBB/BAuA-ETB 2018. Own calculations

4.3 Relation to Technology Use and Digital Potentials

A look at the use and potential of technology confirms previous assumptions about the leading role of autonomous front runners (Table 5): 70% of the employees can be assigned to the high-tech or medium–high-tech areas of manufacturing. This is not surprising, as high technological intensity also goes hand in hand with higher demands and, ultimately, qualification requirements. In the clusters of the potentials and passive balanced, almost half of the employees assigned to them remain in the medium–low-tech or low-tech areas.

Mapping the potential use of AI in the manufacturing sector is more demanding. The exact usage rate of AI in companies is fundamentally difficult to examine due to definitional gray areas. Estimates range from 6.8% in electrical and mechanical engineering (5.1% in automotive, 4.6% in chemicals and pharmaceuticals) (Bundesministerium für Wirtschaft und Energie (BMWi) 2019) to 25% of large companies in the manufacturing sector (Seifert et al. 2018) (overview in Giering (2022)). A more recent study estimates that 9% of all companies in Germany used AI in 2022 (Bitkom 2022). Basically, there are still many employees who, according to their own statements, are not yet or only little affected by digitalization in their daily work, especially in metal processing (64%). In the chemical and pharmaceutical industries, mechanical engineering, and automotive, the picture is more promising, as less than a third of employees report not being affected by digitalization at all or only to a small extent (Institut DGB-Index Gute Arbeit (DGB-Index) 2022). The AI usage rate in

Table 5 Innovation capacity cluster by technology use and potentials

	I (%)	II (%)	III (%)	% of total
Technological intensity				n = 3,839
High technology	6	8	9	8%
Medium–high technology	48	47	61	53%
Medium–low technology	22	21	16	19%
Low technology	24	24	14	20%
AI skills				n = 903
Basic	10	9	31	19%
Advanced	2	2	9	5%
New manufacturing or process technologies				n = 3,771
No	45	62	45	48%
Yes	55	38	55	52%
New computer programs				n = 3,718
No	51	56	33	44%
Yes	49	44	67	56%
New machinery or equipment				n = 3,783
No	41	59	47	47%
Yes	59	41	53	53%

Source BIBB/BAuA-ETB 2018. Own calculations

this study is also in the low single digits for manufacturing (max. 8% for chemicals and pharmaceuticals). More than 90% of employees currently never use AI (Institut DGB-Index Gute Arbeit (DGB-Index) 2022).

Since the BIBB/BAuA-ETB hardly contains any items that document the use of digital tools or even AI, an approximation of the potential AI skills of employees, according to Pfeiffer, is attempted (Pfeiffer 2020). Utilizing the BIBB/BauA items, basic AI skills essentially include information gathering, working with computers, knowledge of mathematics and statistics, and computer application programs. The existence of advanced AI skills is assumed when computer use goes beyond pure application in everyday work. This approach thus centrally presupposes that the use of AI in companies is possible if these skills are available to employees. According to this distinction, Pfeiffer estimates that 4% of all employees already have advanced AI skills. A total of 11% show basic skills in dealing with AI (Pfeiffer 2020).

The mapping of AI skill potential confirms the impression of particularly high innovative capacities among the autonomous front runners. Again, 40% of this cluster has the basic skills to deal with AI. Almost 10% of them even have advanced skills, i.e., skills that go beyond the mere application of technologies (e.g., programming). In the other clusters, there are comparable proportions, with just over 10% of employees having basic skills for working with AI. Although they are just below the averages achieved according to Pfeiffer (2020), they nevertheless provide a positive outlook regarding the use of AI in everyday working life.

Considering the introduction of new production technologies as applied in the BIBB/BAuA-ETB, computer programs, or machines, all clusters are almost exclusively divided into balanced proportions between those who have experienced these organizational changes in the past two years and those who have not. Among the autonomous front runners, it is the slight majority who have experienced changes, especially with regard to new computer programs. The potentials, however, are hardly inferior to this trend. Only the passive balanced appear to have undergone fewer organizational and technical changes, which fundamentally confirms the assumption that there is still hidden potential for innovation here.

5 Summary and Conclusion

The scarce research results on the interplay between workplace autonomy and the use of digital tools and AI so far show an ambivalent relationship: optimistic forecasts prove a positive connection (Giering and Kirchner 2021; Institut DGB-Index Gute Arbeit (DGB-Index) 2022), while others see the autonomy of employees in manufacturing in danger due to increasing external control and standardization (Butollo et al. 2018; Hinojosa and Potau 2017).

To address this existing research gap, the question was asked about the latent structures and linkages between employee autonomy and their innovation capacity. This innovation capacity is essentially operationalized in terms of the creativity- and innovation-friendly characteristics of a workplace and is intended to map the organizational preconditions for digital work. After all, it is well known that digital change is usually incremental and that the manufacturing sector in particular tends to be structurally conservative, i.e. changes tend to take place within given organizational structures without changing them disruptively (Hirsch-Kreinsen 2018).

The explorative analyses of this paper fundamentally show that for many employees, autonomy at work is given to varying degrees, but in any case, to a certain degree. However, when it comes to accessing innovation-friendly resources, the field is already divided into those employees who have resources at their disposal but also work in pressure-filled organizational scenarios and those who find less conflictual working environments.

Of course, an explorative approach such as the one adopted in this study leaves room for interpretation. The results mentioned should therefore also be regarded as *suggestions* for conclusions. These could be validated in the future by analyzing other waves of BIBB/BAuA-ETB data (2006, 2012, and soon 2024). It is also plausible to compare the methodical approach on the basis of other sectors and occupational groups, such as the service sector.

However, the cluster analysis undertaken points to three groups of employees for further specification: the potentials, passive balanced, and autonomous front runners. Each of these groups faces a specific field of resources and requirements. In combination with Hirsch-Kreinsen's distinction between pioneer companies, followers, and laggards regarding the digital transformation (Hirsch-Kreinsen 2018), the autonomous front runners are presumably to be assigned to the pioneer companies that have relied early and systematically on digital technologies in production. Here, the description of manufacturing as a producer and developer of digital technologies is most accurate. These tend to be large, technology-intensive companies from the mechanical engineering, automotive, electrical, and process industries. This group shows that, in addition to the given work resources, the demands are also particularly high, especially with regard to the complexity of the work and the pressure to perform. This clearly shows the relationship along the lines of Karasek's job demand-control model (Karasek 1979), where high demands and given autonomy lead to particularly satisfying jobs. However, the given scope for action must always exceed the demands for this balance to be fruitful. In order to avoid crushing the highly innovative employees in this conflict, it is urgently advisable to take a close look at the workload situation of the autonomous front runners.

The employees among the potentials are probably more likely to be wait-and-see followers. Digital technology is used to a limited extent to improve running processes. SMEs from the metal and process industries are particularly represented here (Lerch et al. 2017). These employees have fewer work resources available in the sense of autonomy, and the demands in terms of target expectations are relatively high. If these employees were given more autonomy, they might be able to move more quickly towards the front-running companies. They still have the highest potential and opportunities for expansion in terms of innovation capacity. Companies need to ask themselves how these potentials can be utilized for them. For example, various organizational practices are available to strengthen the autonomy of employees: participative management, alternative working arrangements, or autonomous working groups (Gagné and Bhave 2011).

Finally, it remains to be discussed to what extent the passive balanced employees can be found among the waiting laggard companies with regard to digitalization. As a rule, they are uncertain about which path to take in the direction of digitalization, which is why the digital possibilities are still rather rudimentary. They tend to be small and medium-sized companies that manufacture less complex products and have not yet felt any lasting pressure to innovate. Yet, the conditions for the employees here are fundamentally encouraging. They are able to work autonomously and are less subject to organizational or interactional pressure and dependencies. The question to be discussed from a company perspective is how innovation-friendly resources can be activated in everyday work. Ultimately, production workers in particular often have the necessary competence to deal with complex situations and come directly "from the production process". They are thus ideally suited to accompany innovations (Pfeiffer 2018).

References

Bakker, A.B., Demerouti, E.: The job demands-resources model: state of the art. J. Manag. Psychol. **22**(3), 309–328 (2007). https://doi.org/10.1108/02683940710733115

Bitkom: KI gilt in der deutschen Wirtschaft als Zukunftstechnologie—wird aber selten genutzt, Berlin (2022)

Breaugh, J.A.: The measurement of work autonomy. Hum. Relat. **38**(6), 551–570 (1985). https://doi.org/10.1177/001872678503800604

Bundesministerium für Wirtschaft und Energie (BMWi): Einsatz von Künstlicher Intelligenz in der Deutschen Wirtschaft. Stand der KI-Nutzung im Jahr 2019, Berlin (2020)

Butollo, F., Jürgens, U., Krzywdzinski, M.: Von Lean Production zur Industrie 4.0. Mehr Autonomie für die Beschäftigten? Arbeits- und Industriesoziologische Studien **11**(2), 75–90 (2018). https://doi.org/10.21241/ssoar.64864

Carayon, P.: Human factors of complex sociotechnical systems. Appl. Ergon. **37**(4), 525–535 (2006). https://doi.org/10.1016/j.apergo.2006.04.011

Demerouti, E., Bakker, A.B., Nachreiner, F., Schaufeli, W.B.: The job demands-resources model of burnout. J. Appl. Psychol. **86**(3), 499–512 (2001). https://doi.org/10.1037/0021-9010.86.3.499

Dispan, J., Schwarz-Kocher, M.: Digitalisierung im Maschinenbau. Entwicklungstrends, Herausforderungen, Beschäftigungswirkungen, Gestaltungsfelder im Maschinen- und Anlagenbau. Working Paper Forschungsförderung. Nr. 94. Hans-Böckler-Stiftung (HBS) (ed.), Düsseldorf (2018)

Fabrigar, L.R., Wegener, D.T., MacCallum, R.C., Strahan, E.J.: Evaluating the use of exploratory factor analysis in psychological research. Psychol. Methods **4**(3), 272–299 (1999). https://doi.org/10.1037/1082-989X.4.3.272

Gagné, M., Bhave, D.: Autonomy in the workplace: an essential ingredient to employee engagement and well-being in every culture. In: Chirkov, V.I., Ryan, R.M., Sheldon, K.M. (eds.) Human Autonomy in Cross-Cultural Context. Perspectives on the Psychology of Agency, Freedom, and Well-Being. Cross-Cultural Advancements in Positive Psychology, pp. 163–187. Springer, Dordrecht (2011)

Giering, O.: Künstliche Intelligenz und Arbeit: Betrachtungen zwischen Prognose und betrieblicher Realität. Zeitschrift Für Arbeitswissenschaft **76**, 50–64 (2022). https://doi.org/10.1007/s41449-021-00289-0

Giering, O., Kirchner, S.: Künstliche Intelligenz am Arbeitsplatz. Forschungsstand, Konzepte und empirische Zusammenhänge zu Autonomie. Soziale Welt **72**(4), 551–588 (2021). https://doi.org/10.5771/0038-6073-2021-4-551

Hackman, J.R., Lawler, E.E.: Employee reactions to job characteristics. J. Appl. Psychol. **55**(3), 259–286 (1971). https://doi.org/10.1037/h0031152

Hackman, J.R., Oldham, G.R.: Development of the job diagnostic survey. J. Appl. Psychol. **60**(2), 159–170 (1975). https://doi.org/10.1037/h0076546

Hinojosa, C., Potau, X.: advanced industrial robotics: taking human—robot collaboration to the next level. In: Impact of game-changing technologies in European manufacturing. Working Paper. The Future of Manufacturing in Europe (FOME) project, Eurofound, European Commission (EC) (eds.) (2017)

Hirsch-Kreinsen, H.: Arbeit 4.0: Pfadabhängigkeit statt Disruption. Soziologisches Arbeitspapier. Nr. 52/2018, Dortmund (2018)

Institut DGB-Index Gute Arbeit (DGB-Index): Report 2022. Digitale Transformation—Veränderungen der Arbeit aus Sicht der Beschäftigten. Ergebnisse des DGB-Index Gute Arbeit 2022, Berlin (2022)

Karasek, R.A.: Job demands, job decision latitude, and mental strain: implications for job redesign. Adm. Sci. q. **24**(2), 285–308 (1979). https://doi.org/10.2307/2392498

Langfred, C.W., Moye, N.A.: Effects of task autonomy on performance: an extended model considering motivational, informational, and structural mechanisms. J. Appl. Psychol. **89**(6), 934–945 (2004). https://doi.org/10.1037/0021-9010.89.6.934

Lerch, C., Jäger, A., Maloca, S.: Wie digital ist Deutschlands Industrie wirklich? Arbeit und Produktivität in der digitalen Produktion. Mitteilungen aus der ISI-Erhebung. Ausgabe 71. Fraunhofer ISI (ed.), Karlsruhe (2017)

Lorenz, E., Valeyre, A.: Organisational innovation, human resource management and labour market structure: a comparison of the EU-15. J. Ind. Relat. **47**(4), 424–442 (2005). https://doi.org/10.1111/j.1472-9296.2005.00183.x

Moldaschl, M.: Herrschaft durch Autonomie - Dezentralisierung und widersprüchliche Arbeitsanforderungen. In: Lutz, B. (ed.) Entwicklungsperspektiven von Arbeit. Ergebnisse aus dem Sonderforschungsbereich 333 der Universität München, pp. 132–164. De Gruyter, Berlin (2001)

Park, R., Searcy, D.: Job autonomy as a predictor of mental well-being: the moderating role of quality-competitive environment. J. Bus. Psychol. **27**(3), 305–316 (2012). https://doi.org/10.1007/s10869-011-9244-3

Pfeiffer, S.: The 'future of employment' on the shop floor: why production jobs are less susceptible to computerization than assumed. Int. J. Res. Vocat. Educ. Train. **5**(3), 208–225 (2018). https://doi.org/10.13152/IJRVET.5.3.4

Pfeiffer, S.: Kontext und KI: Zum Potenzial der Beschäftigten für Künstliche Intelligenz und Machine-Learning. HMD Praxis der Wirtschaftsinformatik **57**(3), 465–479 (2020). https://doi.org/10.1365/S40702-020-00609-8

Pongratz, H.J., Voß, G.G.: From employee to 'entreployee': towards a 'self-entrepreneurial' work force? Concepts Transform. **8**(3), 239–254 (2003). https://doi.org/10.1075/cat.8.3.04pon

Ribbat, M., Weber, C., Tisch, A., Steinmann, B.: Führen und Managen im digitalen Wandel: Anforderungen und Ressourcen. baua: Preprint, Dortmund (2021)

Seifert, I., Bürger, M., Wangler, L., Christmann-Budian, S., Rohde, M., Gabriel, P., Zinke, G.: Potentiale der Künstlichen Intelligenz im Produzierenden Gewerbe in Deutschland. In: Studie im Auftrag des Bundesministeriums für Wirtschaft und Energie (BMWi) im Rahmen der Begleitforschung zum Technologieprogramm PAiCE—platforms|additive manufacturing|imaging|communication|engineering. Institut für Innovation und Technik (iit) (ed.). Berlin (2018)

Shin, I., Jeung, C.-W.: Uncovering the turnover intention of proactive employees: the mediating role of work engagement and the moderated mediating role of job autonomy. Int. J. Environ. Res. Public Health **16**(5), 1–16 (2019). https://doi.org/10.3390/ijerph16050843

Sia, S.K., Appu, A.V.: Work autonomy and workplace creativity: moderating role of task complexity. Glob. Bus. Rev. **16**(5), 772–784 (2015). https://doi.org/10.1177/0972150915591435

de Spiegelaere, S., van Gyes, G., de Witte, H., Niesen, W., van Hootegem, G.: On the relation of job insecurity, job autonomy, innovative work behaviour and the mediating effect of work engagement. Creat. Innov. Manag. **23**(3), 318–330 (2014). https://doi.org/10.1111/caim.12079

Spiegelaere, S.D, van Gyes, G., van Hootegem, G.: Not all autonomy is the same. Different dimensions of job autonomy and their relation to work engagement & innovative work behavior. Hum. Factors Ergon. Manuf. Serv. Ind. **26**(4), 515–527 (2016). https://doi.org/10.1002/hfm.20666

Voydanoff, P.: Implications of work and community demands and resources for work-to-family conflict and facilitation. J. Occup. Health Psychol. **9**(4), 275–285 (2004). https://doi.org/10.1037/1076-8998.9.4.275

Wheatley, D.: Autonomy in paid work and employee subjective well-being. Work. Occup. **44**(3), 296–328 (2017). https://doi.org/10.1177/0730888417697232

Dr. Gina Glock completed her doctorate in sociology of work at the Dresden University of Technology. She is employed at the Federal Institute of Occupational Safety and Health (BauA) in Berlin. Her research focuses on the interplay between digitalization, working conditions, and the shaping of good work. She studied Industrial Engineering and Public Economics in Dresden, Venice, and Berlin.

Development of an Open Innovation Knowledge Platform in the Context of Digital Sovereignty

Juliane Balder and Rainer Stark

Abstract This paper investigates the development of an open innovation platform in the furniture and furnishing industry, emphasizing the significance of digital sovereignty in digital platform design. The study involved various aspects, starting with the identification of stakeholders and their influences on the digital platform. Understanding the roles and interests of different stakeholders is crucial for designing effective governance mechanisms and ensuring inclusivity. Furthermore, an analysis of necessary extensions to existing business models was conducted. An input and output flow analysis elucidates knowledge generation and utilization. Clear requirements, including digital sovereignty, ensure data protection and privacy. The findings revealed that an Open Innovation platform can be conceptualized as a product-service system, combining the functionalities of a product with the administration and moderation services. This perspective facilitated a comprehensive understanding of the platform's interactions with stakeholders and allowed for the integration of digital sovereignty measures from the outset. The approach fosters stakeholder understanding and early integration of digital sovereignty measures and offers insights into designing digital platforms with digital sovereignty in focus. It emphasizes the need for early consideration and the link between research and commercial utilization.

Keywords Digital platform engineering · Open innovation · Circular economy

1 Introduction

The goal of a resource-efficient circular economy (CE) is becoming increasingly important in both industry and politics. The aim of mostfundings arethe development of suitable business models (BM), design concepts and digital technologies to close product cycles and thus contribute to a resource-efficient CE. The project "PERMA"

J. Balder (✉) · R. Stark
Technische Universität Berlin, 10587 Berlin, Germany
e-mail: j.balder@tu-berlin.de

© The Author(s) 2025
U. Schmuntzsch et al. (eds.), *New Digital Work II*,
https://doi.org/10.1007/978-3-031-69994-8_12

(Platform for Efficient Resource Use in the Furniture and Furnishing Industry)[1] picks up on the growing awareness of sustainability and aims to establish resource-efficient BMs in the furniture and furnishing industry (https://innovative-produktkreislaeufe. de/Projekte/PERMA.html).

Simultaneously, digital platforms have emerged as powerful tools for enabling collaboration. They offer digital infrastructure that facilitates stakeholder interaction, knowledge sharing, and value co-creation. However, ensuring digital sovereignty has become a crucial concern in this context. Digital sovereignty encompasses data protection, information control, and privacy preservation, ensuring the autonomy and self-determination of individuals and organizations in the digital realm.

The primary goal of the PERMA joint project is to establish a resource-efficient CE in furniture and contract construction. For this purpose, circular BMs are being developed, the implementation of which is supported by a digital platform. By providing innovative product life cycle and guideline information about cross-manufacturer compatibility, the prototype realisation of this digital platform makes it possible to open new sales markets in the B2B sector according to the three-pillar model of sustainability (economic, ecological, and social). The platform enables extended product life cycles through sustainable and flexible reuse and reutilisation of products, which makes a significant contribution to resource conservation in view of increasingly scarce raw material resources.

During the development of the platform, it was found that there is great potential in stakeholder collaboration. However, there is a lack of effective knowledge sharing and a suitable space for targeted collaboration. Therefore, the PERMA open innovation (OI) platform is developed with a focus on promoting collaboration and OI. This paper presents the investigation and development of the PERMA platform, highlighting key findings and outlining future prospects.

2   State of the Art

2.1   Open Innovation

In times of increasing competition due to higher innovation pressure with simultaneously decreasing R&D budgets, companies are forced to open their innovation process to specifically increase their innovation potential by involving the outside world. This strategic involvement is referred to as the OI approach.

OI is an umbrella term that encompasses several approaches to innovation, including not only the involvement of customers and strangers, but also collaboration with other partners in the innovation process. This enables a combination of internal and external knowledge to generate innovations (Göhring 2017; Hauschildt et al. 2016; Müller-Prothmann and Dörr 2011). OI can arise in different ways; for

---

[1] Funded by the Federal Ministry of Education and Research (BMBF) Germany as part of the research program "Resource Efficient Circular Economy—Innovative Product Cycles".

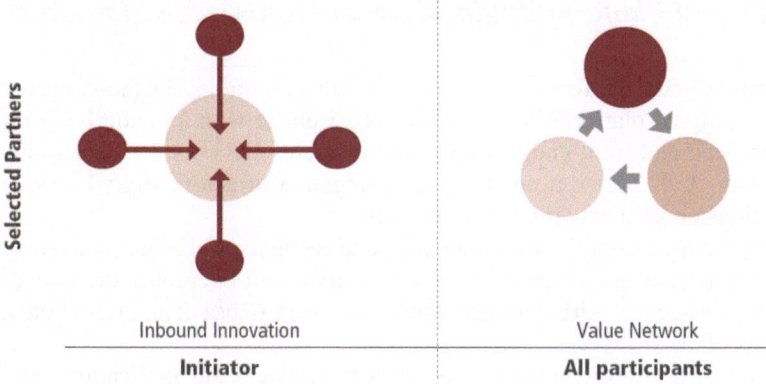

Fig. 1 Comparison of inbound innovation and value network (Own illustration)

the knowledge platform, i.e. inbound innovations and value creation networks are most important (see Fig. 1).

In inbound innovations, such as lead user integration, selected external partners, e.g., customers, suppliers, users or experts, contribute to the company's internal innovation process at various stages. The company has full control over the process and the intellectual property throughout the process (Göhring 2017).

In a value network or coupled process, the companies work together as equals. Each company has control over its part of the BM and the marketing of the products (Chesbrough et al. 2014). The key aspect is a high level of mutual trust between the companies (Bigliardi et al. 2020). Thanks to the complementarity of knowledge and know-how and their combination, it is possible to bring innovation projects to a successful conclusion, e.g., in the case of interprofessional innovation.

A company can pursue several OI approaches at the same time (Myhren et al. 2018). OI approaches follow the assumption that idea development and innovation do not necessarily have to take place in the same place.

Crucial prerequisites for cooperative innovation processes are, that the company can absorb external knowledge and integrate it into its own knowledge and technology base. Also, it can share its own knowledge so that the partner company can benefit from it. Success depends on finding the right partners for fruitful cooperation (Obradović et al. 2021).

Innovation can come to firms in various ways, such as through the licensing of technology developed by other firms, but firms can also contribute to the innovation process of others by making their internal innovations available to other organizations through joint ventures, licensing, spin-offs, etc. (Mortara et al. 2009).

OI requires turning the closed boundaries of a company into a semi-transparent membrane that allows innovation to move easily between the external environment and the company's internal innovation processes (Chiaroni et al. 2011).

2.2 Digital Platform Engineering and Knowledge Platforms

The term "digital platform" can have different definitions. In the context of this chapter, a digital platform is understood as a website with an underlying information back-end and front-end technology that creates and operates web applications (Asadullah et al. 2018). It enables the aggregation of actors, digital services, and transactions under a network effect of users.

Best practice examples are established and recognised sharing platforms as part of the CE framework to close the resource loop and overcome the under-use of technologies such as Airbnb or car-sharing websites (Rosa et al. 2020; Schwanholz and Leipold 2020).

Digital platforms are not only used to share products and applications but also to exchange data and information and facilitate knowledge sharing between companies and networks (both internal and external) (Gawer and Cusumano 2014; Stark 2022a). For example, they encourage consumers to participate in the product development process. In this way, digital platforms also serve as tools for replicating processes along the value chain and as hubs for collaborative exchange, co-development, and innovative ideas that promote social, economic, and environmental value creation for all stakeholders (Rosa et al. 2020; Pialot et al. 2017). At the same time, they enable access to a broad network of users and providers.

Digital platforms can be used for the sale and marketing of products and services. They also promote industrial symbiosis through the resale and retail of used products and waste materials (Halstenberg et al. 2017).

In addition, by using digital platforms, processes can be automated, transaction costs reduced, and efficiency improved. This leads to optimised use of resources and increased productivity.

Overall, three main categories of digital platforms are distinguished: market, operational, and co-creation.

Digital platforms offer an open and participatory environment by providing "a set of common techniques, technologies and interfaces for a diverse group of users to build what they want on a stable substrate". They foster valuable collaborations and interactions between participants (Parker et al. 2016). These developments create new approaches to promote the CE throughout the value chain, including production, transport, consumption, and reuse of resources, or products, even in the after-use phase (Jørgensen and Pedersen 2018).

Knowledge is also disseminated through the creation and maintenance of digital platforms (Stark 2022b). Improved knowledge sharing can impact the CE on multiple levels by raising public awareness, guiding action, and inspiring policy and legislation. Digital solutions can facilitate connections between actors and encourage people to become "active participants in the data economy and co-creators of knowledge" (Hedberg and Šipka 2021).

Despite the advantages of digital platforms, drawbacks need to be recognised. One disadvantage of digital platforms is their dependence on platform operators. Companies are heavily dependent on terms and conditions and technology updates, which can negatively impact their business activities and require extensive adaptation of their processes.

Security and privacy risks are another challenge in the use of digital platforms. The intensive exchange of data and information poses the risk of security breaches and data protection violations. Businesses and users face threats such as data leaks, misuse of personal information, and cyber-attacks, so adequate security measures are essential.

Large digital platforms can have a dominant position in the market, leading to an imbalance of power. Smaller companies and individual providers find it difficult to compete with established platforms and achieve their desired market share. This can limit competitiveness and lead to monopoly formation, which affects the diversity and dynamism of the market (Berg and Wilts 2019).

3　Description of the Open Innovation Platform's Objective and Needs

The extension of the PERMA platform through OI aims to provide improved access to solution information by networking internal and external stakeholders. This offers a competitive advantage through shortened iteration cycles, especially in product development. These stakeholders are manufacturers, planners, and service providers, in the following also referenced as *internal market actors*. In the context of PERMA, this means that the platform participants jointly create innovations within the framework of a value creation network for the furniture industry. This approach will serve as a prototypically implemented component of the platform.

In particular, this approach will be used for the community-oriented design of furniture, with a focus on the standardisation of connectivity solutions. Due to changing application and usage scenarios, objects also require constructive solutions at the transport level to permanently minimise wear and tear, for example when users change or when they are transported to the warehouse. The knowledge gained is to be transferred to the community through the platform in the form of a dynamic knowledge library and interactive discussion forums. In addition, the community should be encouraged to try out, evaluate and further develop new solutions by offering novel products and services. Manufacturing companies gain core functionality to track their inventory and gain insights into the location, condition, and purpose of the object. This information can be used for continuous product improvement (feedback for design) as well as for optimising maintenance cycles for long-life products and directly correlates with the already prototypically developed revenue platform of the internal market actors.

The revenue platform enables the generation of multiple life cycles for furniture, while the OI platform provides space for development, training, exchange, and networking. There is an overlap of product life cycles on both platforms, leading to sustainable use of resources and holistic development in the furniture sector.

To ensure quality, the recyclable label, which has already been developed in its basic features, is to be further developed at the product and component level in order to continuously assess resource efficiency. This serves both as an incentive for furniture manufacturers, assembly, and refurbishment service providers participating in the platform and as a basis for decision-making for the further life history of an object through a transparent condition assessment.

Considering the original project intention, three basic approaches to OI emerge for the project:

- A growing participatory community as a driver for novel developments.
- A live inventory management for permanent traceability of manufacturer-owned products.
- Increased product transparency as a digital file and basis for trust.

Overall, through the OI approach, the PERMA project aims to create a comprehensive platform framework within which not only existing objects are traded, but which also enables innovation through interactive and cross-stakeholder design.

4 Importance of Digital Sovereignty

In the context of digital platforms and OI, digital sovereignty plays a crucial role. The European Union (EU) has outlined several criteria for online platforms to ensure digital sovereignty and protect users' rights. These criteria include transparency reports, consideration of fundamental rights in terms of use, cooperation with national authorities, provision of contact points and legal representatives, reporting obligations, complaint and redress mechanisms, and transparency of recommendation systems and online advertising.

Intellectual property (IP), data protection and data security are fundamental aspects of digital sovereignty. IP (like tangible property) can be bought, sold, given away, leased, and exchanged, although certain limits apply. In essence, the property's owner can prevent others from using the property and potentially acting in the ways described above by creating a legal framework. (Bogers et al. 2012).

However, the protection of knowledge—especially technical knowledge— depends on its embodiment, as this determines how it can be transferred and protected. For example, it is challenging to identify and transfer tacit knowledge (as opposed to explicit or codified knowledge). It is difficult to put implicit knowledge into words and to determine its value, which complicates commercial transfer (Cowan 2000; Polanyi 1967). Multi-partner R&D cooperations require complex regulations to determine which results of the cooperation can be used by whom and in what way during and after the conclusion of the cooperation. Certain access rights to IP may be agreed on, and after the end of the cooperation, the internal intellectual property may become freely available (Roh et al. 2021).

PERMA, as a digital platform, must ensure the security and protection of the users' data from unauthorized access. Clear privacy policies should be in place to inform users about the collection, usage, and accessibility of their data.

User control is another essential element of digital sovereignty. PERMA should empower users to have full control over their data sharing and privacy settings. Customizable privacy settings and control over data access are vital for user autonomy.

Transparency is key to building trust and ensuring digital sovereignty. PERMA should be transparent about its data collection practices, usage policies, and who can access the data. Regular transparency reports can help users understand how their data is being used and who has access to it.

Interoperability is also crucial for digital sovereignty. PERMA should allow users to share their data between different platforms and services, adhering to interoperability standards. This enables users to have more control over their data and reduces dependence on a single provider.

As a European platform, PERMA is committed to upholding EU core values, including individual data ownership, self-determination of digital assets, and privacy as a fundamental right. The EU asserts sovereignty over identity, data and information, software, and operations. Ensuring users' control over their digital identities, protecting and utilising user data transparently, and ensuring the secure operation of applications and workloads without reliance on specific provider software are key aspects of digital sovereignty.

The collaborative approach of PERMA aligns with the EU's objectives of upholding European values and empowering citizens to have greater control over their digital assets. Digital sovereignty is crucial for cooperative platforms like PERMA, as it enhances user trust and expands the user base. By prioritizing digital sovereignty, PERMA can foster a secure and trusted environment for stakeholders, contributing to its long-term success in promoting resource-efficient CE in the furniture and furnishing industry.

In order to ensure the basic features of digital sovereignty, the categories shown in Table 1 according to Schauf and Neuburger (2021) are applied.

5 Methodical Approach

The steps shown in Fig. 2 provide a basic framework for the development of an OI platform and serve as a starting point for future considerations, adaptations, and optimisations. The following section explains these steps in detail.

Table 1 Digital sovereignty criteria in relation to platforms (Schauf and Neuburger 2021)

Degree categories	Low level (=high dependence)	Medium level (=some dependence)	High level (=no dependence)
Data	Controlled data access and usage by platform operator	Users have complete control over data access and deletion by users	Users have full data control, incl. reading, modifying, deleting, and choosing storage location
Distribution and federation.	The platform can only be operated by a commercial operator	Instances of the platform can be operated under self-control	Instances of the platform can be operated independently without central organization consent
Interfaces	No interfaces available or only proprietary interfaces	Support for many open standards and interfaces	Access to all data and functions via open, freely usable interfaces with open-source reference implementation
Software for technical platform	Software only available to the platform operator	Software can be used independently under self-control/another operator	Source code can be modified and utilized independently
Hardware	Technical solution purchased outside the EU by operator	Existing non-EU solutions can be complemented by EU developments	All technical components can be developed and maintained in the EU
Technical control	No control/migration options; one single provider	Users can migrate important parts and establish self-operated solutions	Users have control over all components (source code, hardware, etc.)

(continued)

Table 1 (continued)

Degree categories	Low level (=high dependence)	Medium level (=some dependence)	High level (=no dependence)
Competencies	Limited understanding of principles, requirements, technology; focus on applications	Basic understanding of principles, requirements, technology; competent in assessing existing players	Competencies in understanding principles, requirements, technology, and evaluating existing players are present
Jurisdiction	Platform subject to non-EU law	Platform subject to non-EU law with reliable contracts ensuring compliance with EU standards	Platform located in Germany or the European Union, subject solely to this jurisdiction

Fig. 2 Methodical approach (Own illustration)

1. Research & Analysis
> OI PERMA Definition
> Qualitative Interviews

2. General platform conceptualization
> Adapting business models
> Stakeholder analysis
> Definition of required tools

3. Process design
> BPMN 2.0 processes

4. Prototypical implementation
> Tools & IT Systems
> Virtual artefacts

5.1 Research and Analysis

Extensive research was conducted to understand OI in the context of the furniture and furnishing industry. Literature sources, case studies and experiences of other platforms were analysed. Opportunities, challenges and impacts of OI were identified. Particular attention was paid to digital sovereignty, intellectual property protection and the examination of best practice examples. The insights gained form the basis for the design of an OI platform.

Definition of OI in the context of digital platforms

The next step was to define OI in the context of the Platform to ensure that all stakeholders and actors have a common understanding of it.

OI is a systematic exchange with external stakeholders to improve innovation processes and includes opening up the innovation process to external stakeholders.

In the context of the PERMA Platform, OI takes place in the community area, where competences and curated knowledge are communicated and exchanged through participation and interaction of users in modelled spaces.

OI can be understood as an extension of the PERMA internal market and aims at the networking and knowledge transfer of the internal market actors. On the one hand, the knowledge area serves the collective collection, management and sharing of curated knowledge on the CE in the furniture and furnishings industry.

On the other hand, the focus is on networking economic, scientific as well as political-social actors in order to stimulate cooperation and research projects, to convey competences and to exchange knowledge. Through interaction and bidirectional exchange between science and practice, as well as the possibility of sharing knowledge that is not self-generated, a targeted increase in knowledge takes place. OI is implemented on the platform via various media (blogs, wiki, forums, media library).

Qualitative interviews

The qualitative interviews are conducted to clarify the needs and expectations of the potential users of the OI platform regarding the functionalities and design of the platform to create a user-oriented and trustworthy platform.

The first step was to define the interview partners from business and academia and to ensure that they had relevant insights and experience in the field of OI research. The objectives and requirements for the interviews were then defined to ensure that the data collected would provide the desired information and meet the research objectives.

An interview guide was prepared for conducting the interviews.

The interviews were conducted in an open and exploratory style to allow the interviewees to share their personal perspectives, experiences, and insights. A total of eight interviews were conducted.

After the interviews were completed, the first step was a quantitative analysis of the data. Here, the data was coded using nominal scales and a frequency analysis was conducted. The open questions were categorised for an efficient and methodically controlled evaluation according to the method of qualitative content analysis. Inductive category formation according to Kuckartz and Rädiker (2022) was used. This procedure allows statements to be made on specific thematic aspects. The choice of inductive categorisation is based on the exploratory nature of the survey, since the range of answers is unknown (Kuckartz 2009). However, it should be noted that subjective deductive categorisation already takes place through the engagement with the topic. The procedure can be described as follows: First, the answer of participant 1 was analysed. Each topic mentioned was given a new category. In this way, all answers were systematically recorded. In the case of multiple topic areas, the category was noted in the answer. When looking at the numerical data, it should be noted that several categories can be addressed in a respondent's statement, which affects the principle of multiple mention in open-ended questions (Kuckartz 2009; Gläser and Laudel 2010).

The results of the qualitative interviews provide important insights regarding the design and functionalities of the PERMA platform in the context of OI. The main findings are as follows:

1. Knowledge needs or rather OI: it was identified that there is a need for knowledge and information on OI, as this term is not yet widely used. It is important to provide users with relevant information and training to promote the understanding and application of OI.
2. Quality of the wiki content: Participants in the interviews indicated that wiki contributions should be created by qualified authors who provide background information as well as cite scientific sources. This ensures the reliability and credibility of the information provided.
3. Moderation of the forum: Participants stressed the need for moderation of the forum to ensure appropriate exchange of information and adherence to rules. Furthermore, access rights should be assigned to ensure the protection and privacy of the users.
4. Networking: The platform should offer users the opportunity to network and build relationships with each other. Face-to-face meetings have been identified as a popular channel for knowledge sharing and collaboration.
5. Professional design and imprint: The design of the website was considered important as it influences the professional perception and credibility of the platform. An imprint should also be present to ensure the integrity of the website.
6. Knowledge rating system: The introduction of a knowledge rating system, both in the wiki and the forum, was considered significant. Such a system allows users to evaluate and assess the quality and relevance of the information provided.

By taking these results into account, PERMA can create a user-oriented and trustworthy platform that meets the requirements in the context of OI.

5.2   General Concept of the OI Platform

Based on the findings of the analysis, a general concept of the platform was developed. The BM of the platform and the platform operator were adapted, and features and criteria of digital sovereignty were integrated. A stakeholder analysis was conducted to identify the relevant actors and understand their needs and expectations.

Adapting business models to open innovation

The successful development of a platform requires a comprehensive identification and knowledge of the required OI BM components. To achieve this, important features of the OI BM were identified using the Triple-Layer-Canvas. The application of the Triple-Layer-Canvas provides a structured method to capture these OI BM features and integrate them into the platform design. Characteristics include, for example, expanding the partner pool and creating closer proximity to customers to build a diverse and engaged community, fostering dialogue and knowledge transfer between stakeholders to enable shared learning and exchange of ideas, reducing risk, and making savings in development by leveraging resources and expertise from external partners, and addressing licensing issues to establish the legal framework for sharing knowledge and innovation.

Stakeholder analysis

Stakeholder analysis aims to determine the characteristics of stakeholders based on their power, legitimacy, and urgency in order to subsequently evaluate the relationship and coalition behaviour in an organisational hierarchy. Following (Freeman 1984), the stakeholders (see Table 2) are classified according to their ability to influence and their influence on others. The collected data are combined in a relevance matrix (see Fig. 3).

For the knowledge platform it is important to analyse the stakeholders to know if one stakeholder holds a specific power on the platform. It helps to create a balance of power and an equal standing by implementing specific tools and possible constains.

The stakeholders (see Table 2) were determined, and the sources of power identified. To define the individual sources of power, (Morgan 1986) description of the potential for influence and power was used, which describes the ability of individuals and groups to influence others in their interests.

To define the weighting of the sources of power, a weighting factor between 0 and 1 was determined for each source of power. To determine the values and power

Table 2 Display of the identified stakeholders of the knowledge platform

Stakeholder type (excl. internal market actors)	Roles
Manufacturer (MA)	Enabling circular development of products
Planner (PL)	Conceptualization, implementation and execution of circular solutions
Educational Institutions (EI) University	Publication/research topics, industry placement, educational mission
Research Institutions (RI) e.g., Fraunhofer, Leibniz, etc.	Publication/provision of research topics
Service Provider (SP)	Enabling sustainable service products
Industry Associations (IA) e.g., VDM etc.	Community building intermediaries and industry advocates
Chambers (CH) e.g., Chamber of Trades, IHK	Updating examination and teaching content
Economic Develop. Institutions (EDI) e.g., WFBB	Consultants, intermediaries, market development
Journalists (JO)	Networking
Standardization Institutes (SI)	Providers of guidance and expertise pool
General Public (GP)	Observers
Multiplipliers (MU)	Influence on stakeholders outside of the platform

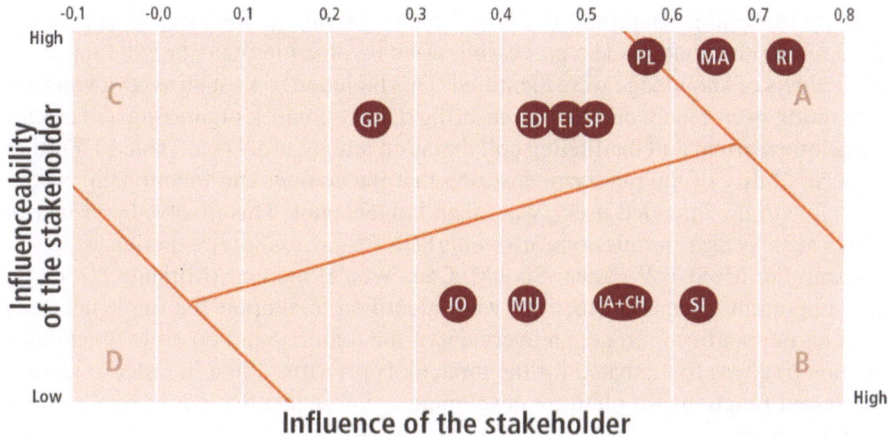

Fig. 3 Power factors of knowledge actors within the framework of strategic positioning, for list of stakeholders please see Table 2 above (Own illustration)

positions of the individual stakeholders, their power factors are analysed for each stakeholder. The extent to which the stakeholder possesses the various sources of power and how the stakeholder can be influenced is examined.

The result for the OI platform is the relevance matrix shown in Fig. 3. Research institutions and manufacturers are so-called game changers (type A) or key stakeholders. This means that they have a great influence on the company but can equally be influenced by the company. The planner is categorised as a game maker as well as standardisation institutions, chambers and industry associations, multipliers, and journalists as jokers (type B). Jokers have a high degree of influence but are difficult to be influenced themselves. The distribution of power lies with the stakeholder. Service providers, educational institutions, economic development agencies and interested parties can be categorised as so-called laws (type C). They have little influence on the company, but are strongly influenced by it. None of the stakeholders was identified as an insignificant marginal figure (type D) (Müller-Stewens and Lechner 2005).

Definition of required tools

Based on an analysis to understand the perspectives, needs and expectations of the results of the research, stakeholder analysis and interviews, input and output flows of knowledge on and from the platform were identified and categorised by purpose, e.g., promotion of research projects, enlightenment, exchange with practitioners, advice and assistance, and network building. The form and nature of the input and output sources were further examined and assigned to subcategories. This step allowed for a more detailed understanding of the different types of knowledge sources and their characteristics. For each input and output stream, specific content, material forms (e.g., text, images, videos), and functions that would facilitate the effective flow of knowledge on the platform were developed and defined. This included determining the types of information or resources that would be shared, the format in which they would be presented, and the functionalities required for users to access and interact with the content. The processing tasks involved in managing the input and output flows of knowledge were identified. This included tasks such as reviewing and moderating user-generated content, ensuring quality control, organizing and categorizing information, and facilitating collaboration among users (see Table 3). Further, the accessibility of the platform, ensuring that the content and functionalities were accessible to the intended users, was taken into account. This involved considering factors such as user permissions, user interface design, and user experience.

Using the MoSCoW (Must, Should, Can, Won't) method (Miranda 2011), the most important tools and functions were identified to support the implementation of OI on the platform. To get an overview of the required and possible functions, a function tree was first created for the three tools prioritised first in order to identify all relevant functions for platform development. Subsequently, lists of requirements were drawn up.

At the same time, existing software tools for knowledge management and community building were evaluated to determine which were already available and could be implemented in the prototypes, for example a wiki. Information sharing and collaboration on the platform were designed to consider the needs and requirements of all stakeholders. In addition, potential third-party tools that could relieve the editorial work were considered.

5.3   Process Design and Prototypical Implementation

To facilitate the programming of the platform the Engineering Operating System (EOS) was used. EOS puts human engineering activities at the centre and describes three interaction activities: "processes and organization", "tools and IT systems" and "virtual and physical artifacts" (Stark 2022a; Randermann et al. 2021).

In the process of developing the platform, the same approach was used as in the development of the revenue platform. Here, the seven steps from the developed process design according to Balder et al. (2023) was applied. First, the current actual processes of knowledge sharing of the individual stakeholders and the main target processes were identified. Second, all dependencies and correlations between the target processes of the individual stakeholders were identified. Third, the process profiles and the identified dependencies were visualized. Forth, the designed profiles were classified and assigned as a core, support or management processes of the platform. Fifth, the processes were displayed in a process map in the form of a flow chart and, sixth, divided into sub processes. Seventh, the defined sub-processes were described, formulated as user stories and, eighth, formalized in Business Process Model and Notation (Balder et al. 2023).

These processes serve as a guide for programming and ensure iterative, efficient and error-free development of the platform. By defining the processes in a clear and understandable way, development time can be shortened, and programming can be made more effective.

Table 3 Selected input- and output stream for identified knowledge clusters (this/next page)

	Existing and desired input streams				
Catego-ries	Purpose	Subcatego-ries of input streams	What? What specific knowledge/content?	What? What shape is the material?	How? Which Tools bring input to the platform?
Focus Research	Promote research projects	Provision of research topics	Tenders	Text, image, PDF	Upload option
			event offer	Text, image, PDF	Upload option
			Job offers	Text, image, PDF	Upload option
			research reports	Text, image, PDF	Up+Down-load option
	Enlighten-ment	Presentation of current political guidelines, guidelines, trends	Political requirements	Text, image, PDF	Up+Down-load option
		Access to sources, methodologies, advice	Discussion and controversy about current methodologies	Text, image, PDF	contact form
			Questions about FP, process management	Text, image, PDF	Upload option
			networking opportunity	Text, image, PDF	Chat & call function
Focus on Design and Production	Exchange from practice	Exchange and sharing of design strategies	sustainability reports	Text, image, PDF	Up+Down-load option
			corporate strategies	Text, image, PDF	Up+Down-load option
		Representation of knowledge about own products	Design strategies and methodologies	Text, image, PDF, video	Up+Down-load option
			Guide for recyclable products	Text, Image, PDFs, Video, CAD, Drawings	Up+Down-load option
Focus Skills and Competencies	Mediation to save resources	Experience	Exchange of niche knowledge	Text, image, PDF	Upload option
			networking opportunity	Text, image, PDF	Chat & call function
			Exchange of niche knowledge	Text, image, PDF	Up+Down-load option

(continued)

Table 3 (continued)

Processing		Output streams			
Who? Processing	How? Processing	Subcategories of output streams	What is the output?	How? Which Tools generate output?	Access for whom?
editorial staff	Comparison with requirements	Provision of information about events, offers, tenders	Hints, search, offer	bulletin board	semi-public
moderation	Comparison with requirements		Notices of interesting and relevant events	calendar	public
editorial staff	Comparison with requirements		Hints, search, offer	bulletin board	semi-public
Editing, authorization access for external	Read, Write, access via invite link		Resource Saving Research Report	media library	semi-public
editorial staff	Checking the sources	Clarification of systemic and bureaucratic requirements, action recommendations	Removal of implementation hurdles	blog	semi-public
moderation	Comparison with requirements	Recommendations for action and direct exchange with actors from research	Opinions, news, contributions to discussions	Forum	semi-public
editorial staff	Checking the sources		Info, field reports, competent contributions, answers, references to the how-to research	Forum	semi-public
automatic logic	Comparison with requirements		direct contact	P.O. Box	closed
Editing + authorization access for external	Comparison with requirements	Knowledge transfer and enlightenment from practice	Enlightenment, attention, PR	media library	semi-public
editorial staff	Comparison with requirements		Enlightenment, attention, PR	media library	semi-public
editorial staff	Comparison with requirements		Resource Saving Presentation, read-only.	media library	semi-public
editorial staff	Comparison with requirements		Education about ways of thinking about cycles in an economic context	media library	semi-public
editorial staff	Comparison with requirements	Knowledge transfer and education about niche knowledge from practice	specific knowledge sharing	Forum	semi-public
automatic logic	Comparison with requirements		Personal meeting/ direct contact	P.O. Box	closed
editorial staff	Comparison with requirements		specific knowledge sharing	blog	semi-public

Based on the general concept, the information technology infrastructure and the defined requirements, a prototypical implementation of the essential functions of the platform was carried out. For this, both the information and data flow were modelled. This also includes the design of the first content of the platform such as recommendations for action, blog and wiki contributions and scientific papers.

In this way, the feasibility of the platform was tested and valuable feedback was obtained from users. These findings can be used in the further development and optimisation of the platform. Result documents will be created and communication with possible stakeholders (publications, trade fairs, etc.) will be initiated. Finally, a legal review of the platform is carried out.

6   Implementation of Digital Sovereignty

By considering these aspects of digital sovereignty at different stages of the platform's development, it was possible to ensure that the platform supports data protection, information control and privacy.

In the qualitative interviews, aspects of digital sovereignty were emphasised through asking participants about their concerns, expectations and requirements regarding data protection and control over their data. This allowed for a better understanding and consideration of users' needs and preferences in relation to digital sovereignty for designing the platform. The interviews revealed that the platform is perceived as serious and secure if it has an appealing general appearance on the one hand and a meaningful imprint and comprehensible sources on the other.

Within the framework of the rough conception of the platform, digital sovereignty was taken into account in the stakeholder analysis, among other aspects. Here, the power relations over data and data flows and their effects on the platform design and control were considered.

Data protection and the platform operator's approach to data and breaches were also integrated in the definition of the BM and the tasks of the platform operator. This includes defining governance rules, ensuring data security, transparency about data processing practices and protecting user privacy.

The definition of the tools needed, functions and requirements were defined to support data protection, data encryption, secure data transfer and other aspects of digital sovereignty.

Digital sovereignty played a central role in the evaluation of third-party tools. It was important to ensure that the selected tools met data protection standards, securely encrypted data, and ensured compliance requirements. Through careful evaluation and selection of third-party tools, potential risks to digital sovereignty were minimised.

Finally, processes were defined for the platform to implement data management policies, provide transparency on data flows, establish mechanisms for user consent over their data, and put in place procedures for data breach notification and data security.

Incorporating digital sovereignty from the beginning can be challenging, especially when developing something new and with limited resources. By considering digital sovereignty as a fundamental principle from the outset, potential risks can be minimised and a solid foundation for a trustworthy and successful platform can be established.

Regarding the criteria presented in Table 1, the following assessment can be made:

- Data

All categories of digital sovereignty are applicable, depending on the data. The location of the data is fixed and enables a clear allocation, especially when specifying the data. With a low level of digital sovereignty, there is less control over one's own data e.g., data uploaded in the mediacenter. An example of a medium level of data sovereignty would be profile data. A high level of digital sovereignty enables the user to view and use a comprehensive product catalogue directly from the manufacturer.

- Distribution and Federation

Since the PERMA OI platform is to be run as a cooperative and thus as an operating community, the middle level is estimated. However, this decision cannot be made universally for other OI platforms.

- Interfaces

In the current state, the exchange of files only takes place via export and import, as there are no interfaces. This leads to an exclusively proprietary output and thus to a low level of digital sovereignty. An intended increase to at least a medium level is currently not realisable within the framework of the research project.

- Software for Technical Platform

The use of open source and the release of the source code of the operating company are still pending, as the decision depends on the economic usability of the platform. This will be determined outside the research project.

- Hardware

As planned, the server should be positioned in the EU and thus a high level should be achieved. Currently, the prototype is hosted by Google VM in Ireland.

- Technical control

The operating organisation has the availability of technical control. If the cooperative model is used as planned, then medium level is achieved.

- Competencies

The assessment of competencies is difficult at the time of development when the end users are not yet clearly enough defined, as this is highly dependent on the future operators and users. Since, in addition to the technically experienced internal market players, there will be a scientific community, a medium to high level of competence can be assumed.

- Jurisdiction

The jurisdiction is high level, as the platform will be located in Germany.

7 Conclusion and Outlook

Important results were achieved within the framework of the investigation of the PERMA OI platform. The identification of stakeholders and their influences on the platform, the analysis of the necessary extensions of BMs, and the definition of OI in the context of the platform were carried out. Consideration of the platform's input/output flows in terms of knowledge generated and the definition of clear requirements, including digital sovereignty, were also components of the development.

As Balder et al. (2023) already stated can this OI platform also be understood as a product-service system. The product in this case can represent the platform itself and its functions, the service includes, among other things, the moderation and administration of the platform. This perspective enabled a comprehensive view of the platform and its interaction with the stakeholders. Furthermore, it was found that the consideration of digital sovereignty is of great importance from the very beginning. By involving all relevant aspects of digital sovereignty at an early stage, it was possible to ensure that the platform offered the necessary safeguards and controls.

The development of the PERMA OI platform also showed that a special focus must be placed on certain factors in research projects. The subsequent economic exploitation plays a decisive role here, as some aspects of digital sovereignty can only be fully considered in this context. Nevertheless, it is important to consider the possibilities and interdependencies between the individual categories already during development. In this way, potential limitations and challenges can be identified at an early stage and appropriate measures can be taken.

Overall, the results of the development of the PERMA OI platform have shown that consideration of digital sovereignty is essential from the outset and that there is a close link between research projects and subsequent commercial exploitation.

The next steps are the practical implementation, the prototype and the development of a process guide. Here, the prototype is implemented and evaluated to check its functionality and effectiveness. Through tests and evaluations, the prototype can be further developed and optimised. It should be noted that during the research project there were constraints in the form of a small research team and limited resources both in terms of time and material. These constraints required the definition of different stages of development and the prioritisation of the feasible and verifiable aspects of the platform.

In conclusion, the results and outlook suggest that the PERMA OI platform provides important insights into the design and development of digital platforms focused on digital sovereignty. The next steps will help to implement, evaluate and optimise the platform in a practical way in order to maximise the benefits for stakeholders and to meet the requirements of digital sovereignty.

References

Asadullah, A., Faik, I., Kankanhalli, A.: Digital platforms: a review and future directions. In: PACIS (2018)

Balder, J., Mathi, C., Hagedorn, L., Stark, R.: Digital platform engineering to enable circular-economy core mechanism (2023)

Berg, H., Wilts, H.: Digital platforms as market places for the circular economy—requirements and challenges. Nachhaltigkeits Manag. Forum **27**(1), 1–9 (2019). https://doi.org/10.1007/s00550-018-0468-9

Bigliardi, B., Ferraro, G., Filippelli, S., Galati, F.: The influence of open innovation on firm performance. Int. J. Eng. Bus. Manag. (2020). https://doi.org/10.1177/1847979020969545

Bogers, M., Bekkers, R., Granstrand, O.: Intellectual property and licensing strategies in open collaborative innovation. In: de Pablos Heredero, C., López, D. (eds.) Open Innovation in Firms and Public Administrations, pp. 37–58. IGI Global (2012)

Chesbrough, H., Vanhaverbeke, W., West, J.: New Frontiers in Open Innovation. Oxford University Press (2014)

Chiaroni, D., Chiesa, V., Frattini, F.: The open innovation journey: how firms dynamically implement the emerging innovation management paradigm. Technovation **31**(1), 34–43 (2011). https://doi.org/10.1016/j.technovation.2009.08.007

Cowan, R.: The explicit economics of knowledge codification and tacitness. Ind. Corp. Chang. **9**(2), 211–253 (2000). https://doi.org/10.1093/icc/9.2.211

Freeman, R.E.: Strategic Management : A Stakeholder Approach. Pitman, Boston [u.a.] (1984)

Gawer, A., Cusumano, M.A.: Industry platforms and ecosystem innovation. J. Prod. Innov. Manag.innov. Manag. **31**(3), 417–433 (2014). https://doi.org/10.1111/jpim.12105

Gläser, J., Laudel, G.: Experteninterviews und qualitative Inhaltsanalyse: Lehrbuch. Wiesbaden, VS Verlag (2010)

Göhring, A.: Analyse und Vergleich von Innovationsansätzen. Masterthesis, Hochschule Pforzheim (2017)

Halstenberg, F.A., Lindow, K., Stark, R.: Utilization of product lifecycle data from PLM systems in platforms for industrial symbiosis. Proc. Manuf. (Seliger, G., Kohl, H., Oosthuizen, G.A. (eds.)) Elsevier, pp. 369–376 (2017). http://www.sciencedirect.com/science/article/pii/S2351978917300537

Hauschildt, J., Salomo, S., Schultz, C.D., Kock, A.: Innovations Management, 6th edn. Verlag Franz Vahlen, München (2016)

Hedberg, A., Šipka, S.: Toward a circular economy: the role of digitalization. One Earth **4**(6), 783–785 (2021). https://doi.org/10.1016/j.oneear.2021.05.020

Jørgensen, S., Pedersen, L.J.T.: RESTART Sustainable Business Model Innovation. Springer International Publishing, Cham (2018)

Kuckartz, U.: Evaluation online: Internetgestützte Befragung in der Praxis. VS Verlag für Sozialwissenschaften, Wiesbaden (2009)

Kuckartz, U., Rädiker, S.: Qualitative Inhaltsanalyse: Methoden, Praxis, Computerunterstützung, 5th ed. Beltz Juventa, Weinheim, Basel (2022)

Miranda, E.: Time boxing planning. SIGSOFT Softw. Eng. Notes **36**(6), 1–5 (2011). https://doi.org/10.1145/2047414.2047428

Morgan, G.: Images of Organization. Sage, Bristol (1986)

Mortara, L., Napp, J.J., Slacik, I., Minshall, T.: How to Implement Open Innovation: lessons from studying large multinational companies (2009)

Müller-Prothmann, T., Dörr, N.: Innovationsmanagement: Strategien, Methoden und Werkzeuge für systematische Innovationsprozesse, 2nd edn. Hanser, München (2011)

Müller-Stewens, G., Lechner, C.: Strategiesches Management: Wie strategische Initiativen zum Wandel führen, 3rd edn. Schäffer-Poeschel, Stuttgart (2005)

Myhren, P., Witell, L., Gustafsson, A., Gebauer, H.: Incremental and radical open service innovation. JSM **32**(2), 101–112 (2018). https://doi.org/10.1108/JSM-04-2016-0161

Obradović, T., Vlačić, B., Dabić, M.: Open innovation in the manufacturing industry: a review and research agenda. Technovation (2021). https://doi.org/10.1016/j.technovation.2021.102221

Parker, G., van Alstyne, M., Choudary, S.P.: Platform Revolution: How Networked Markets are Transforming the Economy and How to Make Them Work for You, 1st ed. W.W. Norton & Company, New York [etc.] (2016)

Pialot, O., Millet, D., Bisiaux, J.: "Upgradable PSS": clarifying a new concept of sustainable consumption/production based on upgradablility. J. Clean. Prod. **141**, 538–550 (2017). https://doi.org/10.1016/j.jclepro.2016.08.161

Polanyi, M.: The Tacit Knowledge Dimension. Routledge & Kegan Paul, London (1967)

Randermann, M., Blüher, T., Stark, R., Jochem, R.: Reifegradmodell für die kollaborative Produktentwicklung (2021)

ReziProk, PERMA. https://innovative-produktkreislaeufe.de/Projekte/PERMA.html. Accessed 4 Nov 2022

Roh, T., Lee, K., Yang, J.Y.: How do intellectual property rights and government support drive a firm's green innovation? The mediating role of open innovation. J. Clean. Prod. **317**, 128422 (2021). https://doi.org/10.1016/j.jclepro.2021.128422

Rosa, P., Sassanelli, C., Urbinati, A., Chiaroni, D., Terzi, S.: Assessing relations between circular economy and industry 4.0: a systematic literature review. Int. J. Prod. Res. **58**(6), 1662–1687 (2020). https://doi.org/10.1080/00207543.2019.1680896

Schauf, T., Neuburger, R.: Digitale Souveränität im Kontext plattformbasierter Ökosysteme (2021)

Schwanholz, J., Leipold, S.: Sharing for a circular economy? An analysis of digital sharing platforms' principles and business models. J. Clean. Prod. **269**, 122327 (2020). https://doi.org/10.1016/j.jclepro.2020.122327

Stark, R.: The set-up of virtual product creation in industry—best practices, error modes and innovation speed. In: Stark, R. (ed.) Virtual Product Creation in Industry, pp. 75–111. Springer, Berlin Heidelberg (2022a)

Stark, R.: Major technology 10: artificial intelligence (AI) in virtual product creation. In: Stark, R. (ed.) Virtual Product Creation in Industry, pp. 381–401. Springer, Berlin, Heidelberg (2022b)

Juliane Balder has been a research associate at the Chair of industrial information technology of the Technical University (TU) Berlin since 2021. Holding a Master's degree in Industrial Engineering and Management with a specialization in Mechanical Engineering from TU Berlin, she completed her studies in 2020, including semesters at Aalto University in Finland and Universitat Politècnica de València in Spain. In her current role, her research delves into the realms of New Work, collaboration dynamics, and sustainability within the domain of product development. Adept in areas like innovation and knowledge management as well as process design and control. She leads a dynamic research project focussing on the circular economy, digital platforms and collaboration in product development. Simultaneously, she is responsible for pivotal teaching modules at the chair.

Prof. Dr. Rainer Stark an experienced expert in the field of industrial infor-mation technology, studied mechanical engineering at Ruhr University Bochum and Texas A&M University. His academic career includes a doctorate in design engineering/CAD at Saarland University, focussing on a mathematical model for tolerances in (3D) CAD systems. Following his research activities, he moved to Ford-Werke AG in Cologne, where he worked on methods for virtual product development. Promoted to Technical Specialist, he led global initiatives at Ford Motor Company from 2002 to 2008. Prof. Stark joined the TU Berlin as Professor of Industrial Information Technology in 2008. He was Head of Virtual Product Creation at the Fraunhofer Institute for Production Systems and Design Technology (IPK) until 2021 and Man-aging Director of the Institute for Machine Tools and Factory Management (IWF) from 2011 to 2013. He is a member of renowned organizations such as acatech, WiGeP, Design Society and CIRP and sits on the boards of ProSTEP iViP, VDI Prod-uct Development and Product Management and the acatech Industry 4.0 research committee.

Shaping Transformation: Becoming a Changemaker

Simone Kauffeld⊙ and Ann-Kathleen Berg⊙

Abstract This chapter introduces the pivotal role of changemakers in steering personal and organizational development through the execution of transfer projects. Participants undergo a competence-based training program, comprising five modules, which equips them with knowledge from both inside and beyond their organizations to facilitate change and navigate the complexity of organizational transformation processes. The training program modules feature diverse learning formats, including work-integrated learning and digital tools, ensuring learning transfer between the program and the workplace. Ultimately, the program contributes significantly to participants' competence development by creating a robust network among them. The discussion also emphasizes the necessity for leadership involvement in support of the participants' new role.

Keywords Transformation · Changemaker · Personal development · Organizational development

1 Transformation

Modern-day Germany's economy stands at the edge of transformation, challenged by the dynamic interplay of four fundamental forces—digitalization, decarbonization, deglobalization, and demographic shifts—collectively termed the "Four Ds of Transformation" (Demary et al. 2021). Digitalization, featuring the adoption of digital technologies to transform services and businesses, goes beyond technological change in its reinvention of business models and the emergence of a new paradigm. This force demands continuous learning and adaptation, which necessitates upskilling the workforce to master evolving technologies and practices. Decarbonization describes

S. Kauffeld (✉) · A.-K. Berg
Industrial/Organizational and Social Psychology, Technische Universität Braunschweig, Universitätsplatz 2, 38106 Braunschweig, Germany
e-mail: s.kauffeld@tu-braunschweig.de

A.-K. Berg
e-mail: a.berg@tu-braunschweig.de

© The Author(s) 2025
U. Schmuntzsch et al. (eds.), *New Digital Work II*,
https://doi.org/10.1007/978-3-031-69994-8_13

the shift toward more sustainable and low-carbon economic growth. Thus, the urgent need to reduce global carbon emissions mandates Germany's businesses to align their strategies with green technologies and sustainable practices, in turn necessitating an environmentally literate and technologically skilled workforce. Deglobalization represents an economic trend characterized by becoming more regionally focused and less integrated on a global scale. As unpredictability in the global context increasingly frequent, businesses more than before must enhance their ability to cultivate resilience, navigate changes in international trade, and understand local markets. Lastly, demographic changes reflect the shifting age characteristic of the German population. In the face of an aging workforce, businesses must foster an inclusive culture that values age diversity, encourages knowledge transfer across generations, and integrates varied talent.

Training and development programs tailored to these "Four Ds" are essential for businesses to equip their workforce with the skills needed to thrive amid these changes (Berg et al. 2023). These include developing partnerships with tech firms and universities for digital upskilling, creating platforms for understanding sustainable business practices, fostering skills for local market understanding, and promoting lifelong learning for sustainable competence development. Accordingly, this chapter presents a competence-based training program, "Changemaker," which addresses key facets of contemporary learning and transfer designs. Figure 1 illustrates the five modules of the program, which was developed to ensure that organizations can navigate transformational change successfully with their employee's aid. In brief, the six-month program combines formal and informal learning formats, empowering participants to develop competencies for supporting transformation processes while teaching them to act sustainably as "changemakers" in their own work contexts. The program's formal training spans four content modules divided into learning and application days. During the six month of the program, professionals collaborate in pairs to implement a company-specific transfer project. The goal envisions professionals applying the training content to their daily work by autonomously and continuously implementing their own concrete projects, ensuring both personal *and* organizational development. The (optionally digital) workshop-based format is conducted by 4A-SIDE GmbH in cooperation with "Demografieagentur für die Wirtschaft" and with the participation of the Lower Saxony Research Center for Automotive Engineering (NFF). Training program development was funded by the European Social Fund and the Ministry of State of Lower Saxony. Additionally, the program has been adapted as a transfer format and launched in other funded projects

Execution of transfer projects during the training

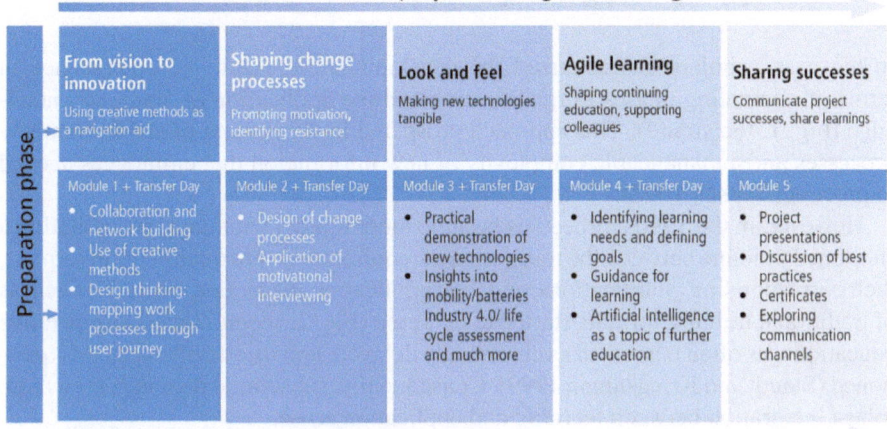

Fig. 1 Modular structure of the "changemaker" training program. All rights reserved 4A-SIDE (own illustration, cf. Kauffeld and Berg 2022)

(e.g., REPASE,[1] ROBUST,[2] KREIS,[3] MIAMy[4]), or as in-house training programs. Participants engaging in (or having completed) the program are now "changemakers" for digitalization, systems engineering (e.g., REPASE), resilient value chains (ROBUST), cultural change, the circular economy (KREIS), and artificial intelligence (MIAMy). The training program has been applied to both SMEs and large organization. To date, a total of 65 participants have been qualified as changemakers.

This discussion is intended to provide a full understanding of the considerations and suggestions related to the training design behind the competence-based training program. First, we outline the developments within the competence literature that combine learning and application. We then go on to explain how new work demands require considering individualized learning paths to ensure learning transfer and explore the best approach to integrating learning and organizational goals. Finally, we present the key aspects of the concept behind being a "changemaker."

---

[1] These projects are funded by the German Federal Ministry of Education and Research (BMBF) and managed by the Project Management Agency Forschungszentrum Karlsruhe (PTKA) (grant no. 02J19B140, 02J21C176, 02L22C100).

[2] These projects are funded by the German Federal Ministry of Education and Research (BMBF) and managed by the Project Management Agency Forschungszentrum Karlsruhe (PTKA) (grant no. 02J19B140, 02J21C176, 02L22C100).

[3] These projects are funded by the German Federal Ministry of Education and Research (BMBF) and managed by the Project Management Agency Forschungszentrum Karlsruhe (PTKA) (grant no. 02J19B140, 02J21C176, 02L22C100).

[4] This project is funded by the German Federal Ministry Federal Ministry for Economic Affairs and Climate Action (BMWK) and managed by Project Management Agency VDI/VDE Inno-vation + Technik GmbH.

2 Building and Using Competencies

In the past, scholars characterized the development of employee competence in terms of alternating phases of qualification and the application of acquired knowledge (Fig. 1, top strand). This approach prospered in the context of stable company processes and a manageable complexity of new information that employees needed to process (Baitsch 1998).

However, in the 1990s, processes became more rapidly changing, necessitating a shortened timeline between the translation of requirements into qualification formats, their corresponding qualifications, and their application in practice. The effectiveness of traditional training formats started to decrease (Fig. 2, second strand). Imperative education was often perceived as chronically delayed, and its effectiveness was questioned (Staudt and Kriegsmann 1999). Consequently, substantial demands for a more robust integration between learning and application arose.

The requirement of a strong connection between learning and application is echoed in the concept of competence, underscoring the successful handling of new tasks at work (Kauffeld and Paulsen 2018). Currently, new workplace demands are met, for example, by integrating learning into real-world work assignments (Fig. 2, third strand). Thus, learning occurs within the context of application and usage, serving to minimizing the common implementation issues and frictional losses that often arise in the wake of training programs.

Nowadays, learning processes are consistently embedded within work routines and increasingly supported in an online format. Rather than maintaining a static set of qualifications, such formats allow the necessary competencies to be readily available through just-in-time approaches, such as self-learning, job rotation, and peer instruction. The concept of "learning on demand" ensures employees' instant access to learning materials, comprising resources such as wikis, blog posts, and digital learning libraries, right from their desks. For example, online tutorials featuring insights from organizational experts can be accessed exactly as needed.

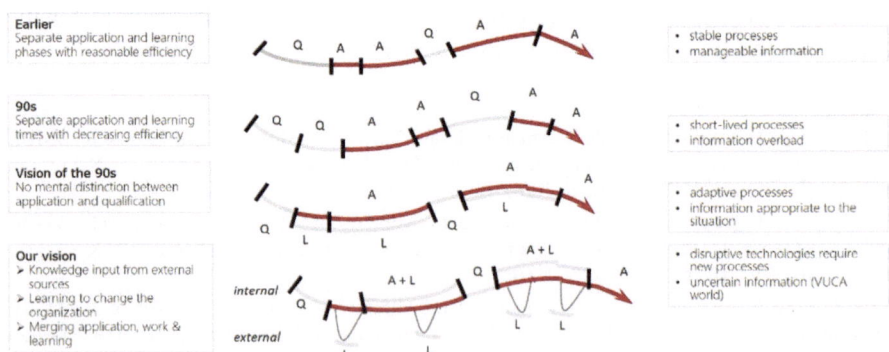

Fig. 2 Qualification, application, and learning within and beyond a company in the past and future. (Own illustration, based on Baitsch 1998; Kauffeld and Paulsen 2018)

Overall, these developments demonstrate that the learning environment is a crucial factor for effective learning. In addition to the organizational framework, which should include the provision of adequate learning time and access to learning units, a learning-friendly work design considers the potential learning opportunities inherent in work tasks, such as feedback, a holistic view of tasks, and the diversity of the tasks themselves.

2.1 Guiding on Learning Paths

In the new learning environments that have emerged, assisting employees as they traverse their unique learning paths can boost their success. Learning paths encompass individual learning goals and topics that necessitate cultivating and maintaining an employee's motivation to learn and change. Various learning opportunities increasingly interconnect and contribute toward a tailored learning journey for each employee (Poell 2017; Kauffeld and Paulsen 2018).

In this vein, Richter et al. (2020) suggest that employees should be guided on their individual learning journeys. Thus, potential steps might include defining learning objectives, identifying relevant learning units, encouraging learning transfer, and reflecting upon both the learning progress and the application of the acquired knowledge. Along these lines, their Richter et al. (2020) research findings also demonstrated that participants in formal training continued to learn autonomously if they perceived the initial training as satisfactory.

Establishing such guidelines increases the probability that employees will attain the desired competence development and incorporate their new knowledge into the organizational context. As an added bonus, this approach strongly supports employees' employees' individual career development.

2.2 Connecting Learning and Organizational Goals

Steering employees on their individual learning paths goes a step beyond simply providing them with learning units. However, even this extra step may not suffice during periods of disruptive change when numerous work processes are being redesigned. When addressing new and unfamiliar disruptive technologies and complex forces (Fig. 2, fourth strand), relying solely on work-integrated learning is likely to fall short, as competence development primarily relies on internal organizational knowledge. In this case, considering input from outside the organization is crucial. In fact, such external input is now necessary to generate innovation and drive organizational development. In this light, an approach entailing "stewing in one's own juices" is no longer conducive for organizational transformation.

Consequently, linking learning to organizational goals has become equally crucial, allowing for immediate organizational benefits to arise from individual learning. This means that, individual learning goals must contribute to organizational objectives, and the implementation of the learned material must be specifically tailored toward organizational and employees' needs. This emphasis can be attained through the design of the learning transfer system (Table 1), which determines the extent to which learning transfer (from training into organizational practices) is achievable via both individual and social contextual factors (e.g., Holton and Baldwin 2003; Kauffeld et al. 2008; Kauffeld and Paulsen 2018).

In this context, the anticipated goal is that employees will be equipped to design the future structure of their own organizations.

Table 1 Success factors and measures for optimization (Kauffeld 2016; Kauffeld and Paulsen 2018)

	Success factors		Optimization actions (Examples)
Specific participant characteristics	Motivation to transfer	Direction, intensity, and duration of effort to make skills and knowledge learned in training usable in the work environment	Planning concrete steps for implementation in the training Scheduling transfer day after a few weeks Accepting takeover of a transfer project
General participant characteristics	General self-efficacy belief	Belief that one is generally capable of changing one's performance if one wishes	Providing positive feedback Adapting the learning content to the abilities of the participants
	Enhancing performance through effort	Anticipation that investing effort in learning transfer will result in improved job performance	Having managers and colleagues report on successful training experiences Offering comparison with another group that has already completed the training
	Outcome expectancy	Expectation that changes in work performance will lead to desirable outcomes for the participant	Providing signals from management that what has been learned will be rewarded Selecting the right staff members for the training

(continued)

Table 1 (continued)

	Success factors		Optimization actions (Examples)
Training features	Transfer design	Training design that offers opportunities for practical application and training exercises that prepare participants for real job demands	Including realistic exercises (e.g., using participants' personal examples) Anticipating resistance and developing and acting out possible solutions in the training session Applying interval training that alternates between between learning and application phases
	Training–work match	Training content that aligns with job requirements	Aligning with a competency model and competence-based design Analyzing the organization and employees' tasks
Specific work environment characteristics	Expectation clarity	The training participant's understanding of what to anticipate	Defining training objectives, content, and procedures and providing participants with written information beforehand, possibly using a competency model
	Personal transfer capacity	The training participant's possession of the time and capacity to apply newly acquired knowledge	Allocating time for reflection on the practical application of newly gained insights
	Ability to apply knowledge	Accessibility of materials, tools, budgets, etc., for implementing what has been learned	Supplying work equipment
	Positive consequences upon application	Favorable consequences as a result of applying the training at work	Highlighting achievements Establishing recognition and reward system
	Negative consequences of non-application	Negative consequences as a result of refraining from applying the training content at work	Encouraging evaluation by supervisors following successful implementation of the training content

(continued)

Table 1 (continued)

	Success factors		Optimization actions (Examples)
	Sanctions by supervisors	Participants' perception of adverse responses from their supervisors when they put their acquired knowledge into practice	Involving managers in needs analysis (e.g., via competency models)
	Supervisor support	Supervisors' provision of encouragement and reinforcement for on-the-job learning	Identifying the need for individual employee training by supervisors Defining learning objectives with employees prior to the training Creating implementation agreements
	Peer support	Colleagues' encouragement and reinforcement of on-thejob learning	Cultivating a shared learning interest Assigning the training participant to a team
General work environment characteristics	Work group's readiness for change	Existing norms within the group that support using skills and knowledge	Including training for the entire group Conducting workshop on group norms
	Feedback	Formal and informal evaluations of an individual's job performance	Scheduling regular employee evaluations Providing 360° feedback

3 Considerations and Suggestions for Implementing Competence-Based Training Programs

Having outlined the factors that are relevant for competence development in order to meet contemporary organizational challenges, we will now dive into the key aspects of the concept behind becoming a "changemaker" (see Fig. 1).

3.1 Leveraging Transfer Projects

Transfer projects serve as a critical connection between the training received and its practical application. These projects are the key to successful learning transfer between training content and work-based learning and may ensure that the organization reaps the benefits of further training (Kortsch and Kauffeld 2019).

As a first step, a specific project (e.g., providing a digital adaptation of customer requests in production, digitizing the complaints and quality assurance process, or introducing a digital time tracking system) must be defined in collaboration with the company management before formal training begins. The learning content and transfer project must align with the organization's strategic goals. In addition, resources should be allocated for the designated project, and expectations for learning outcomes and implementation results should be set as metrics for success measurement. This process requires an explicit agreement between the management and participants, with both parties committed to the project's success (Table 1). The information learned and successfully implemented depends on participants' motivation and willingness to transfer the acquired knowledge, the nature of the training itself (transfer design, work–training congruence, training atmosphere), and the work environment. Several studies (e.g., Kauffeld 2016) have highlighted factors in the work environment as determining the level of successful transfer into "everyday working life." Notably, support from colleagues and supervisors, as well as opportunities to apply knowledge, play vital roles.

3.2 Applying Design Thinking Methods and Motivational Interviewing

Implementing new projects (similar to the transfer project incorporated into the program) that are expected to rebuild structures and transform work processes is generally accompanied by a change process affecting both the training participant and their colleagues. Change initiatives are rarely universally embraced. Passive or even active resistance must be anticipated, considering the rarity of universally embraced change initiatives. This resistance will inevitably impact the training participants while executing their transfer projects.

One viable approach to effectively addressing these circumstances and instigating changes through the transfer project involves using design thinking methods (Endrejat and Kauffeld 2017). This user-centered methodology helps participants delineate the problem space in their projects. Data gathered from internal organizational interviews can help participants identify needs and benefits for both colleagues and the organization at large.

In terms of effectively managing resistance in such situations, a second promising approach involves motivational interviewing (Endrejat et al. 2015; Klonek and Kauffeld 2012). The principles of motivational interviewing (originally designed for effective communication with addiction patients) lend themselves to be successfully adapted to the organizational context. This approach enhances the potential for change by exploring and eliciting change motives within a supportive atmosphere. Participants engage in practical exercises, including motivational interviewing, employing case studies and a digital learning tool that they can directly apply to their transfer projects.

3.3 Providing Insights into New Technologies and Agile Learning

Previous experiences with the process of selecting transfer projects revealed that organizations often choose projects that projects that yield the implementation of new digitalized work structures. Therefore, integrating technological content into the training program is crucial, allowing participants to familiarize themselves with new digital technologies and gain experience with such new approaches as, for example, life cycle assessment, Internet of Things, Industry 4.0, augmented reality, artificial intelligence, specialized software programs, and systems engineering. Such opportunities will help them select and adapt the content best suited to their organization. External experts in the new digital technologies or methods can provide valuable assistance and project-specific support.

It is also noteworthy that introducing new technologies, revamping procedures, or redesigning entire work processes requires others in addition to the training participants to learn and adapt since colleagues who are also impacted by the changes must learn as well. Therefore, training participants should also acquire methods to guide their colleagues during the transition period and support the latter's learning process. Agile learning is integrated into the workflow, allowing for rapid testing, reflection, and adjustment.

3.4 Learning in Networks Within and Outside the Organization

Transformation and change measures in companies can be advanced and accelerated through strong networking and social exchange. Therefore, employees participate in pairs and work together on a shared transfer project. This collaboration encourages learning while additionally fostering mutual social support during project implementation. The opportunities for consultation and reflection characteristic of this setup thus enhance the effectiveness of the project implementation.

Encouraging networking among the participants in the learning or training group is also essential. In this context, the method of collegial case consultation serves as an example of an ideal tool (Tietze 2010). This approach leverages the shared expertise of the participants in their project work, enabling them to learn from each other's experiences. Thus, participants can gain insights into overcoming obstacles and their evolving roles as "changemakers."

Additionally, the integration of alumni of previous cohorts as role models and mentors further reinforces this network, ensuring the diffusion of innovative practices throughout the organization. In line with the diffusion of innovation theory, following Everett Rogers (2003), such alumni are integrated into the training, drawing upon their experiences, and providing a guiding role. As early adopters of change, these alumni can inspire and influence others in their learning journey, accelerating the acceptance and implementation of innovative practices and technologies in the organization.

Research indicates that these types of networks, which exhibit a decentralized structure and significant clustering, increase efficient knowledge transfer and innovation diffusion, all essential for driving change. Strategies such as project funding and the establishment of collaborative project structures have proven to enhance the transfer of complex knowledge and facilitate organizational transformation (Stasewitsch et al. 2022a). Accordingly, companies should actively foster these collaborative networks and provide necessary structural and financial support. In addition to accelerating innovation and learning, this proactive approach also allows employees to become transformation agents, effectively driving organizational change (Stasewitsch et al. 2022b).

3.5   Supporting Reflections with Digital Tools

A critical component of accompanying employees in the training process alongside their transfer projects and supporting them throughout their learning and implementation journey (Fig. 1) is reflection (e.g., Kauffeld and Paulsen 2018). This activity occurs in the training sessions with the trainers, as well as with the help of a digital tool referred to as process pilot®. The tool's integrated outcome-oriented training evaluation measures the effectiveness of the training at various levels, such as reaction, learning, and behavior (Kirkpatrick and Kirkpatrick 2006). This function provides a scientifically grounded overview, highlighting potential areas for further development in specific measures. Process-related aspects influencing the improvement of the competence-based training approach are captured, including factors that are conducive or hindering, encompassing participant-related aspects (e.g., motivation), training-related aspects (e.g., practical relevance), and aspects within the work environment (e.g., support from supervisors) (Baldwin and Ford 1988). Additionally, the tool enables swift feedback exchanges, fostering a positive learning and feedback culture within the organization. This type of comprehensive digitalized tool is crucial in efficiently managing transfer projects, facilitating seamless support, ensuring that all necessary information is available at the right time, and encouraging reflection and practical application of the acquired knowledge. The tool also identifies clear responsibilities by allowing specific tasks to be assigned to individuals, enhancing project management.

3.6   Presentation of Successes

The training program culminates in a final forum, which provides a platform for recognizing the achievements of implementation projects and providing company-wide notice. Projects are presented in terms of the initial situation, the process of development and implementation, the impact, and the lessons learned. This forum disseminates information and ideas while also providing recognition for the participants, thus reinforcing their self-efficacy. One example of an ideal conclusion to the final forum is a certificate presentation. Depending on the company, other communication channels can also be called upon to disseminate the transfer projects.

3.7 Conclusion

In conclusion, the evolution and transformation of organizations in this rapidly digitalizing era clearly require efforts that go beyond work-integrated learning. The introduction of disruptive technologies necessitates the design of new processes and an infusion of external knowledge to expand the potential scope of actions at both the individual and organizational levels.

Designing an effective learning transfer system and building networks alongside with integrating digital tools are promising factors in this journey of transformational change. These components support employees along their individual learning paths, foster readiness for change, and leverage personnel development as a catalyst for organizational development. Digital tools can facilitate efficient task assignment, along with encouraging reflection and the practical application of knowledge. On the human level, the active involvement of leadership in understanding and calculating the economic benefits of these change initiatives and promoting an atmosphere of growth and recognition is essential. The leaders' oversight directly affects the overall success of the organization's change strategies.

The integration of external knowledge is especially vital in adapting to new technologies and methodologies that lie outside the existing competency framework of the organization. Insights from outside of the company pave the way for innovative thinking and the capability to manage technological disruptions effectively.

In essence, the organizational transformation is a comprehensive process, centered around the entity's people and facilitated by strategic learning, networking systems, and digital tools. This success involves empowering employees to become change-makers by equipping them with the necessary tools and strategies, along with fostering a conducive environment for learning, sharing, and application of knowledge. This holistic approach goes beyond merely increasing the organization's adaptability in the face of disruption by also laying a foundation for sustainable growth and development (Berg et al. 2023).

References

Baitsch, C.: Lernen im Prozeß der Arbeit - zum Stand der internationalen Forschung. In: ABWF (eds.) Kompetenzentwicklung, vol. 98: Forschungsstand und Forschungsperspektiven, pp. 269–337 (1998)

Baldwin, T.T., Ford, J.K.: Transfer of training: a review and directions for future research. Pers. Psychol. **41**(1), 63–105 (1988)

Berg, A.K., Schulte, E.M., Schultz, A., Kauffeld, S.: Transformationsprozesse gestalten und evaluieren: Wirkmodelle als Ansatz um strategische Personalentwicklung mit Organisationsentwicklung zu verbinden. Gruppe. Interaktion. Organisation. Zeitschrift für Angewandte Organisationspsychologie (GIO), **54**(3), 371–389 (2023)

Demary, V., Matthes, J., Plünnecke, A., Schaefer, T.: Gleichzeitig: Wie vier Disruptionen die deutsche Wirtschaft verändern. Herausforderungen und Lösungen. IW-Studien-Schriften zur Wirtschaftspolitik aus dem Institut der deutschen Wirtschaft (2021)

Endrejat, P.C., Kauffeld, S.: Wie könnten wir Organisationsentwicklungen partizipativ gestalten?. Gruppe. Interaktion. Organisation. Zeitschrift für Angewandte Organisationspsychologie (GIO) **48**(2), 143–154 (2017)

Endrejat, P.C., Klonek, F.E., Kauffeld, S.: A psychology perspective of energy consumption in organisations: the value of participatory interventions. Indoor Built Environ. **24**, 937–949 (2015)

Holton III, E.F., Baldwin, T.T.: Making transfer happen: an action perspective on learning transfer systems. In: Holton III, E.F., Baldwin, T.T. (eds.). Improving Learning Transfer in Organisations, pp. 3–15. Jossey-Bass, San Francisco, CA (2003)

Kauffeld, S., Bates, R., Holton III, E.F., Müller, A.C.: Das deutsche Lerntransfer-System-Inventar (GLTSI): Psychometrische Überprüfung der deutschsprachigen Version. Zeitschrift Für Personalpsychologie **7**, 50–69 (2008)

Kauffeld, S.: Nachhaltige Personalentwicklung und Weiterbildung. Betriebliche Seminare und Trainings entwickeln, Erfolge messen, Transfer sichern, vol. 2, Springer, Berlin (2018)

Kauffeld, S., Berg, A.-K.: Wie Mitarbeitende Veränderungsprozesse in Unternehmen gestalten können. Personalquarterly **74**, 18–25 (2022)

Kauffeld, S., Paulsen, H: Kompetenzmanagement in Unternehmen. Kompetenzen beschreiben, messen, entwickeln und nutzen, Springer, Stuttgart (2018)

Karwehl, L.J., Kauffeld, S.: Traditional and new ways in competence management: application of HR analytics in competence management. Gruppe Interaktion Organisation. Zeitschrift für Angewandte Organisationspsychologie (GIO) **52**(1), 7–24 (2021)

Kirkpatrick, D., Kirkpatrick, J.: Evaluating Training Programs: The Four Levels. Berrett-Koehler Publishers (2006)

Klonek, F.E., Kauffeld, S.: „Muss, kann... oder will ich was verändern?" Welche Chancen bietet die Motivierende Gesprächsführung in Organisationen. Wirtschaftspsychologie **14**(4), 58–71 (2012)

Kortsch, T., Kauffeld, S.: Validation of a German version of the dimensions of the learning organization questionnaire (DLOQ) in German craft companies. Zeitschrift Für Arbeits-und Organisationspsychologie **63**(1), 15–31 (2019)

Poell, R.F.: Time to 'flip' the training transfer tradition: employees create learning paths strategically. Hum. Resour. Dev. Q.resour. Dev. q. **28**(1), 15–19 (2017)

Richter, S., Kortsch, T., Kauffeld, S.: Understanding learning spillover: The major role of reflection in the formal-informal learning interaction within different cultural value settings. J. Work. Learn. **32**(7), 513–532 (2020)

Rogers, E.M.: Diffusion of Innovations. Free Press, New York (2003)

Stasewitsch, E., Dokuka, S., Kauffeld, S.: Promoting educational innovations and change through networks between higher education teachers. Tert. Educ. Manag.. Educ. Manag. **28**(1), 61–79 (2022a)

Stasewitsch, E., Barthauer, L., Kauffeld, S.: Knowledge transfer in a two-mode network between higher education teachers and their innovative teaching projects. J. Learn. Anal. **9**(1), 93–110 (2022b)

Staudt, E., Kriegsmann, B.: Weiterbildung: Ein Mythos zerbricht. Der Widerspruch zwischen überzogenen Erwartungen und Misserfolgen der Weiterbildung. In: ABWF (eds.) Kompetenzentwicklung '99. Aspekte einer neuen Kultur, pp 17–95 (1999)

Tietze, K.O.: Wirkprozesse und personenbezogene Wirkungen von kollegialer Beratung, Wiesbaden (2010)

Prof. Dr. Simone Kauffeld holds the chair of Industrial/Organizational Psychology at Technische Universität Braunschweig. Her research focuses on competence management, teamwork, counseling, and leadership. To make HR concepts practically accessible, she co-founded 4A-SIDE GmbH in 2008, dedicated to developing and implementing digitalized HR tools.

Ann-Kathleen Berg, M.Sc., has served as a research associate at the Department of Industrial/Organizational and Social Psychology, Technical University (TU) Braunschweig, since 2019. Specializing in team dynamics and transformation processes, she brings a wealth of expertise to her research endeavours. Additionally, since 2020, Ann-Kathleen has taken on the roles of project manager and HR consultant at 4A-SIDE GmbH, where she actively supports professionals in their journey towards becoming effective agents of change.

Digital Care-Pathways and Inter-Organizational Systems: A Perspective on Digital Sovereignty Along Shared Responsibilities

Katharina Dassel and Maxie Lutze

Abstract The concept of digital sovereignty has gained attention on levels spanning from individual information sovereignty to digital sovereignty of organizations or the state level, concerning, for example, digital infrastructures. Taking into account the reality of digital work however, information and information systems are practically managed and worked with across organizations and thus, interdependencies challenge a classical notion of a sovereign authority that builds on supreme authority over a certain domain. We argue that for the conceptual understanding to allow for the realities of practical interdependencies, a perspective of sovereignty is needed that considers the shared responsibilities that emerge in these inter-organizational systems settings. We apply this idea to the context of digital care pathways for a practical exploration of the phenomenon in healthcare and suggest a conceptual understanding of inter-organizational, shared digital sovereignty which is informed by shared responsibility. We discuss open questions for further research.

Keywords Inter-organizational systems · Digital care pathways · Shared digital sovereignty

1 Introduction

Change in the healthcare sector of Germany is fueled by structural problems such as high workload, lack of qualified personnel, or challenges in meeting the needs of patients and their families including care quality.

The vision of new, digitally supported care pathways brings hope that such digital transformation can help to tackle the change and spur innovative and efficient ways

K. Dassel (✉) · M. Lutze
Institute for Innovation and Technology (Iit), Steinplatz 1, 10623 Berlin, Germany
e-mail: dassel@iit-berlin.de

M. Lutze
e-mail: lutze@iit-berlin.de

U. Schmuntzsch et al. (eds.), *New Digital Work II*,
https://doi.org/10.1007/978-3-031-69994-8_14

of organizing work and quality of care. Already, the healthcare sector is characterized by a high interdependence of tasks between organizations and professions. In light of the vision of digital care pathways (see Sect. 2), the inter-sectoral and inter-organizational management of information systems gains increasing importance to effectively address problems such as a high workload and mismatches in patient care. Thus, organizations in the healthcare sector not only need to manage the digital transformation inside the boundaries of their organizations, but also have to coordinate and negotiate issues of digital ways of working, such as data usage and designing processes, in cooperation. These issues have been recognized as key characteristics of inter-organizational information systems being shared information systems among a group of companies, based on electronic data interchange (Reimers et al. 2004). It shows that organizations in healthcare no longer only need to be mere sovereigns of their own digital transformation, but that the power and scope to shape digital work needs to be negotiated and distributed among organizations as well.

The concept of sovereignty has gained importance in the digital sphere, broadly spanning the levels of the individual citizen and social movements (e.g., discussing issues around digital participation and competencies), organizations (e.g., discussing implementation of new technologies or sovereignty in developing technologies) and the state (e.g., control of cyberspace) (Wittpahl 2017; Couture and Toupin 2019). Along the different perspectives, sovereignty is highly woven into the narratives of European political discourse (Lambach and Oppermann 2023). Across all different units of analysis, the notion of sovereignty, when referring to the digital sphere, generally means some kind of (collective) ability to control or exercise power within a defined digital space. However, it has been widely discussed that the interpretations of digital sovereignty can significantly differ, revealing a somehow contested concept (Couture and Toupin 2019; Glaze et al. 2023).

While the core idea of sovereignty provides a useful lens to spur valuable discussion, especially in light of a digital sphere characterized by interdependent information systems and users, this paper assumes a "de facto" sovereignty is hard to achieve on a practical level. To not render discussions around digital sovereignty somewhat theoretical, we aim to elaborate on aspects that highlight aspects specifically relevant to information exchange in the healthcare sector.

We argue that now on the verge of digital work and interdependent systems, we need a different perspective on the concept of sovereignty to understand how it can be useful to include a perspective of inter-organizational systems, for example along digital care pathways. Accordingly, we hypothesize that a new aspect of digital sovereignty emerges that describes the abilities to manage and negotiate spheres of power and responsibilities along inter-organizational information systems. We pose the following research question: *How can we understand digital sovereignty along inter-organizational information systems and in interdependent settings such as digital care pathways?*

In this paper, we thus explore how a perspective of inter-organizational information systems and shared (digital) responsibilities on digital sovereignty can help to spur discussions that emerge on a practical level when considering the inter-organizational digital transformation in healthcare. To approach this concept, we leverage the argument that a scope of power and authority over a certain digital domain also entails the responsibility to cater to this domain. Accordingly, we consider the idea of digital sovereignty from the perspective of shared, digital responsibilities, to shed light on inter-organizational, digital work in the healthcare sector.

Methodologically, we explore the first parts of the process of concept definition by Podsakoff et al. (2016). We start by introducing the practical problem of sovereignty in the context of digital care pathways, characterized by inter-professional, inter-sectoral and inter-organizational digital work in Sect. 2. In Sect. 3, while building on the roots of the concept of digital sovereignty, we explore what we can learn about an understanding of shared digital sovereignty through the perspective of digital care pathways and inter-organizational systems in Germany's healthcare system, after which we pose open questions for further research and conclude in Sect. 6.

Thus, we take a first step to contributing a first conceptual understanding of shared, digital sovereignty and offer a new perspective on managing the interdependencies of digital transformation in the healthcare sector. We offer a new view and understanding of the emergence of digital sovereignty in interdependent contexts such as healthcare and digital care pathways.

2 Issues of Inter-Sectoral and Inter-Professional Cooperation as Issues of Inter-Organizational Work

Continuous and inter-sectoral care is crucial in order to avoid deterioration and hospitalization in the face of rising patient numbers with increased treatment needs and complex treatments (SVR 2018). The coronavirus pandemic has made it particularly clear how essential well-functioning and transparent care and information chains are for patients. The German healthcare system is self-administrating and is operated by many institutions and players. Since 2007, the self-administration has been obliged to establish integrated, inter-sectoral forms of care (Gerlinger 2017). However, healthcare in Germany is highly fragmented. The sectoral separation is accompanied by considerable interface problems, meaning that prevention, outpatient and inpatient treatment, rehabilitation, and care are largely separated from each other, as are healthcare professions, disciplines, and service provider institutions. The separation of sectors goes hand in hand with different remuneration structures and responsibilities, which makes it difficult to provide coordinated and continuous care for patients and focuses care on remuneration regulations and service provision rather than on medical and nursing requirements. Inadequate coordination and cooperation along these interfaces mean that existing potential in healthcare is not exploited and treatment outcomes fall short of their potential.

Overcoming this sectoral boundary is being driven by the implementation of integrated care services outside of standard care and the introduction of discharge management, which is somewhat loosening up care at the interface between the outpatient and inpatient sectors. However, the fundamental division of care areas that pervades the legal framework of statutory health insurance and its downstream institutions remains in place. In the current coalition agreement,[1] concrete ideas for overcoming the sector boundaries are mentioned, such as hybrid Diagnosis Related Groups (DRG), outpatient-inpatient care and remuneration concepts, inter-sector planning, pilot and community nurse approaches. Approaches are needed that could help to ensure that inter-sectoral integrated care does not remain the exception, but becomes the rule, with shared, mediating responsibility for care between outpatient and inpatient care.

The lack of inter-sectoral, inter-disciplinary, and inter-professional care organization reflects on the service provider side and shows its effects in a fragmentation of patient care, communication problems, loss of information, and inefficient processes between the various service providers. In particular, the lack of information sharing leads to confusion regarding the responsibilities and roles of care providers and inadequate care, which has an unfavorable impact on health outcomes (Schneider et al. 2016; Frandsen et al. 2015). Multiple examinations and redundant diagnostic procedures are associated with this, as information is not shared seamlessly between different facilities, innovation is inhibited and the billing systems and bureaucratic processes within sector boundaries can lead to additional administrative burdens for service providers. The effects of separation can also be seen on the organizational and directly at employee level, where employees report that ineffective use of systems across professions affects their ability to care effectively which leads to strain and quitting (Dassel et al. 2024). Last but not least, the sector-based financing system according to the cost recovery principle contributes to the fact that healthcare providers maximize the quantity of services instead of the quality of services and minimize internal costs, e.g. through low-paid employees, in order to improve their financial situation.

In the current healthcare system, doctors are located at the highest hierarchical level. Other professional groups are seen as supporting actors. The different roles of health professions in multi-professional teams are often explored in pilot projects (Hoster and Majjouti 2022). Multi-professional teams are comparatively rare in Germany, such as community health nurses or advanced practice nurses, who take on coordination tasks in the patient's care continuum, compensate for breaks in the care process and support patients in navigating the healthcare system.

When it comes to actual care, citizens often experience the existing sector boundaries. Chronic patients in particular would benefit from integrated care that is inter-sectoral, interdisciplinary, and inter-professional. Up to now, they tend to find themselves in the role of "letter carriers" of the healthcare system, because they have to distribute information between service providers who do not exchange information with each other. In many cases, this leads to unnecessary health measures because the necessary information is not available or financial disincentives encourage this.

---

[1] https://www.bundesregierung.de/breg-de/service/gesetzesvorhaben/koalitionsvertrag-2021-199 0800.

The concept of patient pathways is an approach for improving care processes that are designed across sectors in order to contribute to more patient-centered care in organizational terms. These are structured and multidisciplinary approaches, which describe the sequence of steps or interventions for the assessment, treatment, and aftercare of patients with specific conditions or following specific medical procedures. Patient pathways also serve to standardize and streamline care and are designed to ensure that patients receive evidence-based, high-quality care in a consistent and efficient manner.

In the course of digitalization, analog pathways are becoming digital or, in some cases, hybrid care pathways. "A digital [..] (care) pathway combines a patient engagement platform with care workflows to seamlessly navigate patients through various care experiences, such as telehealth visits, remote monitoring, assessment of patient-reported outcomes, and appropriate escalations to the care team" (Shapiro et al. 2023, 2). Wisniowski und Kurscheid (2023) point out that digital patient pathways have so far been designed primarily for the inpatient sector. Accordingly, existing platforms often reflect the separation of sectors and hinder continuous care.

For cross-sector, cross-specialty, and cross-institutional treatment along a treatment pathway, it is necessary that all relevant data and information is available and that all service providers involved in the treatment process have access to it. This basis should be available in future with the electronic patient record, which will be created as an opt-out file by the health insurance funds for all insured persons from 2024 (BMG 2021). With regard to data exchange, the responsibility and control of health data lies with the insured persons, which is why they will be a central player in inter-sectoral data exchange in the future.

Various reports (Braeske et al. 2020; Lutze et al. 2021; Amelung et al. 2013) show that there is still a lot of potential for digitalization in Germany. To date, there are no established methods for comprehensive digital maturity assessment among service providers in Germany, who face a challenging task with regard to digital patient pathways. Setting up a state-of-the-art hardware and software infrastructure, connecting to the telematics infrastructure and using its applications are just one part of this. Service provider and care-specific as well as systemic and general knowledge of digital healthcare services, offers and processes must also be acquired as part of digital competence.

According to the ENISA report,[2] increasing cyberattacks also show the Europe-wide need for adequate strategies for IT security measures. The latest laws in Germany that have an impact on this are the Digitization Acceleration Act and the Health Data Use Act, which are associated with personnel and financial costs for healthcare organizations and institutions.

Since the introduction of the General Data Protection Regulation (GDPR), there have been equal requirements in place within the EU regarding the handling of data protection. Compared to other EU countries, this is sometimes interpreted more restrictively by state and federal law in Germany. In general, the German healthcare system is characterized by federal structures (cooperation between the federal and

---

[2] https://www.enisa.europa.eu/publications/health-threat-landscape.

state governments), which means that the GDPR is interpreted differently. The state hospital laws or nursing home laws also provide for corresponding regulations on digitalization in the healthcare sector.

Also still relatively new in Germany's digital healthcare system are consistent interoperability developments such as the commitment to standards FHIR by HL7[3] and joining SNOMED CT,[4] which are an indispensable prerequisite for cross-sector, cross-specialty, and cross-institutional care. The interoperability of systems must be guaranteed in the form of technical, semantic, syntactic, and organizational interoperability in order to derive meaningful, well-founded data-based findings and make decisions.

In the healthcare context, it has been shown that the effective use of digital systems across organizations will have a major impact on realizing the potential of digital transformation and integrated care, such as to relief work strain. It is also apparent that there is a need to use efficient and effective systems across all stakeholders to solve issues of inter-professional and inter-sectoral communication and cooperation which can be considered an inter-organizational (systems) issue.

3 Towards Inter-Organizational, Digital Sovereignty in Practice: Employing Shared, Corporate Digital Responsibilities

3.1 Digital Sovereignty and Inter-Organizational Health Information Systems

The idea of sovereignty was transferred to the digital sphere and gained popularity especially in European countries under the umbrella term of "digital sovereignty" (Friedrichsen and Bisa 2016; Wittpahl 2017). While discussed by various academic disciplines, most discussions span the domains of the state or supranational institution, the organizational or the individual (Fries et al. 2023; Hartmann 2021). Under the umbrella term of digital sovereignty, frameworks exist for the organizational domain, that offer guidance on aspects like data governance and control, infrastructure control, cybersecurity, risk management, and compliance or competencies to help organizations navigate the complexities of digital governance (Madiega 2020). Organizations often tailor these frameworks to their specific needs and regulatory environments.

---

[3] FHIR (Fast Healthcare Interoperability Resources) supports data exchange between software systems in the healthcare sector. The standard is defined by the international organization HL7 (HL7 Deutschland e.V.).

[4] SNOMED CT stands for Systematized Nomenclature of Medicine and is a comprehensive clinical healthcare terminology that is used to encode meanings of health information in a standardized way (BfArM).

However, actors across those domains are not isolated and individuals, organizations, and nations, while formulating claims to sovereignty in some form of autonomy, freedom or power over something, always find themselves in interdependence. The interdependent nature that is relevant to consider becomes especially apparent in light of networked systems and eco-systems in healthcare, which build on the exchange of information (Deutscher Ethikrat 2017). Similarly, and as our healthcare case of digital care pathways shows, healthcare organizations are increasingly dependent on cooperation and need to organize work in eco-systems, as they depend on information systems as services or underlying information infrastructures that work across organizations. The issues of inter-sectoral and inter-professional collaboration in Sect. 2 showed that to seize the potential of digital transformation, health information systems need to be managed successfully on the inter-organizational level, i.e. across actors and organizations.

To understand digital sovereignty in the sphere of digital care pathways we thus argue that there is a need to take into account the interdependencies that arise between actors. Recognizing this allows to shed light on new aspects of the notion.

3.2 Attributes of Shared Digital Sovereignty in a Digital Care Pathway

We take the exemplary context as a case to investigate how issues of digital sovereignty come into play in interdependent environments. Following, we explore possible attributes of digital sovereignty in healthcare organizations using an example of digital care pathways in order to gain insight into the practical significance of digital sovereignty in inter-organizational settings. We built on prominent work that investigates attributes of digital sovereignty (cf. Friedrichsen and Bisa 2016; Wittpahl 2017; Fries et al. 2023; Hartmann 2020; Hummel et al. 2021; Stubbe 2017; Bogenstahl und Zinke 2017) to explore those aspects that are relevant to the inter-organizational, shared perspective.

We summarize the aspects of shared digital sovereignty in Table 1: Along a digital care pathway, *roles and responsibilities* need to be agreed upon across care providers, especially to establish trusting relationships with other providers and patients. On a patient-centered pathway, this can be considered an issue of shared digital sovereignty. While all the issues discussed are known areas of *data governance* for organizations, in digital, interdependent settings, like across a digital care pathway, to govern data in a sovereign way becomes a shared responsibility. Exploring a classic domain of digital sovereignty as issues of *infrastructure, soft-and hardware* in the context of digital care pathways shows that issues of digital sovereignty need to be managed in a shared way, especially when it comes to aspects of *interoperability*. Aspects of *IT- and cybersecurity* now also concern inter-organizational systems and need to be governed in a sense of shared sovereignty. *Transformational capabilities* come to light as it becomes clear that healthcare providers do not merely need to manage their own organization and organizational culture towards being adaptive to change but have to consider the strengthening of collaboration across organizations through the digital care pathway a joint effort. Also, responsible action is an essentially shared part of digital sovereignty, as in interdependent settings also compliance measures need to be considered together and are a shared responsibility. The shared perspective also applies to another key characteristic of digital sovereignty, and looking at digital care pathways it becomes clear that *digital competencies* are also inter-organizational issues and thus, a relevant attribute of shared sovereignty.

There are different settings in which digital care pathways can be used, e.g. in the care of people with multiple sclerosis, children with complex chronic neurological diseases, and in the treatment of chronic wounds. While digital care pathways are generally discussed as an approach for high-quality care, their introduction is not considered reasonable in all areas with regard to standardization, for example in the case of rare diseases (Olsen et al. 2021). With the further development of the telematics infrastructure, more and more design options for inter-sector, digitally supported care pathways are also being opened up in Germany. In order to explore the impact of these possibilities on the digital sovereignty of healthcare organizations, we take a look at the issues that concern aspects of digital sovereignty when it comes to digital care pathways.

To examine the aspects of shared digital sovereignty, we ask two questions to shed light on the degree of possible dependencies or degrees of freedom. First: How can a healthcare organization address the respective care concern? Second: What characterizes the digital interdependence?

Roles and responsibilities: Healthcare involves a multidisciplinary approach, with various professionals providing sequential or parallel treatments. Establishing digital care pathways requires defining objectives within legal frameworks and clarifying roles across healthcare sectors. In chronic wound care for example, ensuring data sharing is a shared responsibility among providers in a patient-centered design. While individual processes may be isolated, providers must view themselves as part of a larger care system. It is crucial to envision wound care as interconnected steps forming a patient pathway. In digital settings, avoiding overarching responsibility is not possible, given the platform's purpose of efficient data collection. To address interdependent issues of digital sovereignty, the focus should be on continuous care processes with a shared goal rather than isolated tasks.

Table 1 Aspects of shared digital sovereignty in inter-organizational systems

Aspects of shared, digital sovereignty
1. Roles and responsibilites
2. Data exchange, access and use
3. IT infrastructure, hardware and software, interoperability
4. IT and cybersecurity
5. Transformation capabilities
6. Ethics and responsible practices, trustworthy algorithms and sustainability
7. Competencies

Data exchange, access, and use: A requirement for high-quality treatment is the flow of information between the practitioners. The current exchange of information about patients is often incomplete or non-existent and is often only briefly communicated in doctor's letters. The essence of a digital care pathway from this perspective is effective data management, integral for quality-oriented healthcare. Data sovereignty in healthcare organizations entails compliance with data protection laws, including GDPR, and secure encryption methods. It is vital to establish user authorizations, clear ownership areas, and defined responsibilities within and across organizations. Both personal and non-personal data require careful handling for fair cooperation in the data ecosystem. Protocols for addressing data breaches, including notification to authorities and subjects, must be legally compliant. Continuous monitoring through access logs and security assessments ensures ongoing compliance.

While these sovereignty requirements are central to the use of different digital systems and must be implemented by the organization itself, corresponding security measures and access regulations via corresponding organization- or person-related authorization (security module card and health professional card) are provided within the telematics infrastructure, which must be implemented at an organizational level.

In cross-sectoral (wound) care, ensuring data sovereignty involves reliable data entry by all service providers so they are available at the relevant touchpoints. A shared understanding among healthcare professionals about the exchanged data's relevance to patient care is crucial. Service providers must agree on measurement parameters to contribute to the patient pathway and improve outcomes. Patients can selectively grant access to their wound transfer form data. Hence, it is of high importance that they understand the data's relevance to their personal health. The opt-out principle, allowing insured persons to prevent data use, depends on the relationship between patient and healthcare professional and the powers granted to insured persons regarding the electronic health record (EHR) use. The Digital Health Act (DigiG)[5] mandates health insurance funds to introduce and communicate EHR usage options (BMG 2023). Empowerment approaches for diverse insured groups may be necessary (e.g. older people or people with low digital literacy or concerns of data misuse) to ensure that all actors are contributing to the shared responsibility of exchanging data for digital sovereignty. Also, health insurance companies and other stakeholders play a central role in accessing data in digital cross-sector care pathways, alongside service providers in wound care.

IT infrastructure, hardware and software, interoperability: Digital sovereignty in healthcare involves maintaining control over software and hardware to protect patient data and reduce dependence on external entities. Collaborating with transparent manufacturers minimizes the risk of compromised hardware, preserving healthcare infrastructure integrity. Transparent supply chains are crucial, and control over hardware resilience is vital. Implementing backup systems and disaster recovery plans ensures uninterrupted service during disruptions. Selecting software with stringent security standards protects patient confidentiality. Customizing patient

---

[5] „Gesetz zur Beschleunigung der Digitalisierung des Gesundheitswesens " (Digital-Gesetz—DigiG).

or practice management software to specific workflows enhances efficiency and reduces dependence on generic solutions. Co-creation approaches, where healthcare organizations actively contribute to development, drive innovation and address professional needs, improving digital tool effectiveness. Solutions like open source systems provide transparency, control over source code, and customization, reducing dependency on proprietary systems.

Data exchange standards such as HL7 FHIR are crucial for seamless healthcare communication and interoperability. To enhance digital sovereignty, healthcare organizations should minimize reliance on a single vendor by selecting interoperable solutions. This flexibility facilitates vendor switchovers, reducing "vendor lock-in" effects. Controlling vendor relationships is essential, involving careful selection and negotiated contracts ensuring data ownership, security, and regulatory compliance. Telematics infrastructure addresses hardware and software prerequisites for interoperability and security. This is an issue of shared responsibility: External digital patient pathways require aligned software selection and technical/semantic interoperability. Cooperative efforts among healthcare organizations can reduce costs, enhance negotiation power, and encourage favorable offers from providers. Shared sovereignty involves joint assurance of hardware and software function, achieved through regular updates and agreed-upon requirements and conditions of use.

IT and cybersecurity: Healthcare organizations need robust cybersecurity measures to protect their IT infrastructure from cyber threats. This includes the implementation of firewalls, encryption, intrusion detection systems, and regular security audits. Training employees in cybersecurity best practices is critical to avoid human error that can lead to security breaches. Cybersecurity agreements among organizations of the health care sector can contribute to a mutual understanding. They are essential for safeguarding patient confidentially, ensuring the integrity of healthcare data and system availability. Mutual protocols help mitigating risks, rules for incident response and access control and approaches for ongoing employee training concerning data protection and privacy regulations.

Transformation capabilities: Digital sovereignty for healthcare organizations hinges on their transformational capability, the ability to adapt and innovate. This involves a patient-centered approach in designing processes, optimizing workflows, and fostering interdisciplinary collaboration. Seamless integration of data across systems is crucial, prioritizing real-time information access for healthcare professionals. Telemedicine necessitates processes to support remote consultations, monitoring, and follow-up to enhance accessibility and continuity.

Evaluation of clinical outcomes is paramount to ensure quality care, using metrics like patient outcomes, complication rates, and adherence to guidelines. Patient satisfaction, measured through feedback surveys, gauges the impact of process changes. Ensuring continuity of care involves tracking metrics such as readmission rates and patient-reported experiences with care transitions. Interoperability metrics, assessing the extent and speed of information exchange, are vital for data integration success. Leadership plays a central role, driving a patient-centered culture and collaboration. This involves cultivating empathy, effective communication, and shared decision-making among healthcare professionals. Breaking down silos and fostering a culture of teamwork and open communication improves care coordination. Digital literacy among healthcare professionals is also essential. Training programs for adapting to new technologies promote continuous learning and innovation. Ethical leadership emphasizes patient privacy, data security, and compliance with ethical standards when using digital technologies.

We conclude, that achieving digital sovereignty in healthcare requires effective process design, meticulous change measurement, and a culture valuing patient-centricity, collaboration, and ethical leadership. This, however, can not be sufficiently achieved by one healthcare organization alone, rather organizations along a care pathway need to collaborate and create a shared understanding of these aspects. This holistic strategy ensures digital transformation enhances operational efficiency while prioritizing the well-being and satisfaction of professionals and patients.

Ethics and Responsible Practices, Trustworthy Algorithms, and Sustainability:Digital sovereignty in healthcare hinges on trustworthy algorithms, emphasizing responsible technology and data use. This involves ensuring accurate, fair clinical decision support algorithms, with transparent processes fostering trust among healthcare professionals, patients, and regulators. Trustworthy algorithms adapt, evolving with medical knowledge and changes in healthcare practices for sustainable digital health solutions.

Shared sovereignty emerges through collaborative efforts where healthcare institutions work with tech developers and researchers. This extends to algorithm development, data-sharing initiatives, and adherence to industry standards. Initiatives for shared responsibility include for example:

Standards adherence: Collaborative efforts ensure adherence to industry standards in algorithm development, with stakeholders updating these standards collectively.

Educational programs: Joint initiatives promote awareness and understanding of trustworthy algorithms among healthcare professionals, policymakers, and the public.

Inclusive decision-making: Active involvement of patients and the public in algorithm development ensures diverse perspectives for algorithms meeting broader needs.

Collaborative reviews: Healthcare organizations, tech providers, and regulators work together on reviews; ensuring algorithms meet standards and ethical guidelines.

Ethics committees: Healthcare organizations establish committees with diverse representation for ethical evaluations of algorithm use in patient care.

Another relevant aspect to consider from the perspective of shared digital sovereignty is that of *sustainability*. In healthcare, digital sovereignty's sustainability involves enhancing capabilities for the long term while considering environmental, economic, and social factors. Healthcare organizations must address global issues such as climate change and resource scarcity to ensure resilient digital infrastructure. To achieve environmental sustainability, healthcare organizations should adopt energy-efficient hardware, optimize software code for efficiency, and embrace technologies that minimize power consumption. Choosing suppliers with environmentally friendly practices, promoting responsible recycling of equipment, and transitioning to paperless processes through electronic health records are essential steps.

Conducting life cycle assessments for digital technologies helps understand their environmental impact, aiding in informed decision-making for resource management. Moreover, healthcare organizations need to meet legal requirements, reporting on their environmental impact and sustainability efforts.

From this ethical perspective shared digital sovereignty relies on trustworthy algorithms and shared responsibility with regard to collaboration, adherence to standards, education, inclusive decision-making, and ethical evaluations. This understanding of shared digital sovereignty can contribute to a resilient and ethical healthcare ecosystem. Moreover, sustainability of digital sovereignty in healthcare involves strategic decisions that align with environmental responsibility and long-term resilience, emphasizing shared responsibility among healthcare organizations.

Competencies: In their pursuit of digital sovereignty, healthcare organizations must cultivate a range of competencies to effectively navigate the dynamic landscape of digital healthcare. This is not just about technical skills, but also a deep understanding of patient needs, ethical considerations, and the strategic vision to lead organizations into a digitally-enabled future. Skilled professionals are central to shaping a healthcare landscape in which digital technologies improve patient care while upholding the principles of safety, interoperability, and sustainability. The following overview lists competencies that are relevant to consider as organizations collaborate along a digital care pathway:

A proficient professional in cross-sector care pathways possesses interdisciplinary knowledge to facilitate seamless collaboration across diverse healthcare areas. Strategic thinking is applied to design and implement digital solutions aligning with organizational goals and patient-centered outcomes. Deep understanding of patient needs integrates patient perspectives into personalized digital solutions supported by clinical knowledge for optimal patient care. Technological competence guides decisions on digital system introduction and integration. Data literacy ensures responsible data utilization to enhance patient outcomes and organizational efficiency. Systemic thinking optimizes digital infrastructure, while assessing vendors aligns with digital sovereignty goals. Security measures, compliance with regulations, and data governance practices safeguard patient data. Expertise in data-driven decision-making and ethical standards contribute to a secure digital ecosystem. Interoperability knowledge enables seamless collaboration, forming a comprehensive skill set for effective management of cross-sector care pathways.

We, therefore, propose that along with a shared understanding of digital sovereignty, a shared digital responsibility also emerges for the various domains to cater to the interdependent realms of power. A notion of shared digital sovereignty can thus refer to collaborative arrangements where multiple entities, such as healthcare organizations, agree to jointly manage and govern certain aspects of their digital activities. Understanding shared sovereignty in digital relationships requires a balance between collaboration and the protection of individual or organizational interests, with a focus on developing inclusive and adaptable governance frameworks that foster mutual benefits and responsible digital practices.

4 Open Questions, Further Research, and Conclusion

In this paper, the perspective of shared digital sovereignty is proposed to advance our understanding of the use of inter-organizational information systems (e.g., along digital care pathways) and to address, in particular, the relational and interdependent nature of digital, information-based work across organizations in healthcare. Subsequently, we address the need for further research around two key themes. *First*, we highlight the limitations and the need for further research around the conceptual development of shared digital sovereignty. *Second*, while we emphasize the contribution of leveraging the notion to advance inter-professional and—organizational work along health information systems, we highlight the need for further work to investigate how the idea of shared digital sovereignty can be negotiated and achieved in practice.

First, it becomes clear that this paper explores one specific case example methodologically—that of digital care pathways in the German healthcare system. The conclusions drawn are thus limited to a specific context and are only a first step to understanding a concept. More research is needed that includes issues that take a broader range of contexts into account, such as different care pathways and different health care infrastructure and regulations. Only then, all attributes relevant to the understanding of shared digital sovereignty can be depicted.

Second, exploring the idea of sovereignty in a digital sphere through the lens of inter-organizational, digital sovereignty highlights that managing the networked realities comes with inextricably linked spheres of power and authority. Hence, in order to perform agency over "core domains of digital sovereignty", the links to domains of other actors need to be negotiated, managed, and taken into account. Discussions around corporate sovereignty have recognized that sovereignty, understood as power of corporations, is "intimately related" to business ethics and to extending the growing responsibility of corporations to use their spheres of power in line with ethical and sustainable societal norms. This shows that the idea of corporate or organizational digital sovereignty in general also closely ties not only to exploring spheres of influence and responsibility, but to the ability to enforce claims of power as well. To address this, we contribute the proposition of attributes relevant to an understanding of shared digital sovereignty. On a theoretical level, this understanding can offer an integrative perspective on digital sovereignty. On a practical level, considering the healthcare sector, the interest groups have different abilities to enforce claims of power, which especially in Germany has grown historically. Therefore, the question remains whether and how this perspective can be leveraged to advance and manage digital transformation. Especially when addressing shared responsibilities practically, a focus on care pathways is beneficial for designing healthcare with people and patients in mind, and it is crucial to align the incentive structures of participating organizations and stakeholders with the core concept of placing the patient at the center of healthcare (Wisniowski and Kurscheid 2023).

It will thus be a fruitful avenue for future research to investigate questions of how shared digital sovereignty can be achieved and practiced in interdependent contexts such as healthcare. What do healthcare organizations and service providers need to consider in future digital care, and how can shared responsibilities be addressed to achieve a shared digital sovereignty? Some strategies to navigate interdependence might involve issues of establishing data governance policies and strategies that include the telematics infrastructure. Additionally, efforts may include establishing shared compliance frameworks, maintaining transparency in digital operations, and establishing accountability mechanisms. For example, this can be achieved through extended reporting mechanisms under the umbrella of corporate social responsibility or corporate digital responsibility. Finally, in interdependent relationships, shared digital sovereignty should be seen as a dynamic concept that requires ongoing communication, collaboration, and adaptation to balance independence with cooperation effectively.

References

Amelung, V., Jensen, S., Krauth, C., Wolf, S.: Pay-for-performance: Märchen oder Chance einer qualitätsorientierten Vergütung? G+G Wissenschaft **13**(2), 7–15 (2013). https://www.wido.de/fileadmin/Dateien/Dokumente/Publikationen_Produkte/GGW/wido_ggwaufs1_0313_Amelung.pdf. Last accessed 22 Jan 2024

BfArM: Snomed CT. https://www.bfarm.de/DE/Kodiersysteme/Terminologien/SNOMED-CT/_node.html. Last accessed 17 Jan 2024

BMG: Die elektronische Patientenakte (ePA). https://www.bundesgesundheitsministerium.de/elektronische-patientenakte (2021). Last accessed 22 Jan 2024

BMG: Gesetz zur Beschleunigung der Digitalisierung des Gesundheitswesens (Digital-Gesetz-DigiG). https://www.bundesgesundheitsministerium.de/ministerium/gesetze-und-verordnungen/guv-20-lp/digig (2023). Last accessed 24 Nov 2023

Bogenstahl, C., Zinke, G.: Digitale Souveränität - ein mehrdimensionales Handlungskonzept für die deutsche Wirtschaft. In: Wittpahl, V. (Hg.): Digitale Souveränität. Bürger, Unternehmen, Staat. Springer (iit-Themenband), Heidelberg, S. 65–82 (2017)

Braeske, G., Pflug, C., Tisch, T., Wentz, L., Pörschmann-Schreiber, U., Kulas, H.: Umfrage zum Technikeinsatz in Pflegeeinrichtungen. Sachbericht für das Bundesministerium für Gesundheit. In: IGES Institut GmbH (2020)

Couture, S., Toupin, S.: What does the notion of "sovereignty" mean when referring to the digital? New Media Soc. **21**(10), 2305–2322 (2019). https://doi.org/10.1177/1461444819865984

Dassel, K., Busch, A., Lutze, M.: Gütekriterien von Pflegesoftware. Grundlage für Entlastung von Fachpersonal in der stationären Langzeitpflege? Hg. v. Bertelsmann Stiftung (2024)

Deutscher Ethikrat: Big Data und Gesundheit—Datensouveränität als informationelle Freiheitsgestaltung. https://www.ethikrat.org/fileadmin/Publikationen/Stellungnahmen/deutsch/stellungnahme-big-data-und-gesundheit.pdf (2017). Last accessed 17 Jan 2024

Frandsen, B.R., Joynt, K.E., Rebitzer, J.B., Jha, A.K.: Care fragmentation, quality, and costs among chronically ill patients. Am. J. Manag. Care **21**(5), 355–362 (2015)

Friedrichsen, M., Bisa, P.-J.: Digitale Souveränität. Springer Fachmedien Wiesbaden, Wiesbaden (2016)

Fries, I., Greiner, M., Hofmeier, M., Hrestic, R., Lechner, U., Wendeborn, T.: Towards a layer model for digital sovereignty: a holistic approach. In: Hämmerli, B., Helmbrecht, U., Hommel, W., Kunczik, L., Pickl, S. (Hg.): Critical Information Infrastructures Security. 17th International Conference, CRITIS 2022, Munich, Germany, 14–16 September 2022, Revised Selected Papers,

Bd. 13723. 1st ed (2023). Springer Nature Switzerland, Cham, Imprint Springer (Lecture Notes in Computer Science, 13723), S. 119–139 (2023)

Gerlinger, T.: Gesetzliche Regelungen zur Integration von Versorgungsstrukturen. Hg. v. Bundeszentrale für politische Bildung (bpb). https://www.bpb.de/themen/gesundheit/ges undheitspolitik/255488/gesetzliche-regelungen-zur-integration-von-versorgungsstrukturen/ (2017). Last accessed 17 Jan 2024

Glasze, G., Cattaruzza, A., Douzet, F., Dammann, F., Bertran, M.G., Bômont, C., et al.: Contested spatialities of digital sovereignty. Geopolitics **28**(2), 919–958 (2023). https://doi.org/10.1080/14650045.2022.2050070

Hartmann, E.A.: Digitale Souveränität in der Wirtschaft. Gegenstandsbereiche, Konzepte und Merkmale. In: Hartmann, E.A. (Hg.) Digitalisierung souverän gestalten. Innovative Impulse im Maschinenbau. Springer, Berlin, Heidelberg, S. 1–16 (2021)

HL7 Deutschland e.V. https://hl7.de/themen/hl7-fhir-mobile-kommunikation-und-mehr/warum-fhir/. Last accessed 17 Jan 2024

Hoster, B., Majjouti, K.: Das Potenzial der künstlichen Intelligenz. In: Die Schwester Der Pfleger (12), S. 32 (2022)

Hummel, P., Braun, M., Tretter, M., Dabrock, P.: Data sovereignty: a review. Big Data Soc. **8**(1), 205395172098201 (2021). https://doi.org/10.1177/2053951720982012.

Lambach, D., Oppermann, K.: Narratives of digital sovereignty in German political discourse. Governance **36**(3), 693–709 (2023). https://doi.org/10.1111/gove.12690

Lutze, M., Trauzettel, F., Busch-Heizmann, A., Bovenschulte, M.: Potenziale einer Pflege 4.0. Hg. v. Bertelsmann Stiftung. Gütersloh. https://www.bertelsmann-stiftung.de/fileadmin/files/user_u pload/Pflege_4.0_final.pdf (2021). Last accessed 22 Mar 2021

Madiega, T.: Digital sovereignty for Europe. Hg. v. European Parliamentary Research Service (EPRS). European Parliamentary Research Service (EPRS). https://www.europarl.europa.eu/RegData/etudes/BRIE/2020/651992/EPRS_BRI(2020)651992_EN.pdf (2020). Last accessed 22 Jan 2024

Olsen, C.F., Bergland, A., Bye, A., Debesay, J., Langaas, A.G.: Crossing knowledge boundaries: health care providers' perceptions and experiences of what is important to achieve more person-centered patient pathways for older people. BMC Health Serv. Res. **21**(1), 310 (2021). https://doi.org/10.1186/s12913-021-06312-8.

Podsakoff, P.M.; MacKenzie, S.B., Podsakoff, N.P.: Recommendations for creating better concept definitions in the organizational, behavioral, and social sciences. Organ. Res. Methods **19**(2), 159–203 (2016). https://doi.org/10.1177/1094428115624965

Reimers, K., Johnston, R., Klein, S.: The shaping of inter-organisational information systems: main design considerations of an international comparative research project (2004)

Schneider, A., Donnachie, E., Tauscher, M., Gerlach, R., Maier, W., Mielck, A., et al.: Costs of coordinated versus uncoordinated care in Germany: results of a routine data analysis in Bavaria. BMJ Open **6** (6), e011621 (2016). https://doi.org/10.1136/bmjopen-2016-011621

Shapiro, M., Renly, S., Maiorano, A., Young, J., Medina, E., Neinstein, A., Odisho, A.Y.: Digital health at enterprise scale: evaluation framework for selecting patient-facing software in a digital-first health system. JMIR Form. Res. **7**, e43009 (2023). https://doi.org/10.2196/43009

Stubbe, J.: Von digitaler zu soziodigitaler Souveränität. In: Wittpahl, V. (Hg.) Digitale Souveränität. Bürger, Unternehmen, Staat. Springer (iit-Themenband), Heidelberg, S. 43–60 (2017). http://www.springer.com/de/book/9783662557884. Accessed 19 Dec 2024

SVR: Bedarfsgerechte Steuerung der Gesundheitsversorgung. https://www.svr-gesundheit.de/filead min/Gutachten/Gutachten_2018/Gutachten_2018.pdf (2018). Last accessed 17 Jan 2024

Wisniowski, N., Kurscheid, C.: Digitale Patientenpfade: Zukunftsmodell für intersektorale Versorgung. In: kma Klinik Management aktuell 28 (02/03), S. 108–109 (2023). https://doi.org/10.1055/s-0043-1768326

Wittpahl, V. (Hg.): Digitale Souveränität. Bürger, Unternehmen, Staat. Institut für Innovation und Technik in der VDI/VDE Innovation + Technik GmbH (iit). Springer (iit-Themenband), Heidelberg (2017)

Dr. Katharina Dassel is a research associate at the Institute for Innovation and Technology (iit) and scientific consultant in the group 'Demographic and Socio-digital Change'. At the iit, she worked on projects such as 'Nursing Software 4.0' funded by the Bertelsmann Stiftung among others. With a background in Digital Business and Information Systems, Dr. Katharina Dassel received her doctoral degree from the University of Münster where her research focused on information privacy descision-making and data-driven innovation in healthcare. Her work was published in the Journal of Business Research and conferences in the field of Information Systems.

Maxie Lutze, Research Associate at the Institute for Innovation and Technology (iit) and Head of 'Demographic and Socio-digital Change' at the VDI/VDE Innovation + Technik GmbH, specializes in digital healthcare innovations and policy strategies for societal challenges. With a background in computer science and human factors, Ms. Lutze, a key expert, advises federal and state ministries and innovation policy organizations. Notably, she led a study on patient-centric evaluation of assistive technologies funded by the National Association of Statutory Health Insurance and spearheaded a Bertelsmann Stiftung-funded study on innovative technologies' impact on challenges and job satisfaction for long-term care professionals.

Edge Computing for Digital Sovereignty in the Data Economy

Nils Jahnke⊕, Marieke Rohde⊕, and Tom Kraus⊕

Abstract Edge computing is a pivotal technology for value creation in the data economy. Edge computing moves computation and storage closer to the point of data creation, offering manifold advantages for data-driven applications. These benefits include reduced latency in data analytics and the local analysis of confidential or privacy-relevant data. Yet, particularly small and medium enterprises struggle to design and implement edge computing applications. This chapter addresses this issue by providing guidance for enterprises embarking on their own edge computing endeavors. Drawing on an analysis of ten cases of edge computing adoption across diverse industries in Germany, this contribution elaborates on the most prevalent potentials of edge computing for early adopters. As essential characteristics for participation in the data economy it puts extra emphasis on the aspects of digital sovereignty and data sovereignty of the enterprise. Further, current challenges in the implementation of edge computing applications are discussed. Based on the experiences of the analyzed cases, it identifies important areas of action for enterprises initiating their own edge computing endeavors and provides recommendations for action.

Keywords Edge computing · Data economy · Digital sovereignty · Data sovereignty

N. Jahnke (✉)
Fraunhofer ISST, Speicherstraße 6, 44147 Dortmund, Germany
e-mail: nils.jahnke@isst.fraunhofer.de

TU Braunschweig, Rebenring 58, 38092 Braunschweig, Germany

M. Rohde · T. Kraus
Institute for Innovation and Technology (iit), Steinplatz 1, 10623 Berlin, Germany

© The Author(s) 2025
U. Schmuntzsch et al. (eds.), *New Digital Work II*,
https://doi.org/10.1007/978-3-031-69994-8_15

1   Introduction

A thriving data economy is a pillar for future innovation, economic growth, and social prosperity in Europe (European Comission 2019). In the data economy, data is a strategic asset for value generation: Digital processing, analysis, and transformation can turn data into actionable information for operational excellence, strategic decision-making, and new products, services, and business models (Legner et al. 2020; Koutroumpis et al. 2020).

Small and medium-sized enterprises (SMEs) are the backbone of Europe's economy. Yet, only few SMEs have access to all resources necessary for generating value from data on their own, which includes access to relevant data along the value chain as well as human expertise in data analysis and the implementation of data-driven solutions. To take full advantage of the business opportunities provided by the data economy, SMEs typically must collaborate with business partners and data technology providers on developing and deploying data-driven services together.

In some cases, partnerships with cloud service providers are a suitable approach for leveraging the potential of company-internal data, as they offer computation, storage, data services, and consulting on the base of a pay-as-you-go payment model (Demchenko et al. 2018). However, cloud services have both technical and strategic disadvantages. Technically, services running in cloud environments often cannot provide reliable and robust functionality with respect to real-time requirements, especially in the presence of network outages (Linux Foundation 2022). Hence, the technical requirements of domains such as manufacturing, mobility, and energy cannot be fulfilled. Strategically, using cloud services can diminish the autonomy and digital sovereignty of data holders. Firstly, potentially confidential data leaves the data holder's premises when it is uploaded to the cloud leading to risks of disclosing trade secrets or violating legal regulations (e.g., GDPR). Secondly, cloud providers tend to secure their position in the market by using proprietary data formats and interfaces that lead to poor portability and interoperability (vendor lock-in). This prevents data holders from using or sharing their data assets for other purposes and unlocking their full potential.

Edge computing technologies can mitigate both kinds of limitations. Edge computing is generally understood as the provision of computing power and storage close to the boundary (edge) between the digital and physical worlds, with interaction at the boundary occurring through sensors and (possibly) actuators (Linux Foundation 2022; ISO 2020). Moving computation and storage closer to the place of data creation can benefit an enterprise's digital sovereignty in several ways:

Firstly, edge computing can enable or improve data-driven applications that are currently limited by the existing network infrastructure. This concerns, on the one hand, applications in sectors with strict latency requirements such as manufacturing and mobility. In the manufacturing sector, high processing latencies can inhibit the use of smart services, e.g., in applications for adaptive control of machine tools. The existing internet infrastructure merely works on a "best-effort" basis and therefore cannot guarantee sufficiently small processing latencies for such applications

(Ahmed et al. 2017). Similarly, applications in sectors that are prone to network outages and bandwidth instability, such as agriculture, forestry, and mining, can be enabled through the use of edge computing. The resulting increase in both the range of possible data-driven applications and the resilience and performance of such applications can increase an organization's socio-technical digital sovereignty as it is defined by Hartmann (Hartmann 2022): novel data from sensors and other edge systems can increase the *transparency* of digitally assisted business processes; the resilient functioning of edge computing systems in real-time, even in the presence of connectivity hiccups, can increase the *efficiency* of business processes; finally, the mere possibility of implementing edge systems in application areas where cloud computing solutions are not viable constitutes an increase in an organization's options for technology use (*divergence*).

Secondly, as sensitive data processed on edge devices does not leave the data holders premises, edge computing can also benefit an enterprise's *data sovereignty*, which is an important aspect of digital sovereignty: One typical approach in edge computing is to process confidential or privacy-relevant information at the edge while only transferring less confidential processing results or anonymized data to third parties (Varghese et al. 2021). This can be beneficial in industries with strict regulatory requirements, and when implementing solutions that rely on sensitive internal company data. Edge computing, as part of edge-cloud applications, improves digital sovereignty by simultaneously promoting privacy, scalability and performance, which were previously difficult to combine (Kortum et al. 2023).

Apart from digital sovereignty, edge computing also benefits environmental sustainability. Studies show that the use of fossil fuels can be reduced by about 20 percent through the implementation of respective frameworks (European Commission 2023). The main reason for the higher energy efficiency in the referenced publication is the possibility to delete raw data after analysis at the edge is carried out, thus reducing data transfers and network usage. This can subsequently also reduce cloud computing expenses of an enterprise (Marcham 2021). Sustainability benefits of edge computing also result from the enablement of sustainability applications in many application domains, such as the intelligent control of fossil fuel consuming systems (e.g. heating systems) (Kortum et al. 2023).

In summary, edge computing plays a crucial role for developing novel applications in the data economy by supporting the increased demands towards local data processing and thereby complementing centralized cloud computing infrastructures (Marcham 2021).

This chapter is intended as a practical orientation guide to edge computing for decision-makers in organizations, particularly SMEs. It draws on the practical experience of edge computing practitioners from different economical sectors in Germany. The work presented is based on a qualitative study commissioned by the German Ministry of Economic Affairs and Climate Action (BMWK) as part of the supporting measures for the technology program "Edge data economy". It starts with an overview of relevant edge computing concepts (Sect. 2), then illustrates concrete application potentials with respect to digital sovereignty and challenges in implementing edge computing applications (Sect. 3). It concludes with an overview on fields of action when implementing first edge computing applications (Sect. 4).

2 Edge Computing Technologies

This section provides a brief summary of edge computing technologies and is restricted to the most basic concepts, definitions and design options.

Even though the general concept of edge computing is widely used, there is still no established operational definition of edge computing in academia or practice. However, some characteristics of edge computing are frequently used in descriptions: Edge computing is a computational paradigm that moves storage and computation closer to the edge of the network; it has a distributed nature; it must be complemented by cloud computing processes.

Along the continuum from the network edge to the cloud, organizations can leverage different options for deploying edge computing resources, either individually or in combination. These options differ in characteristics such as distance to data creation, computing power and storage, or scalability. A report by leading industry players in Europe (3DS Outscale et al. 2021) distinguishes the design options of edge computing in the edge-to-cloud-continuum as follows (cf. Fig. 1):

- *On-device edge*: Storage and computation take place within the data creating object. Objects supporting such on-device edge computing range from machine tools to cars and mobile phones. Hence, the devices may or may not be mobile. Typically, on-device edge computing only provides limited computation and storage capacity.
- *On-premise edge*: Storage and computation take place locally in a specific facility, such as a factory or office building. Hence, the resources are typically provided for a single entity. On-premise edge computing can scale storage and computation up to a certain degree. On-premise edge computing infrastructure can be physically accessed by the user (as indicated by the orange box), just like on-device edge computing.
- *Near-edge and Far-edge*: Storage and computation take place in centralized computing facilities, located within specific zones or regions close to the user. This enables use cases with high demands for latency, bandwidth, and scalability close to the user. Such infrastructure is typically used by multiple entities and can be integrated into the existing infrastructure of telecommunication service providers.
- *(Public) Cloud*: Storage and computation take place in centralized computing facilities, which can be located anywhere, e.g., overseas. This enables rapid on-demand access to network, computation and storage resources. Core characteristics are elasticity, scalability, and accessibility, as well as robustness due to server redundancy. Resources are shared between multiple entities. In the public cloud model, the cloud infrastructure is open for use by the general public. Natural tasks

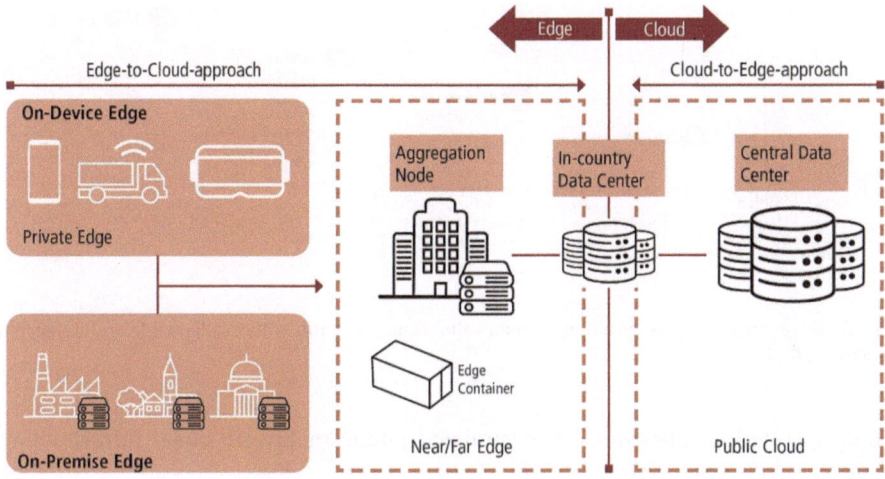

Fig. 1 Edge-to-cloud-continuum, adapted from 3DS Outscale et al. (2021)

carried out in the cloud layer are e.g. the orchestration of computing processes (Fig. 1).

In practice, computation processes at the edge and in the cloud go hand-in-hand (ISO 2020). The use of cloud-resources and edge-resources is orchestrated in line with the application demands. Cloud services offer flexibility and scalability. They are used for storage, expensive computations and orchestration. By contrast, computation processes at the edge are used for sub-tasks that either require real-time capacity or resiliency or involve the processing of confidential data that should not be transferred to a cloud-environment. Processing raw data on edge devices and passing on just the computation results decreases the amount of data transmitted and thus reduces the overall storage requirements of hierarchical edge-cloud-systems.

Several edge computing-related concepts exist. They include fog computing, mist computing and cloudlets, amongst others. *Fog computing* was initially coined by Cisco. It mainly describes a horizontal architecture that distributes computing, storage, control, and networking functions along the continuum from cloud to object (Jain and Mohapatra 2019). In this sense, the "horizontal architecture" refers to the support of a diverse range of use cases and includes cloud, edge and things, amongst others (Chiang et al. 2017). However, practitioners state that the term has recently fallen out of favor due to conceptual ambiguity (Linux Foundation 2020). *Mist computing* describes a computing paradigm, where computation and storage is distributed amongst smart devices at the extreme edge, essentially fostering autonomy and self-organization of systems (Preden et al. 2015). The term *cloudlet* is mainly used in academia. As indicated by the suffix "let" cloudlets refer to small-scale public or private cloud nodes deployed at the infrastructure edge. Cloudlets offer elastically-allocated computation, data storage and network resources and are

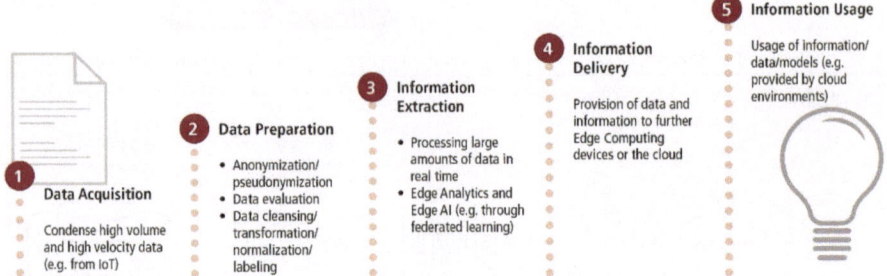

Fig. 2 Edge computing usage along the data value chain, own illustration adapted from Jahnke and Niehoff (2024)

often operated as extensions of centralized public or private cloud infrastructure (Linux Foundation 2020).

Edge computing can also serve as an enabler for other data-driven technologies with a high economic potential, such as artificial intelligence (AI), robotics, and the Internet of Things (IoT). Particularly, in strictly regulated domains such as healthcare, but also for applications requiring customer acceptance the additional digital souvereignity provided by edge computing can be crucial for the technology uptake.

The data economy is characterized by the creation of value from data through the use of digital technologies. For data value creation, the data undergoes five consecutive processing steps that turn the technical asset data, first, into task-relevant information and then, when the information is used for decision-making in an operational context, into value (Jahnke and Niehoff 2024). This chain of steps is called the data value chain (Fig. 2). Edge computing can support all steps of the data value chain:

1. *Data Acquisition* refers to the phase where digital data is recorded and collected. Edge computing can be used during data acquisition to collect high volume and high velocity data, for instance, from sensor networks in industry 4.0 applications or from cameras in robotics or visual monitoring applications. The use of edge computing can provide higher bandwidth between the point of data capture and the point of data collection, reducing disruptions and data loss compared to cloud computing applications.

2. The *Data Preparation* phase refers to bottom-up preparatory data processing steps such as data-cleansing, data transformation, and filtering. Such pre-processing can be performed at the edge to improve efficiency of a data-driven system or to ensure data protection. Edge computing can be used to anonymize or pseudonymize sensitive data locally and before it is sent to the cloud or other stakeholders for further processing. To this end, potentially sensitive data can undergo pre-evaluation to identify its sensitive characteristics.

3. In the *Information Extraction* phase, data from different sources are combined and processed, e.g., using AI or analytics procedures, to extract information that is relevant to achieving a certain goal, such as identifying machine failure to initiate maintenance procedures. Information extraction can be performed at the

edge to ensure real-time system performance. Approaches to extract information while ensuring privacy by means of edge computing comprise e.g., so-called "compute-to-data" approaches (where analysis processes take place at the edge and only task-relevant extracted information is passed on to processes in the cloud) or federated learning (where a machine learning model is trained by collecting results from local entities rather than sending all training data to a central location).

4. The *Information Delivery* phase refers to the processes by which information is made available for use, either by humans or by machines, e.g., by providing programming interfaces, by writing it to a database, by including it in a report, or by visualizing it on a screen. Edge computing can serve as a relay point for information delivery, distributing selected information from cloud services to certain devices and users, and vice-versa.

5. In the *Information Usage* phase, information is used for decision-making in an operational context and achieves its business impact. Edge Computing can support information usage in applications where network connections on site are unreliable or where real-time system performance is of the essence, as it is the case, for instance, in driverless transport systems or autonomously controlled production or farming machinery.

While these general application potentials and principal limitations of edge computing are well-documented, deploying edge computing solutions in concrete application scenarios often demands customization or even research and development efforts. Especially SMEs should therefore carefully weigh the costs and benefits of edge computing solutions. The following Sect. 3 draws upon the experience of current applied research projects from different industrial sectors in Germany to report on the potentials and challenges for the practical implementation of edge computing systems. Section 4 illustrates relevant fields of action for the development of first edge computing applications in the enterprise.

3 Potentials and Challenges from an Early Adopters Perspective

The following section illustrates the application potentials of edge computing applications in different industrial sectors from an early adopters' view, with specific emphasis on digital sovereignty, i.e. socio-digital sovereignty (Hartmann 2022) and data sovereignty (Hellmeier and von Scherenberg 2023). In addition, the challenges for their implementation are reported.

Table 1 Summary of the examined cases of edge computing adoption

Industries	Tasks
– Agriculture and food – Energy – Healthcare – Manufacturing and processing – Smart living – Water management	– Condition monitoring – Energy monitoring – Safety and surveillance – Predictive maintenance – Production planning – System control – Process automation – Carbon footprint modelling – Quality management – Document analysis

3.1 Methodology and Context

The results in this and the following section are based on an online survey conducted in August and September 2023 amongst representatives from enterprises and scientific institutions funded through the technology program "Edge data economy" by the German Federal Ministry for Economic Affairs and Climate Action (BMWK). The technology program aims to promote the development and testing of edge computing in different industries, the transfer of state-of-the-art technology to SMEs, and sovereign data exchange and use.[1] The early adopters of edge computing funded in the program and participating in the survey rated a preselection of potentials and challenges of edge computing that were most relevant to their respective industry and task (see Table 1) on a four-point Likert scale ranging from "no potential" ("no challenge") to "very high potential" ("very high challenge"). Potentials and challenges were grouped into economic, organizational, technological, and legal factors and rated by group. A total of 30 valid responses were received.

Additionally, the participants were asked to name further challenges if they were not included in the preselection provided. For further insights, ten use-cases of edge computing adoption in the technology program were examined in more detail, primarily by analyzing existing documentation. For more methodological details and an in-depth description of each use-case, see Jahnke and Niehoff (2024). Below the opportunities and challenges with the highest average relevance in each of the four groups are presented.

3.2 Potentials of Edge Computing Use

The respondents of the online survey identified *Driving digital transformation* as the main incentive for edge computing adoption. Digital transformation refers to,

---

[1] For more information, see https://www.digitale-technologien.de/DT/Navigation/EN/Programme Projekte/AktuelleTechnologieprogramme/Edge_Datenwirtschaft/edge_dw.html.

possibly disruptive, alterations of economy or society by digital technologies often implying a change in interaction between customers and service providers. Edge computing may foster digital transformation in several ways. On the one hand, edge computing supports the *optimization of existing processes.* On the other hand, edge computing can be leveraged for the *development and marketing of novel products and services.* According to the survey results, both are also key drivers for edge computing adoption.

As an example of the latter, edge computing can be used to implement new service offerings in manufacturing, including condition monitoring and predictive maintenance, by enabling the analysis of large amounts of data in near real time. In this context, also new types of business and billing models based on the use of edge computing are possible. The most prominent example is the "everything-as-a-service" category: By charging only for the actual use of the asset, customers can reduce up-front capital expenditures for capital-intensive assets such as machine tools.

One example of process optimization can be given from production. Participating in bidding processes currently ties up high amount of workforce with often uncertain outcome. Enterprises are therefore turning to AI-based document analysis software to automate the creation of application documents. However, this requires the transfer of confidential documents to cloud servers, which many enterprises are opposed to. Edge computing makes it possible to analyze documents on the organization's premises, using federated machine learning procedures that pass an AI model around from user to user for training instead of uploading each user's confidential internal documents to the cloud for training. As a result, confidential documents are analyzed locally and only the results are shared in the cloud, streamlining the application process for public funding for organizations that are reluctant to expose proprietary knowledge to cloud services.

The possibility to *improve ecological sustainability* by leveraging edge computing is another major incentive for early adopters of edge computing. An example from the domain of agriculture and food retail is the target-oriented analysis of freshness information for vegetables and fruits along the value chain for a more sustainable usage of food products. To this end, freshness data is collected by different parties along the agrifood supply-chain, such as farmers, packagers and wholesalers and shared with retailers. Visual and hyperspectral camera recordings of fresh goods are pre-processed on the edge devices for sensing to extract freshness information. The results are supplemented with freshness-related metadata, such as the storage temperature and the time of harvest, and then shared with retailers, who evaluate this information using AI lifetime models of fruits and vegetables to predict their shelf life more accurately and thus minimize food waste. Similarly, in manufacturing, a product's carbon footprint can be determined by analyzing the production processes at the edge, integrating the results with production planning data from central systems and evaluating them in the cloud using AI models to derive sustainability information (e.g., energy consumption). The results can be used to optimize production processes and to develop new sustainability-based business models as well as to document production sustainability, e.g., for regulatory compliance.

According to the survey results, the opportunity to *improve customer satisfaction* is the most prevalent organizational motivation for edge computing adoption. Typical user requirements are speed, freedom from interruption, privacy protection, and personalization. A practical example of how edge computing can foster customer satisfaction is a use-case from the smart living domain: Commonly, existing heating systems are retrofitted using "smart thermostats" to improve heating efficiency. However, these thermostats do not take the residents' individual preferences into consideration. Therefore, the room temperature is often perceived as uncomfortable. Furthermore, user data is often transferred to cloud services outside Europe when using smart heating solutions, jeopardizing the end-users' sovereignty over their personal data. Edge computing technology deployed on the apartment level to control local heating activity, based on sensor information, external information (e.g., weather conditions), and self-learning AI models can lead to a heating outcome that is perceived as more comfortable by its users without a loss of data sovereignty. By coordinating the local heating systems in each unit with a building's central heating control, overall heating efficiency can be optimized.

Improvements in resilience of the technical infrastructure were rated as the biggest technical incentive for edge computing usage from the perspective of early adopters. Infrastructure resilience refers to the continuity of services in the face of external disruptions, such as network infrastructure failure. Resilience is especially important in applications that are critical for the safety and well-being of people or the natural environment. Edge computing can improve infrastructure resilience by ensuring continuous functioning even in the case of network outages. A use-case in water management illustrates this: In the event of unexpected floodings, e.g., from torrential rains, the conventional reactive control systems for wastewater management frequently discharge wastewater into rivers or lakes and thus damage the natural ecosystem. Current alternatives, i.e., intelligent control centers that use AI models to decide when to discharge wastewater into the environment are typically deployed in the cloud and manage the water system centrally. However, as extreme weather events often go hand in hand with interruptions in network services, the cloud control center may lose the connection to the components of the wastewater system such as retention basins, pumping stations, and sewage treatment plants and centralized control becomes impossible. Establishing an intermediary edge computing layer at a range of important sites for wastewater management makes it possible to evaluate AI models close to the place of data collection and intervention. If the connectivity is sufficient, the system is managed centrally, while in emergency mode, different components are controlled locally by processes computed on the edge devices.

From a legal perspective, the early adopters see the *early consideration of current legislative activities at European level* and the *compliance with domain-specific data protection and data security requirements* as important motivating factors for edge computing adoption. Several regulatory activities in the data economy are currently implemented on the European level, including the EU Data Act. The Data Act intends to give companies the right to use and share any data they generate when using connected devices and the services related to these devices for their own purposes. Amongst other things, this concerns data collected by machinery and IT systems.

Edge computing applications currently developed for the manufacturing domain already prepare for the future opportunity afforded by this regulatory activity. For instance, machine users can decide to give the manufacturers of machine components access to the data created during machine operation, which are currently recorded, stored, and used exclusively by the machine manufacturers. If component manufacturers get access to this data, they can then develop their own data-based services, such as a digital setup, condition monitoring, or predictive maintenance for their components that are built into the machine.

In Germany, domains with particularly high data protection and data security requirements include the energy sector, healthcare, and water management. Due to the legal requirements for respective critical infrastructure, e.g., power plants, hospitals or water treatment plants, cloud services cannot be used for many applications or only with extensive efforts. By processing data on edge devices, services can be partially implemented locally, which reduces the amount of data transferred to the cloud as well as vulnerabilities. Edge computing thus enables enterprises responsible for critical infrastructure to implement novel data and AI based applications, while still making it possible to comply with the strict legal requirements.

These concrete examples of edge computing applications that the surveyed early adopters are currently implementing in their respective sectors illustrate the motivations, design options and potentials of edge computing technologies. They also illustrate the possible benefits for the end-using enterprise's digital sovereignty, be it in terms of keeping sovereignty over the enterprise's internal data assets (Linux Foundation 2020) or their socio-digital sovereignty (Hartmann 2022). The added value can result from an increase in digital innovation possibilities that edge computing provide, from the additional data that edge computing applications generate, and from technically more robust and more efficient solutions through edge computing. The following section will outline the most important challenges that the surveyed researchers are facing in the implementation of their applications.

3.3 Challenges for the Implementation of Edge Computing Applications

According to the survey results, on average, edge computing early adopters see technical and organizational factors as the highest barriers to implementing edge computing applications. Specifically, the early adopters rated the *efficient management and orchestration of edge devices* as the biggest challenge. Device management includes tasks such as configuration, registration, updates, monitoring, deactivation, and removal of edge devices within the overall system. Orchestration refers to the coordination of the hardware and software elements of edge computing applications. Both tasks are notably challenging due to the distributed nature and heterogeneity of components in edge computing.

Missing standards for interoperability and portability are a further significant technical challenge faced by early adopters of edge computing applications. This includes standards being available, but not (yet) widely used. Consequently, integration of different hardware and software components as well as data from different sources requires a lot of effort, which obstructs the formation of edge computing innovation ecosystems. Furthermore, components of edge computing applications cannot be easily switched or replaced, which limits the autonomy of organizations and hence limits digital sovereignty.

To enable the *enforcement of data sovereignty requirements*, several initiatives such as Gaia-X and the International Data Spaces are active in developing organizational and technical mechanisms, which are mainly implemented as open-source software. However, while promising, the early adopters rate the current maturity of these mechanisms as unsatisfactory for their field of application. Hence, they either refrain from using them or only use those parts that are already sufficiently matured.

One big organizational challenge is the recruitment of people with the necessary skills to design and implement edge computing applications. This circumstance is linked to the general *shortage of specialists* affecting the entire IT sector. Limited availability of workforce leads to delays or discontinuation of edge computing projects.

Another major obstacle when implementing edge computing applications are *concerns of different stakeholders within an enterprise regarding the possibility to integrate the edge computing solution into existing processes and workflows*. For a successful implementation of edge computing applications, technical, organizational, and cultural integration into the organization is needed. Data gathered in edge computing applications needs to be integrated into existing enterprise information systems, to make a future reuse possible. This is challenging due to proprietary formats and interfaces associated with existing sub-systems. In addition, edge computing applications often target only specific process subtasks. Complete replacement of existing processes is rarely feasible, resulting in the modification of roles and activities that thus need to be redesigned. This can lead to acceptance problems within the workforce, as employees may feel that their role has been diminished or because they must perform new and additional tasks.

4   Recommendations for Action and Outlook

Based on the most prevalent challenges faced by edge computing early adopters, this section outlines important fields of action for enterprises starting their own edge computing endeavor, to leverage the economic potential of their data and to maintain and expand their enterprise's digital sovereignty.

One aspect to be considered by enterprises in the early stages of edge computing adoption is to *establish the needed edge computing-related competencies and skills*. Developing edge computing applications requires skills in four areas. These areas are (a) hardware, (b) software, (c) communications, and (d) services (Gole et al. 2023).

Additional competencies may be needed in complementary fields such as AI or IoT. As even existing big players on the market mainly focus on one or two of these areas, early adopters are usually not capable of building edge computing solutions on their own. Hence, it is necessary to partner with other organizations to fill the identified skill gaps. Types of collaboration may range from strategic partnerships (including knowledge transfer) to a more loosely coupled fashion by leveraging existing as-a-service offerings. However, for the latter, certain skills must already be available within the organization. The necessary competencies can be developed by hiring skilled workers or by in-house training. Since both hiring and training are time-consuming processes, these tasks should be completed prior to the project start date. Else, project delays may occur.

Ideating or finding first edge computing use cases can be challenging for early adopters of edge computing. It is often unclear where to start and which use cases can deliver potentially high business value for the enterprise. Based on the experiences of the analyzed cases of edge computing adoption, indications of entry points for novel edge computing use cases can be derived.

Edge computing is a pertinent solution for application areas characterized by strict real time requirements and substantial data volumes. It enables use cases that were previously impractical from both technical and financial standpoints. Therefore, areas of the enterprise where real time requirements and extensive data volumes make conventional IT (cloud) infrastructures unsuitable, should be considered as prime candidates for the application of edge computing. One way to identify these areas is to analyze existing value creation processes and the data generated to support them. To this end, the existing value chains and their process steps must be mapped to the information needs. The next step is to identify the data available in each process step and assess its maturity. By matching information needs and available data, it is possible to identify areas where the requirements for low latency or the processing of large data volumes cannot be met with the current IT infrastructure.

While financial performance is often the primary driver for the adoption of new technologies, in the case of edge computing, sustainability improvements should also be taken into consideration as they are becoming increasingly important for attracting customers and investors. Sustainability factors can be considered while analyzing the status-quo and during the evaluation of different design options. To combine indicators of financial performance, sustainability and other metrics, multi-criteria decision-making methods are suitable approaches. Sustainability factors can be derived, for instance, from the United Nations sustainable development goals. Concrete sustainability metrics then need to be deduced based on the specific context. Examples of metrics include total energy consumption or product service life. During the evaluation it is important to take a holistic (systemic) approach, e.g., by considering possible external factors. Analyzing the entire life cycle of a product or service is often a useful starting point.

In contrast to cloud computing services, edge computing solutions (hardware and software) are usually domain specific. While the vendors of these solutions have deep expertise in their own domain, they are often unable to scale their solutions to other domains due to limited resources or market expertise. Thus, existing edge computing applications in adjacent domains should be explored as part of the ideation phase. For example, hardware designed for the specific demands of environment A can also be utilized in environment B if their requirements are similar. Organizations can conduct environmental analysis methods, visit trade fairs and conferences, and exchange with domain experts to identify solutions from related domains. Once relevant solutions have been identified, use case mapping techniques can be used to describe existing applications, outline necessary changes, and design a solution adapted to the domain.

Evaluating and fostering legal compliance of an edge computing application is another important field of action. Several legal questions arise during edge computing projects. As edge computing applications usually process data collected by IoT devices, the edge application under development may be subject to the EU Data Act. Hence, the application may need to provide data access to the application user or other third parties. Other legal issues include securing the freedom to operate, ensuring compliance with applicable data protection regulations, and clarifying liability issues in the use of AI. Failure to comply with these requirements may lead to initially unforeseen costs, project delays or, in the most serious cases, the termination of the innovation project. Thus, legal compliance checks should be included in the early stages of an edge computing project to assess the project's feasibility. However, legal experts are a costly resource. To enable targeted advice and save costs, the edge computing application should already have a certain level of concretization before consultation.

The *support of existing standards and open-source software* is an important factor for interoperability and portability of hardware and data in edge computing applications. Additionally, it can foster the efficient management and orchestration of edge computing devices. All in all, standardization and open-source software allow for extensible edge computing ecosystems. To support the creation of these ecosystems, the market should be examined for suitable standards and open-source software that can be leveraged within the edge computing application project. Popular open-source foundations such as the Eclipse Foundation and the Linux Foundation have recently launched communities for edge computing software development. Additional standards and software are being developed under the umbrella of domain-specific associations such as the OPC Foundation in industrial manufacturing. Participation in standardization initiatives can be valuable for SMEs as it allows them to influence standards and software in a favorable direction and support interoperability with standards already in use.

As with any IT project, *driving integration and adoption of edge computing applications* is critical. While integration merely refers to the technical integration into the existing data infrastructure, adoption often requires cultural and organizational openness. Both can be fostered by involving relevant stakeholders early in the edge computing application project. Early involvement of future users fosters user-centered development, increases acceptance of the application, and creates transparency about possible issues. While potential internal users can be involved throughout the development process, dedicated mechanisms such as co-creation workshops or customer advisory boards are appropriate for external users. Involving senior management helps align the edge computing application with business strategy and emphasizes the importance of the innovation project to the business, resulting in greater stakeholder engagement. Involving IT infrastructure operators ensures that existing development standards (e.g., for data formats or the use of open-source software) are adhered to and that the application is interoperable with the existing enterprise data infrastructure.

To conclude, while the potentials of edge computing for improving digital sovereignty and business success are substantial, implementing edge computing solutions comes with its own set of challenges and corresponding lines of action (cf. Table 2) that early adopters should be aware of. Building edge computing applications is a collaborative endeavor. Such an innovation effort can only be successful when all stakeholders are on board and see their minimum requirements met. Within an enterprise, skills development is essential to digitally empower both individual staff and the organization, and to promote sustainable and robust functioning of edge computing solutions. This will also increase digital sovereignty by decreasing the dependency on external parties, especially cloud and telecommunication service providers that currently hold the biggest share of the edge computing market. Across enterprises and industries, it will be essential to collaborate on standards and open-source software to achieve convergence of edge computing applications within and across domains. This will pave the way for future edge computing applications that involve the sharing of data across organizations for the mutual benefit of all participating organizations, while maintaining digital and data sovereignty.

Table 2 Fields of action and recommendations

Field of action	Recommendations for action
Edge computing-related competencies and skills	– Partner with complementary enterprises – Conduct hiring and in-house training with anticipation
Ideation of first edge computing use cases	– Identify applications with big data volumes and demands for low latency – Consider sustainability and environmental implications – Examine the solutions of adjacent domains
Legal compliance	– Integrate legal expertise in early project stages
Standardization and open-source software	– Scan market for available "products" – Participate in standardization and open-source initiatives
Integration and adoption	– Involve relevant stakeholders at early stages

Acknowledgements Parts of this contribution have been published first in the study "Datenwirtschaft und Edge Computing - Potentiale, Herausforderungen und Handlungsempfehlungen für Unternehmen" available in German language (Jahnke and Niehoff 2024). The study was commissioned by the Federal Ministry for Economic Affairs and Climate Action as part of the accompanying research for the technology program "Edge data economy".

References

Ahmed, E., Ahmed, A., Yaqoob, I., et al.: Bringing computation closer toward the user network: is edge computing the solution? IEEE Commun. Mag. **55**, 138–144 (2017). https://doi.org/10.1109/MCOM.2017.1700120

Chiang, M., Ha, S., Risso, F., et al.: Clarifying fog computing and networking: 10 questions and answers. IEEE Commun. Mag. **55**, 18–20 (2017). https://doi.org/10.1109/MCOM.2017.7901470

Demchenko, Y., Los, W., de Laat, C.: Data as economic goods: definitions, properties, challenges, enabling technologies for future data markets. ITU J.: ICT Discov. **2** (2018)

European Commission Building a Data Economy—Brochure (2019). https://digital-strategy.ec.europa.eu/en/library/building-data-economy-brochure. Accessed 24 Nov 2023

European Commission Study on the Economic Potential of Far Edge Computing in the Future Smart Internet of Things: Final Study Report (2023). https://op.europa.eu/en/publication-detail/-/publication/ff35c457-8f3b-11ee-8aa6-01aa75ed71a1/language-en. Accessed 08 Aug 2023

Gole, J., Zborowska, E., Rotaru, A.: Cloud-Edge-IoT Demand Landscape (2023). https://zenodo.org/records/7821330. Accessed 15 Aug 2023

Hartmann, E.A.: Digitale Souveränität: Soziotechnische Bewertung und Gestaltung von Anwendungen algorithmischer Systeme. In: Hartmann, E.A. (ed.) Digitalisierung souverän gestalten II: Handlungsspielräume in digitalen Wertschöpfungsnetzwerken, 1. Aufl. 2022, pp. 1–13. Springer Berlin Heidelberg, Berlin, Heidelberg (2022)

Hellmeier, M., von Scherenberg, F.: A delimitation of data sovereignty from digital and technological sovereignty. In: Thirty-First European Conference on Information Systems (ECIS 2023) (2023)

ISO ISO/IEC TR 23188:2020: Information Technology—Cloud Computing—Edge Computing Landscape (2020). Accessed 05 Apr 2023

Jahnke, N., Niehoff, N.: Datenwirtschaft und Edge Computing: Potentiale, Herausforderungen und Handlungsempfehlungen für Unternehmen (2024). https://www.digitale-technologien.de/DT/Redaktion/DE/Downloads/Publikation/EDGE-Datenwirtschaft/2023_12_19_Kurzstudie/

Jain, K., Mohapatra, S.: Taxonomy of edge computing: challenges, opportunities, and data reduction methods. In: Al-Turjman, F. (ed.) Edge Computing: From Hype to Reality, pp. 51–69. Springer International Publishing AG, Cham (2019)

Kortum, H., Hagen, S., Eleks, M., et al.: SECAI—sustainable heating through edge-cloud-based AI systems. HMD **60**, 850–871 (2023). https://doi.org/10.1365/s40702-023-00988-8

Koutroumpis, P., Leiponen, A., Thomas, L.D.W.: Markets for data. Ind. Corp. Chang. **29**, 645–660 (2020). https://doi.org/10.1093/icc/dtaa002

Legner, C., Pentek, T., Otto, B.: Accumulating design knowledge with reference models: insights from 12 years' research into data management. JAIS **21**, 735–770 (2020). https://doi.org/10.17705/1jais.00618

Linux Foundation Open Glossary of Edge Computing [v2.1.0] (2020). https://stateoftheedge.com/project/glossary/. Accessed 16 Aug 2023

Linux Foundation State of the Edge 2022 (2022). https://stateoftheedge.com/reports/state-of-the-edge-report-2022/. Accessed 05 Apr 2023

Marcham, A.: Understanding Infrastructure Edge Computing: Concepts, Technologies and Considerations. Wiley, Hoboken, NJ (2021)

Preden, J.S., Tammemae, K., Jantsch, A., et al.: The benefits of self-awareness and attention in fog and mist computing. Computer **48**, 37–45 (2015). https://doi.org/10.1109/MC.2015.207

Varghese, B., de Lara, E., Ding, A., et al.: Revisiting the Arguments for Edge Computing Research (2021)

3DS Outscale, Airbus, Amadeus et al.: European Industrial Technology Roadmap for the Next Generation Cloud-Edge Offering (2021). https://ec.europa.eu/newsroom/repository/document/2021-18/European_CloudEdge_Technology_Investment_Roadmap_for_publication_pMdz85DSw6nqPppq8hE9S9RbB8_76223.pdf. Accessed 04 Apr 2023

Nils Jahnke is a research associate at the Fraunhofer Institute for Software and Systems Engineering ISST in Dortmund, Germany. At Fraunhofer ISST, he is co-leader of the competence field Strategic Data Management. Currently, he works within the accompanying research for the technology programme 'Edge Datenwirtschaft' ('Edge Data Economy') funded by the German Federal Ministry for Economic Affairs and Climate Action (BMWK).

Dr. Marieke Rohde After a first degree in Cognitive Science and a second degree in Evolutionary and Adaptive Systems, Marieke Rohde did a Ph.D. in Artificial Intelligence and Robotics in 2008. She then worked for eight years as an academic researcher to study how humans process novel sensory stimuli for perception and motor control. During her years as an academic, she lived in several German cities as well as in France, the UK and Japan and authored more than 20 highly cited international publications. Afterwards, she co-founded the Berlin-based Affective Signals GmbH, where she developed an online negotiation-training based on intelligent image and sound analysis. In 2018, she took up her work as consultant for Artificial Intelligence and Robotics at the Institute for Innovation and Technology (iit) within VDI/VDE Innovation + Technik GmbH. Her research for German federal ministries helps prepare, accompany and evaluate innovation-political measures. She is especially interested in Big Data, Machine Learning and human–robot-interaction. As part of her job, she also develops data analysis and visualization tools for her company.

Dr. Tom Kraus studied mathematics and computer science with a focus on scientific computing at the Ruprecht-Karls University of Heidelberg and received his Ph.D. in bioscience engineering from the University of Leuven (Belgium) with a thesis on model predictive control and optimization-based state and parameter estimation. For three years, he was responsible for the management of the Scientific Computing programme from Heidelberg University's Institutional Strategy which was funded through the Excellence Initiative, and worked as a research associate at the Heidelberg Interdisciplinary Center for Scientific Computing. Since 2017, Tom Kraus has been working as a consultant at the Institute for Innovation and Technology (iit).

Preserving Digital Sovereignty
in Data-Driven Manufacturing Networks

**Anne Mareike Schlinkert, Leonhard Kunczik, Orlando Hohmeier,
and Michael Kuehne-Schlinkert**

Abstract Leveraging data in collaboration with other organizations requires
resolving a paradox: Organizations of Manufacturing Networks must find a way
to share data while protecting their own intellectual property. Vertical Federated
Learning (VFL) allows collaborative machine learning without compromising data
privacy, as it keeps data local. It facilitates effective, secure data-driven enhancements
across organizations and country borders, enabling more efficient, more sustainable,
and more customer-focused production.

Keywords Data sovereignty · Vertical federated learning · Split learning ·
Predictive quality

1 The Sovereign Data Dilemma

In the age of artificial intelligence, access to data is becoming a key factor
for success in Manufacturing Networks. In these complex ecosystems, no single
entity possesses complete knowledge or information. Manufacturing Networks unite
diverse companies across geographies to collectively enhance value delivery to
shared consumers. The synergies within these networks are instrumental in driving
efficiency, adaptability, and ultimately the competitive edge.

Typically, one entity in the network provides a product or service to the next
entity, and so forth, until the final product reaches the end consumer. This sequence

A. M. Schlinkert (✉) · L. Kunczik · O. Hohmeier · M. Kuehne-Schlinkert
Katulu GmbH, An der Alster 6, 20099 Hamburg, Germany
e-mail: anne@katulu.io; hello@katulu.io

L. Kunczik
e-mail: hello@katulu.io

O. Hohmeier
e-mail: hello@katulu.io

M. Kuehne-Schlinkert
e-mail: hello@katulu.io

U. Schmuntzsch et al. (eds.), *New Digital Work II*,
https://doi.org/10.1007/978-3-031-69994-8_16

of interactions, while organized, often misses opportunities for cross-collaboration or more holistic optimizations.

In a world increasingly driven by data, companies within networks usually operate in data silos, hoarding their data as a proprietary asset. This results in a lack of transparency across the network, inhibiting the potential for collaborative problem-solving and innovation. A classic example is that a supplier's outgoing quality control is usually matched with another round of incoming quality control upon receipt by the customer. This duplication is a wasteful use of resources from a network perspective, as the additional cost is borne by everyone.

Data collaboration is essential in these networks for creating resilient supply chains, maintaining consistent quality, promoting sustainability, and improving resource efficiency. For example, integrating data across supply chains enables a comprehensive understanding of a product's quality throughout its lifecycle, from raw material sourcing and interactions with suppliers to the manufacturing process and logistics. Such data is instrumental in ensuring each production step meets the necessary quality level to fulfill final product quality expectations. The collaborative use and data analysis lead to more informed decisions and optimal resource utilization along the value chain.

The next logical step in enhancing these synergies involves utilizing machine learning to optimize processes through data shared among participants in a Manufacturing Network.

To date, access to data within value networks remains restricted. To enable true data-driven collaboration across organizations, a new approach is needed to resolve a particular dilemma in sharing data with others.

The Sovereign Data Dilemma refers to the challenge of sharing and collaborating with data across organizations in a way that maximizes the collective benefit of the entire network, while concurrently addressing pressing concerns related to data security, privacy, ownership, and competitive advantage.

On the **regulatory front**, there are significant boundaries regarding data sharing due to strict export control regulations such as the US Export Administration Regulations (EAR) and EU Regulation (EC) No 428/2009 (https://www.bis.doc.gov/index.php/regulations/export-administration-regulations-ear; https://www.eumonitor.eu/9353000/1/j4nvk6yhcbpeywk_j9vvik7m1c3gyxp/vitgbgiqqry9#:~:text=It%20sets%20out%20a%20uniform,the%20spread%20of%20nuclear%20weapons) that apply to data from the production of dual-use items, which include products or technologies like high-precision sensors or advanced materials that have both civilian and military applications. Additionally, sharing operational data could inadvertently reveal financially sensitive information, posing a risk for publicly traded companies. For instance, deducing production capacity utilization from timestamps might breach insider trading laws. Moreover, stringent data protection laws, such as the General Data Protection Regulation (GDPR), add another layer of complexity in terms of compliance.

Strategically, the reluctance to share proprietary data within networks is a major barrier. Each organization holds unique processes, innovations, and trade secrets crucial to maintaining its own competitive advantage. Protecting this intellectual property necessitates cautious data-sharing methods to prevent the unintentional exposure of sensitive information.

The **strategic implications** of data sharing are further intensified by geopolitical dynamics, potentially transforming it into a source of business vulnerability. Today, not only does the physical manufacturing footprint need to be carefully evaluated, but also corporate data management strategies must mitigate geopolitical risks. Thus, the act of sharing data is not just a matter of commercial exchange; it becomes a strategic decision with far-reaching implications that go beyond the immediate business interests.

Complicating this issue is the matter of **data ownership**. Sharing data among multiple stakeholders introduces intricate legal and operational challenges concerning who has the rights to control, modify, or delete the data. Without effective data governance frameworks, there is a risk of data being used for purposes that diverge from its original intent, leading to significant commercial and ethical dilemmas.

Cybersecurity is another critical aspect of this challenge. Centralizing sensitive data in a single repository can make it an attractive target for cyberattacks, increasing the risk of significant financial and reputational damage. The necessity for robust cybersecurity measures in centralized data systems is undeniable, yet effectively implementing these measures is increasingly challenging in the face of growing cyber threats.

In light of these considerations, addressing the Sovereign Data Dilemma calls for a forward-thinking strategy that balances the utility of shared data with the imperatives of maintaining data sovereignty and ensuring robust security. Federated Learning (FL) offers a promising approach in this context. More specifically, Vertical Federated Learning (VFL) allows diverse organizations within Manufacturing Networks to collaboratively develop advanced Machine Learning models without the need to exchange raw data. This method not only safeguards sensitive information but also opens up possibilities for groundbreaking innovations in industrial processes. It might be the missing piece to not just optimize but revolutionize industrial processes and establish a new paradigm of efficiency and innovation.

2 The Right Tool for Data Sovereign Collaboration: Federated Learning

In today's business environment, companies commonly share data through direct transfer of raw datasets. These transactions centralize all data in one location, where it is used for machine learning.

The introduction of Federated Learning in 2016 marked a significant departure from these traditional data-sharing practices. Defined by McMahan and Ramage (2017) as a machine learning approach that trains an algorithm across multiple decentralized devices or servers without exchanging data samples, Federated Learning enables multiple devices to "collaboratively learn a shared prediction model while keeping all the training data on device, decoupling the ability to do machine learning from the need to store the data in the cloud" (McMahan and Ramage 2017). At its essence, Federated Learning proposes a reversal of the conventional model: it brings the computational model to the data, rather than centralizing the data.

Vertical Federated Learning (VFL) is a key technology for introducing data-based improvements in Manufacturing Networks. According to Li et al. (2023) it is designed for scenarios where multiple data owners possess "a different subset of features about largely overlapping sets of data sample(s), to jointly train a useful global model." Here, a 'feature' refers to a specific piece of information or attribute related to a data sample, like temperature settings or the speed of an assembly line. Different parties may have distinct sets of such features, depending on their role in the process, like a supplier having material quality data while a manufacturer might have production efficiency metrics. In Manufacturing Networks, such vertically partitioned data usually involves materials or components that undergo various treatment processes along a value chain. In a production process, multiple factories hold distinct information about the same material, at different steps like sourcing, pre-processing, assembly, or testing.

In practice, multiple collaborating data owners make use of split learning methodology, a form of collaborative machine learning where a neural network model is split across different organizations or factories. Unlike traditional models that rely on a centralized dataset, the model is designed to learn from datasets distributed across various participants. Each participant contributes a part of the model, effectively creating a situation where the raw data remains 'vertically partitioned'. One of the key advantages of split learning is its ability to maintain data privacy. Since the raw data does not leave its data owner and only outputs from the specific part of the mode (processed data) are shared, it offers a privacy-preserving approach to collaborative learning.

In the initial phase, a machine learning model is developed tailored to solve a specific business challenge or to enable precise predictions. This is accomplished with an independent Trustee like Katulu that provides the infrastructure to orchestrate this learning process across multiple parties and facilitates cross-organizational predictions. The goal of VFL is to train a global deep neural network that can solve the specific problem while ensuring the data privacy of each participant, also known as 'agent', in the training process.

To achieve this, the global model is split into sub-networks and each participant holds only her part of the global deep neural network. The layers at which the global model was split, and that build the connections between the participants, are referred to as the 'cut layer' (Gupta and Raskar 2018). It allows the outputs from one model segment to be transformed into inputs of another's, and vice versa.

Each of the sub-networks is customized to the specific data of a participant. This segmentation means that each sub-network is trained on a distinct subset of features from the participant's local data and the input from its preceding sub-network in the global model. This ensures that sensitive information does not leave its original location.

To aggregate the local learning, each participant sends their model updates (not the raw data) to the independent Trustee. The Trustee combines these updates to improve the overall model. This aggregated model is then sent back to the participants for further training.

This way, multiple parties collaboratively train a machine learning model while allowing every participant to retain control of their data (Fig. 1).

In this approach, data utilization extends across the entire value chain, with each organization responsible for training a segment of a model relevant to its specific portion of the chain. This collaborative effort enables all parties to gain from the collective insights, enhancing the value of the final product and thereby the Manufacturing Network as a whole.

Simultaneously, each organization maintains control over its intellectual property, adhering to contemporary standards of data protection and privacy. This method is key to Digital Sovereignty, ensuring that each organization's data autonomy is respected and preserved.

For multiple parties along value chains to work together, VFL provides a multifaceted solution to tackle regulatory, cybersecurity, and strategic business challenges:

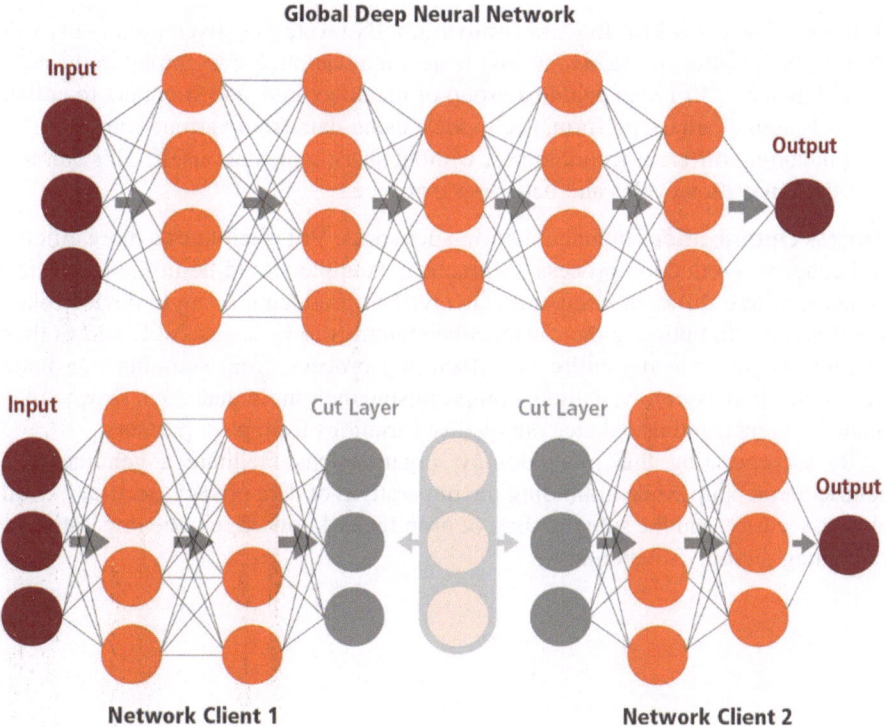

Fig. 1 Visualization of the VFL process between two different participants with the cut layer in between (own illustration)

Regulatory Compliance and Cybersecurity: VFL maintains data localization, drastically cutting down the risks of data breaches and ensuring compliance with strict data protection laws like the GDPR. For example, in a scenario where a European car manufacturer collaborates with an Asian parts supplier, VFL allows them to cooperate without transferring sensitive data across country borders, thus avoiding legal issues. Additionally, by keeping data localized, VFL significantly mitigates the risks linked to centralized storage. For instance, a semiconductor producer using VFL can reduce the likelihood of industrial espionage, as critical data does not leave their own secure network.

Strategic Data Protection and Operational Efficiency: VFL lets companies collaborate in model development without revealing their proprietary information. For instance, a pharmaceutical company could use VFL to enhance drug development processes by learning from data across multiple research labs while keeping their own research data confidential. This ensures that sensitive business information remains within the company, safeguarding its competitive edge. Also, VFL reduces the need for costly data centralization and transfer, thus lowering operational expenses and boosting efficiency. A global manufacturing consortium, for instance, could use VFL to optimize supply chain management without the high costs associated with traditional data-sharing methods.

Enhanced Decision-Making and Innovation: By leveraging diverse data sets, VFL leads to better-informed decisions and fosters innovation. For example, in the automotive industry, VFL can enable a group of manufacturers and suppliers to collaboratively refine battery performance models using data from various sources in the field including different technologies, without sharing the data itself. This can lead to faster innovation cycles and better performance.

Process Optimization: In manufacturing networks, VFL facilitates a more efficient and cohesive production process. A practical example would be in the electronics industry, where different companies involved in producing a single device, like a smartphone, can optimize their respective manufacturing stages. VFL allows them to gain insights into the entire manufacturing process, from sourcing raw materials to the final assembly, without compromising their individual data. Downstream manufacturers can thus balance out quality variations from prior partners.

By incorporating this methodology, organizations facilitate a genuine data-centric partnership while upholding the imperatives of data privacy, accuracy, availability, security, and the safeguarding of proprietary knowledge—the imperative for remaining digitally sovereign.

3 Predictive Quality Control Along the Automotive Painting Process with Vertical Federated Learning

Quality is of paramount importance in manufacturing. Instead of controlling quality at the end of a process, as it is customarily done today, Predictive Quality identifies relevant patterns and correlations that may lead to quality defects while an item is being manufactured. According to Oden Technologies, machine learning is leveraged to "analyze historical and real-time production data to predict the quality of in-process and finished products" (Oden Technologies n.d.). Thus, whether an item will meet a quality range is predicted at the earliest stage possible, allowing mitigation or removal of the item in the process before additional resources are used.

Being able to measure and dynamically adjust different parameters in the full production process is key to enhancing value.

Imagine a painting process of car parts. In its purest form, this process involves two parties, a paint manufacturer providing the paint, and a factory in which the car parts are painted. Usually, these two steps, paint production, and application, are distinct steps within the car painting process, each having separate incoming and outgoing quality control. Although the quality of the final car part depends on the quality of the paint, there is no information exchange between the two parties.

A simple solution would be to share information about the characteristics of each part that is delivered to the next party in the value chain. However, this includes sharing sensitive information that could leak IP of the underlying manufacturing process. Furthermore, each manufacturer of a pre-product could improve its production process by receiving feedback about the end quality of the final product. Thus, in an ideal situation, we would have a two-way information exchange to improve the manufacturing process and the final product.

Using VFL, this vision can be achieved while ensuring that private data never leaves the authority of the data owner. This is depicted below (Fig. 2).

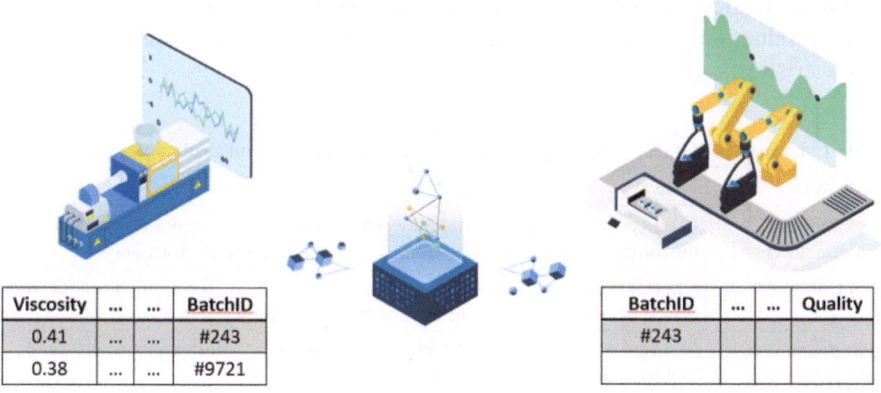

Fig. 2 Visualization of the VFL process between paint production and application (own illustration)

In the example above, we zoomed into a small subprocess within the production and assembly of car parts. However, we can extend the example to the whole Manufacturing Network, which possesses different data related to the same painting process. These are listed in Table 1.

VFL can be used to understand the intricate balance between paint composition, the multi-layered painting process (with paint from multiple suppliers), and equipment operational parameters (from differing equipment manufacturers) to ensure the best possible adherence, finish, and longevity of the paint.

With VFL, each entity can collaboratively train a machine learning model to predict and enhance the overall paint quality without revealing their individual datasets. For instance, data from paint suppliers can be combined with real-time quality control data from software providers to immediately rectify any deviations in the painting process.

Finally, automobile manufacturers gather feedback data from end consumers regarding paint quality, such as how a paint job holds up over time in urban environments or how paint jobs are perceived by human eyes outside the factory doors. Through VFL, this data can be integrated to refine the painting process, ensuring it aligns with consumer preferences and real-world performance.

By employing VFL in the context of automotive painting systems, the industry can foster enhanced collaboration between different stakeholders. This not only ensures optimal paint quality and efficiency but also upholds data privacy and proprietary interests of each party. With VFL, the automotive painting process can be further refined, ensuring vehicles not only look aesthetically pleasing, but also receive a paint job that stands the test of time.

This VFL-enabled collaboration leads to marked improvements in product quality and yield across the automotive chain. It exemplifies how VFL can facilitate industry-wide efficiency, allowing for deep, collaborative insights while safeguarding data privacy and proprietary information.

Moreover, the strategic collaboration on such individually held data offers a pathway to not only reinforce the quality control feedback loops. It can also be used to refine design and material selection in upstream processes, which can lead to the production of paints, equipment, or processes with superior performance characteristics.

Table 1 Examples of data owned by different organizations in the paint application process

Stakeholder	Data types
Automobile manufacturers	Quality specifications, paint adherence metrics, final visual inspections, customer feedback on paint quality, long-term paint performance
Paint suppliers	Chemical composition of the paints, best conditions for application, drying times, possible reactions to environmental variables
Equipment manufacturers	Optimal operational parameters of the equipment, maintenance schedules, technical issues, efficiency metrics
Software providers	Data analytics on quality control, algorithms for optimizing the painting process, prediction models for potential defects

In a similar fashion, privacy-preserving learning extends to many other sectors. Federated Learning is also a key technology for accelerating the maturity of emerging technologies like battery, semiconductor, and medical technology. These areas are characterized by rapid innovation cycles where data-driven insights across organizations are crucial for improving performance, efficiency, and sustainability, and encompassing everything from material sourcing and manufacturing processes to real-world performance and end-of-life recycling. This broad spectrum of data allows manufacturers to swiftly adapt and innovate, leading to more efficient and eco-friendly solutions.

4 Vertical Federated Learning in Practice

In industrial practice, the Katulu Platform stands out with its VFL and AutoFL capabilities, demonstrating significant potential for industries where data privacy and innovative model development are key. Its application in various sectors, such as electronics manufacturing, highlights its ability to handle complex, privacy-sensitive data scenarios effectively.

It provides multiple distributed agents at the factory, production line, or machine level, as well as a central instance, where model updates are aggregated to generate value. The Software Development Kit (SDK) is designed for Data Engineers and Data Scientists to manage aspects of distributed machine learning, like model aggregation, building pipelines, managing data drift, and adding the right amount of privacy. It also allows orchestrating communication between the distributed agents, starting federated learning runs, while keeping communication efficient, secure, and scalable. This is depicted in Fig. 3.

Two case studies demonstrate the utility of VFL in direct comparison with centralized learning:

1. **Sensory Chemical Composition**: The Katulu Platform (http://katulu.io) was applied to a dataset involving gas composition sensory data from a facility in California. In a VFL setup, gas classes were known only to the active party. Using VFL, the Katulu Platform achieved a 94% F1 score with a model comprising 4 hidden deep learning blocks. The performance is similar to central learning.
2. **Retail Store Sales**: The Katulu Platform was used to analyze data from Rossmann, a retail chain, where the supervisory sales data was exclusive to the respective store management. The VFL model developed achieved a 79% F1 score with 2 hidden deep learning blocks, while the central learning model achieved an 81% F1 Score.

These case studies indicate that VFL can reach performances comparable to centralized machine learning settings while preserving data privacy, showcasing its effectiveness in diverse industry applications. There can be a small trade-off in performance between VFL and centralized learning. However, trading a little accuracy provides the opportunity to achieve privacy-preserving learning across the

Fig. 3 Decentralized learning with the Katulu Federated Learning Platform (own illustration)

value chain. This aspect is particularly valuable in scenarios involving multiple stakeholders, who may otherwise be hesitant to collaborate due to data privacy concerns.

5   Conclusion

In an era where data becomes the main driver of innovation, the potential of Federated Learning, a technology already revolutionizing the smartphone industry, extends far beyond mobile consumer devices. It is at the heart of the US National AI R&D Strategic Plan 2023 (https://www.whitehouse.gov/wp-content/uploads/2023/05/Nat ional-Artificial-Intelligence-Research-and-Development-Strategic-Plan-2023-Upd ate.pdf) as it promises to be a catalyst for change across various sectors—from discrete manufacturing and process industries to mobility and healthcare. This technology is pivotal in optimizing systems across factories, organizations, and even country borders, enabling AI-driven advancements without the need to exchange data, with the benefit of confidentiality and sustainability.

Today, sharing data such as design specifications, production data, or internal process details across value networks is essential for keeping pace with market demands and technological advancements.

As these systems encompass entire networks, the role of Federated Learning becomes increasingly crucial. This approach enables organizations, including direct competitors in manufacturing networks, to collaborate and leverage shared insights while maintaining stringent data privacy. It helps tackle the Sovereign Data Dilemma and provides organizations with a tool to leverage data jointly while addressing relevant concerns with regard to IP and strategy.

New federated approaches to ML will be key in an increasingly interconnected world. To date, the biggest challenges across Manufacturing Networks include data availability and data quality. Organizational collaboration is also crucial, demanding clear data usage and contribution agreements between entities. As of today, there are limited commercial mechanisms as to how the shared value is distributed among the Network participants. Furthermore, the continuous management and evaluation of VFL models play a pivotal role. It is crucial to regularly assess and update these models to ensure they remain accurate and relevant to the evolving needs of the industry. This ongoing process of model management guarantees that the insights and outputs provided by FL remain reliable and beneficial for the organization's objectives. To successfully integrate machine learning into industrial operations, robust change management strategies will likely be the biggest leverage. This involves preparing the organization at all levels and providing comprehensive training for this new way of working.

Using data across organizations is about forging a path towards more resource-efficient factories, supply chains resilient to global shifts, smarter and cleaner infrastructure. It is about leveraging the collective power of data to drive innovation in what will be our next industrial revolution—a future where data empowers and protects in equal measure.

References

Export Administration Regulations (EAR) from US. https://www.bis.doc.gov/index.php/regula tions/export-administration-regulations-ear; and Regulation (EC) No. 428/2009 from EU. https://www.eumonitor.eu/9353000/1/j4nvk6yhcbpeywk_j9vvik7m1c3gyxp/vitgbgiqqry9#:~: text=It%20sets%20out%20a%20uniform,the%20spread%20of%20nuclear%20weapons

Gupta, O., Raskar, R.: Distributed learning of deep neural network over multiple agents. J. Netw. Comput. Appl. **116**, 1–8 (2018). https://www.media.mit.edu/projects/distributed-learning-and-collaborative-learning-1/overview/

Katulu Platform: Industrial Scale Federated Learning, Data and Model Engineering Ecosystem. http://katulu.io. www.katulu.io/en/#whitepaper. Accessed 09 Nov 2023

Li, A., Peng, H., Zhang, L., Huang, J., Guo, Q., Yu, H., Liu, Y.: FedSDG-FS: Efficient and Secure Feature Selection for Vertical Federated Learning (2023). https://doi.org/10.48550/arXiv.2302. 10417

McMahan, B., Ramage, D.: Federated Learning: Collaborative Machine Learning Without Centralized Training Data. Google AI Blog (2017). https://blog.research.google/2017/04/federated-lea rning-collaborative.html

Oden Technologies: Predictive Quality in Manufacturing (n.d.). https://oden.io/. Accessed 20 Oct 2023

https://www.whitehouse.gov/wp-content/uploads/2023/05/National-Artificial-Intelligence-Res earch-and-Development-Strategic-Plan-2023-Update.pdf

Anne Mareike Schlinkert is founder and COO at Katulu GmbH, where she is driving the adoption of federated learning in industrial applications. In the past, she has worked in leading industrial OEMs around the world within digital transformation and in redefining business models with data. Anne Mareike holds multiple degrees from leading universities, such a Master of Science CEMS MIM from ESADE Barcelona and Copenhagen Business School.

Dr. Leonhard Kunczik is currently working at the Katulu GmbH as a Data Scientist, where he leads and contributes to different research and industrial projects in the realm of Federated Learning. Before joining Katulu, he worked as a Post-doctoral researcher at the Universität der Bunderswehr München in the Department for Operations Research, where he also finished his doctorate in the area of Quantum Reinforcement Learning.

Orlando Hohmeier currently serves as the Chief Technology Officer (CTO) at Katulu GmbH, where he has been shaping the company's product vision and strategy since September 2020. His efforts are focused on moving AI beyond boundaries to create a better way of gaining value from data, managing project teams, and establishing a product team to realize this vision. Previously, Orlando held positions at D2iQ (formerly Mesosphere), where he managed various teams from Frontend to Networking, Storage, and Data Services teams.

Michael Kuehne-Schlinkert founder and CEO of Katulu GmbH, is a pioneer in Federated Learning (FL) for the industry. As an expert for Privacy Enhancing Technologies like FL, he founded Katulu in 2018 to make artificial intelligence usable in the industry. At Katulu, he is responsible for company building strategic orientation. As a seasoned technologist, he also developed the first non-invasive, AI-based predictive maintenance solution for industrial pumps in 2017. With Katulu he aims to create a sovereign European industry that gains competitive advantages and secures intellectual property through AI, aligning with European values.

A Final Word

Ulrike Schmuntzsch◉, Alexandra Shajek◉,
and Ernst Andreas Hartmann◉

At the end of this book, we would like to make some concluding remarks and give an outlook on further developments of the topic 'Digital Sovereignty', sometimes also referred to as 'Sociodigital Sovereignty', which we have by now covered in four volumes. Since new technologies like AI and other algorithmic systems as well as rapidly changing market conditions have a tremendous impact on companies and organizations as well as on employees at the individual workplace, our publications on Digital Sovereignty address both pillars. The first two edited volumes outlined the concept of Digital Sovereignty (Hartmann 2021, 2022), while the third one focused on the first pillar at the individual level at the workplace (Shajek and Hartmann 2023).

While retaining its fundamentally positive and practical approach, the fourth volume now focuses on the level of companies and other organizations. As in the previous volume, this topic is presented and discussed in a variety of contributions, offering different perspectives with regard to various industries and organizations, as well as from a wide range of different disciplines. In this volume, we have tried to contribute to a deeper understanding of Digital Sovereignty at the organizational level and to provide impulses for actors and decision-makers at the management level to improve the Digital Sovereignty of their company or organization. The topic—Digital Sovereignty—is characterized by such a high degree of variety and complexity that our volume can naturally only provide a few insights into different perspectives. Looking at the very different aspects presented and discussed in the contributions, the following assumptions can perhaps be highlighted at the end of the day:

1. The various facets of Digital Sovereignty are diverse and topics covered in this book reach from industrial engineering (Balder and Stark 2025; Wöstmann et al. 2025) and IT (Braun and Huber 2025; Ganten et al. 2025; Feth et al. 2025; Jahnke et al. 2025; Schlinkert et al. 2025) to law (Straub 2025) as well as to

U. Schmuntzsch (✉) · A. Shajek · E. A. Hartmann
Institute for Innovation and Technology (iit), Steinplatz 1, Berlin, Germany
e-mail: schmuntzsch@iit-berlin.de

© The Author(s) 2025
U. Schmuntzsch et al. (eds.), *New Digital Work II*,
https://doi.org/10.1007/978-3-031-69994-8_17

personnel and organizational development-related domains (Glock 2025; Kauffeld and Berg 2025; Ködding et al. 2025), but also more fundamentally, to the strategic orientation of organizations (Klempert and Ménard 2025) and their external relationships (Coester and Pohlmann 2025; Dassel and Lutze 2025).

2. In practice, the three types of capital described in the introduction are interlinked on the organizational level, meaning that the external relationships of organizations also have an internal impact on the human and structural capital, and vice versa.
3. The organizational level of Digital Sovereignty and the individual level at the workplace are intertwined as well and influence one another.
4. The analysis and the design of Digital Sovereignty at the workplace as well as within the company requires an attitude and a culture of social partnership between management and labor representatives, and, vice versa. It requires the methods and tools for analyzing and designing organizations for Digital Sovereignty, as described here in examples. This can serve well to implement social partnership into the real design of organizations and work systems.

In summary, by examing these various contributions in order to illuminate the topic of Digital Sovereignty on an organizational level from several perspectives, we hopefully have given the one or the other insight and reflection to actors and decision-makers at the management level. Knowing fully well that some topics have been left out, e.g. legal issues referring to the AI-Act or Enterprise Resource Planning (ERP) systems from a perspective of Digital Sovereignty. Other important aspects, such as AI and other algorithmic systems as well as the practical implications of different forms of employee participation, are only touched selectively, but are also important for the development of this topic and should be of broader interest in the future.

Finally, although this series of books has come to an end for the time being, we remain committed to the topic of Digital Sovereignty on both pillars—the individual level at the workplace as well as the organizational level of the company. Having conceived the concept of Digital Sovereignty to be practically applicable, in this volume, we also presented a method and its proof of application in a first industrial use case, demonstrating how to analyze and design Digital Sovereignty on both intertwined levels (Schmuntzsch and Hartmann 2025). With this, a first step has been made for a practical application. Our aim is to provide a tool and further develop it into a complete toolbox in order to integrate and foster Digital Sovereignty in daily business of a company or organization. In this way, we aim to further apply the concept in practice within organizations and at the workplace with different industry partners. Future developments will also give rise to further questions and practical challenges, which we will pursue with interest and which we aim to shape.

Last but not least, we would like to thank all authors for their diverse contributions providing deep insights into their research topics related to various facets of Digital Sovereignty on an organizational level. The publication of this work would not have been possible without their great commitment. We also would like to express our gratitude to our colleagues at the Institute for Innovation and Technology (iit), in particular Désirée Tillack, Alexandra Lescher, and Annelie Hofmann for their excellent support.

Ulrike Schmuntzsch, Alexandra Shajek and Ernst Andreas Hartmann in January 2025.

References

Balder, J., Stark, R.: Development of an open innovation knowledge plat-form in the context of digital sovereignty. In: Schmuntzsch, U., Shajek, A., Hartmann, E.A. (eds.) New Digital Work II: Digital Sovereignty of Companies and Organizations. Springer (2025)

Braun, C., Huber, M.F.: Hybrid AI-driven advances in prognostics and health management within manufacturing environments. In: Schmuntzsch, U., Shajek, A., Hartmann, E.A. (eds.) New Digital Work II: Digital Sovereignty of Companies and Organizations. Springer (2025)

Coester, U., Pohlmann, N.: Why trustworthiness is the cornerstone of digitalization. In: Schmuntzsch, U., Shajek, A., Hartmann E.A. (eds.) New Digital Work II: Digital Sovereignty of Companies and Organizations. Springer (2025)

Dassel, K., Lutze, M.: Digital care-pathways and inter-organizational systems: a perspective on digital sovereignty along shared responsibilities. In: Schmuntzsch, U., Shajek, A., Hartmann, E.A. (eds.) New Digital Work II: Digital Sovereignty of Companies and Organizations. Springer (2025)

Feth, D., Jung, C., Eitel, A.: Concepts for data sovereignty in digital value chains: data cockpits—data usage control—data trustees. In: Schmuntzsch, U., Shajek, A., Hartmann, E.A. (eds.) New Digital Work II: Digital Sovereignty of Companies and Organizations. Springer (2025)

Ganten, P., Seyffarth, M., Kuhlmann, N.: Successful digital transformation in economy and industry requires open source. In: Schmuntzsch, U., Shajek, A., Hartmann, E.A. (eds.) New Digital Work II: Digital Sovereignty of Companies and Organizations. Springer (2025)

Glock, G.: Innovation capacity in manufacturing: a question of autonomy? In: Schmuntzsch, U., Shajek, A., Hartmann, E.A. (eds.) New Digital Work II: Digital Sovereignty of Companies and Organizations. Springer (2025)

Hartmann, E.A. (ed.): Digitalisierung souverän gestalten: Innovative Impulse im Maschinenbau. Springer (2021). https://doi.org/10.1007/978-3-662-62377-0

Hartmann, E.A. (ed.): Digitalisierung souverän gestalten II: Handlungsspielräume in digitalen Wertschöpfungsnetzwerken. Springer (2022). https://doi.org/10.1007/978-3-662-64408-9

Jahnke, N., Rohde, M., Kraus, T.: Edge computing for digital sovereignty in the data economy. In: Schmuntzsch, U., Shajek, A., Hartmann, E.A. (eds.) New Digital Work II: Digital Sovereignty of Companies and Organizations. Springer (2025)

Kauffeld, S., Berg, A.-K.: Shaping transformation: becoming a changemaker. In: Schmuntzsch, U., Shajek, A., Hartmann, E.A. (eds.) New Digital Work II: Digital Sovereignty of Companies and Organizations. Springer (2025)

Klempert, A., Ménard, D.: Wikipedia's atypical organizational model: digital sovereignty 20 years in the making. In: Schmuntzsch, U., Shajek, A., Hartmann, E.A. (eds.) New Digital Work II: Digital Sovereignty of Companies and Organizations. Springer (2025)

Ködding, P., Jahn, M., Koldewey, C., Dumitrescu, R.: Challenges for scenario-based foresight and potentials for digital technologies: insights from practice. In: Schmuntzsch, U., Shajek, A., Hartmann, E.A. (eds.) New Digital Work II: Digital Sovereignty of Companies and Organizations. Springer (2025)

Schlinkert, A.M., Kunczik, L., Hohmeier, O., Kuehne-Schlinkert, M.: Preserving digital sovereignty in data-driven manufacturing networks. In: Schmuntzsch, U., Shajek, A., Hartmann, E.A. (eds.) New Digital Work II: Digital Sovereignty of Companies and Organizations. Springer (2025)

Schmuntzsch, U., Hartmann, E.A.: Analyzing and developing socio-digital sovereignty on individual and organizational levels—a case study. In: Schmuntzsch, U., Shajek, A., Hartmann, E.A. (eds.) New Digital Work II: Digital Sovereignty of Companies and Organizations. Springer (2025)

Shajek, A., Hartmann, E.A. (eds.): New Digital Work: Digital Sovereignty at the Workplace. Springer (2023)

Straub, S.: The European data act and its impact on corporate digital sovereignty. In: Schmuntzsch, U., Shajek, A., Hartmann, E.A. (eds.) New Digital Work II: Digital Sovereignty of Companies and Organizations. Springer (2025)

Wöstmann, R., Schulte, L., Meierhofer, F., Beitinger, G., Deuse, J.: Future challenges of data-driven problem solving in producing companies in context of digital sovereignty and lessons learned from electronics industry. In: Schmuntzsch, U., Shajek, A., Hartmann, E.A. (eds.) New Digital Work II: Digital Sovereignty of Companies and Organizations. Springer (2025)

Dr. Ulrike Schmuntzsch studied 'Business Psychology' at the former University of Applied Science in Lüneburg and 'Human Factors' at the Technical University (TU) Berlin. Focusing on Work and Engineering Psychology, she worked as a research associate at the Chair of Human–Machine Systems at the TU Berlin from 2011 until 2022. As part of several application-oriented research projects with various industry partners, she was responsible for human-centered evaluation and design. Her doctoral thesis, which she completed in 2014, focuses on the development and evaluation of a warning glove as a means of user support during maintenance work in industrial applications. Since 2022, Ulrike Schmuntzsch has been a research associate at the Institute for Innovation and Technology (iit) and a consultant at VDI/VDE Innovation + Technik GmbH. In this role, she is part of the project team 'Digital Sovereignty in Business'.

Dr. Alexandra Shajek is a research associate at the iit and team leader of the 'Education and Work' group at VDI/VDE Innovation + Technik GmbH. Currently, she manages the project 'Federal Report on Young Academics 2025' for a German Federal Ministry. At the iit, she worked on projects such as 'Case studies on the effects of the Covid-19 pandemic on operational transformation processes in organizations' among others. Previously, Alexandra Shajek worked as a visiting researcher at the German Institute for Economic Research (DIW) and, in addition, completed her doctorate in the field of innovation research at Humboldt University of Berlin.

Dr. Ernst Andreas Hartmann After studying psychology, specialising in work and organisational psychology—obtained his doctorate as Dr. rer. nat. at the Faculty of Mechanical Engineering at RWTH Aachen University in 1995. In the 1990s, he worked at the Hochschuldidaktisches Zentrum/Lehrstuhl Informatik im Maschinenbau (University Teaching Centre/Chair of Information Technology in Mechanical Engineering) at RWTH Aachen. In this context, he engaged in projects on academic reform and took part in the development of new forms of academic teaching/learning. Furthermore, he carried out research on the design of man–machine systems, and issues of industrial work organisation. In the mid-1990s, Ernst Andreas Hartmann was an internal consultant for organisation and process development at John Deere Werke Mannheim. In 2002, he qualified as lecturer (habilitation) in psychology and received the 'venia legendi' for Work and Organisational Psychology; since then, he has been a private lecturer for work systems and process design at RWTH Aachen. From 2001 to 2004, he was responsible for the scientific coordination of the programme 'Lernkultur Kompetenzentwicklung' ('competence development and learning cultures') of the German Federal Ministry of Education and Research at the 'Arbeits-gemeinschaft Betriebliche Weiterbildungsforschung ABWF e.V.' ('Association for Research in Continuing Education').

From 2004 to 2016, Ernst Andreas Hartmann was head of the Socio-economic Department at VDI/VDE Innovation + Technik GmbH in Berlin; since 2016, he has been head of the Education, Science, and Humanities Department. Since 2007 he has functioned as one of the directors of the Institute for Innovation and Technology (iit).

Author Index

© VDI/VDE Innovation + Technik GmbH 2025
U. Schmuntzsch et al. (eds.), *New Digital Work II*,
https://doi.org/10.1007/978-3-031-69994-8